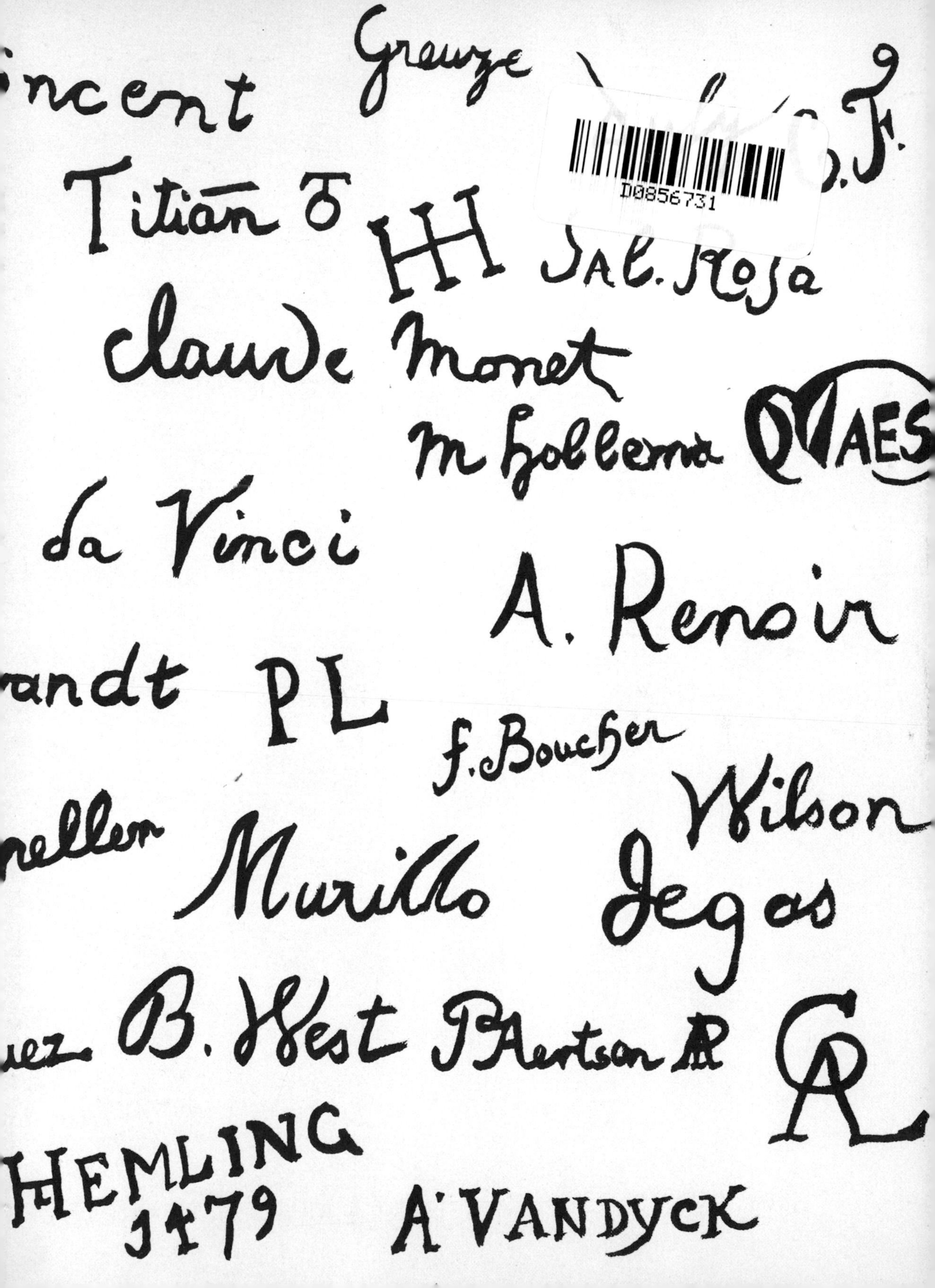

incent Greuze ...... S.F.

Titian ♂

HH SAℓ. ℛOʃa

claude Monet

m hobbema QꞦAES

da Vinci

A. Renoir

andt PL

f. Boucher

Wilson

neller Murillo Jegas

uez B. West PRertson ℝ GꞦL

HEMLING
J179 A·VANDYCK

THE **GUINNESS** BOOK OF
ART
FACTS AND FEATS

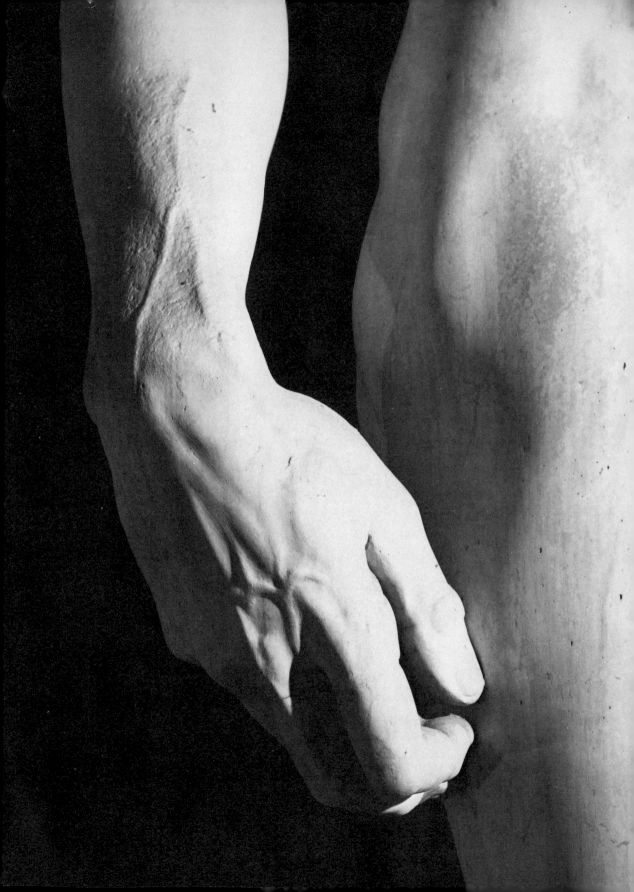

THE **GUINNESS** BOOK OF
ART
FACTS AND FEATS

John FitzMaurice Mills

Guinness Superlatives Limited
2 Cecil Court, London Road, Enfield, Middlesex

© John FitzMaurice Mills
and Guinness Superlatives Ltd, 1978

Published in Great Britain by
Guinness Superlatives Ltd, 2 Cecil Court,
London Road, Enfield, Middlesex

ISBN 0 900424 44 3

Guinness is a registered trade mark of
Arthur Guinness Son & Co. Ltd

Set in 10pt Bembo

Printed and bound by
Jarrold and Sons Ltd, Norwich

Frontispiece: The right hand of the huge white marble statue of 'David' by Michelangelo. (Courtesy Azienda Autonoma di Turismo of Florence, Italy)

Title page illustration: Head of Greek Philosopher in bronze with inlaid eyes, which was salvaged from the sea near Antikythera in the south of the Peloponnese. It was part of a cargo of works of art sunk there in the time of the Roman Emperors. (Courtesy Athens Archaeological Museum)

OTHER GUINNESS SUPERLATIVES TITLES

Facts and Feats Series:

Air Facts and Feats,
3rd ed.
John W R Taylor, Michael J H Taylor and David Mondey

Rail Facts and Feats,
2nd ed.
John Marshall

Tank Facts and Feats,
2nd ed.
Kenneth Macksey

Yachting Facts and Feats
Peter Johnson

Plant Facts and Feats
William G Duncalf

Structures—Bridges, Towers, Tunnels, Dams . . .
John H Stephens

Car Facts and Feats,
2nd ed.
edited by Anthony Harding

Business World
Henry Button and Andrew Lampert

Music Facts and Feats
Bob and Celia Dearling with Brian Rust

Animal Facts and Feats,
2nd ed.
Gerald L Wood

Weather Facts and Feats
Ingrid Holford

Astronomy Facts and Feats
Patrick Moore

Guide Series:

Guide to Saltwater Angling
Brian Harris

Guide to Mountain Animals
R P Bille

Guide to Underwater Life
C Petron and J B Lozet

Guide to Motorcycling,
2nd ed.
Christian Lacombe

Guide to French Country Cooking
Christian Roland Délu

Guide to Bicycling
J Durry and J B Wadley

Guide to Water Skiing
David Nations OBE and Kevin Desmond

Other titles:

The Guinness Guide to Feminine Achievements
Joan and Kenneth Macksey

The Guinness Book of Names
Leslie Dunkling

Battle Dress
Frederick Wilkinson

History of Land Warfare
Kenneth Macksey

History of Sea Warfare
Lt.-Cmdr. Gervis Frere-Cook and Kenneth Macksey

History of Air Warfare
David Brown, Christopher Shores and Kenneth Macksey

The Guinness Book of Answers, *2nd ed.*
edited by Norris D McWhirter

The Guinness Book of Records,
edited by Norris D McWhirter

The Guinness Book of 1952
Kenneth Macksey

The Guinness Book of 1953
Kenneth Macksey

100 years of Wimbledon
Lance Tingay

Book of World Autographs
Ray Rawlins

British Hit Singles
Jo and Tim Rice

Foreword

A work of art is a window into the sight of another. Visual art is a force for understanding which is not necessarily dependent on learning. The use of colour, line and form, together build an image that excites deep-running emotions.

Reasons for liking or disliking works of art reflect almost every facet of our personal experience. The humble craftsmen or geniuses from the past and today have in many cases thrown their all, mentally and physically, into their creation. They ask no more than that we shall look at what they have done with an eye free from bigotry.

There are no rules that can be set down for the appreciation of art. Every work is as individual as the artist creating it and the person observing it. Thus seemingly the infinite quality of art can be demonstrated. The multiples of works, creators and views go beyond assessment.

It is arguable that art is not a necessity to our existence. In the manner of food, drink and clothing it is not. Yet within us there are intangible desires for beauty, colour, harmony, decoration, visual enrichment, that cannot be denied if a whole personality is to be attained.

The story of art has told the historian much, more lucidly and exactly than the written word: how those before us looked, dressed, behaved; the things they used, they lived in; customs, transport, sport and amusements. In a second almost, theirs is thrown into ours. The artist has been connected with propaganda, glorification, symbolism, cen-sure, to an extent that has caused him to play an intense part with religion, politics and social growth.

Whatever the subject the painter or sculptor works on, the realization is an abstraction. It is a distillation by the mind and hand, resulting from the impact of the matter before him on his sensitivity and ability to translate the inspiration.

To understand and know the problems of the artist, to glimpse the astounding array of materials used, the complex variety of methods and techniques, the rules and theories that have accumulated over the long period of creativity is to be better able to appreciate and enjoy what has come into being.

Fashion is perhaps one of the most insidious will-o'-the-wisps. It flashes up with often little warning, casts around its influences and then is gone; or it can repeat itself with predictable cycles. But in the process of knowing the artist and joining in the intimate excitement and joy, fashion can be a set of blinkers. By its nature it constricts to a personal limited outlook and defeats the intention of the infinite involvement and under-standing of all the aspects that are available and should be for every one from all time and today.

What follows in the naming of materials and the explaining of methods, theories and individual artistic experiences may, it is hoped, clarify and point a way through the intricate situation in which the artist moves to bring his work to light.

Contents

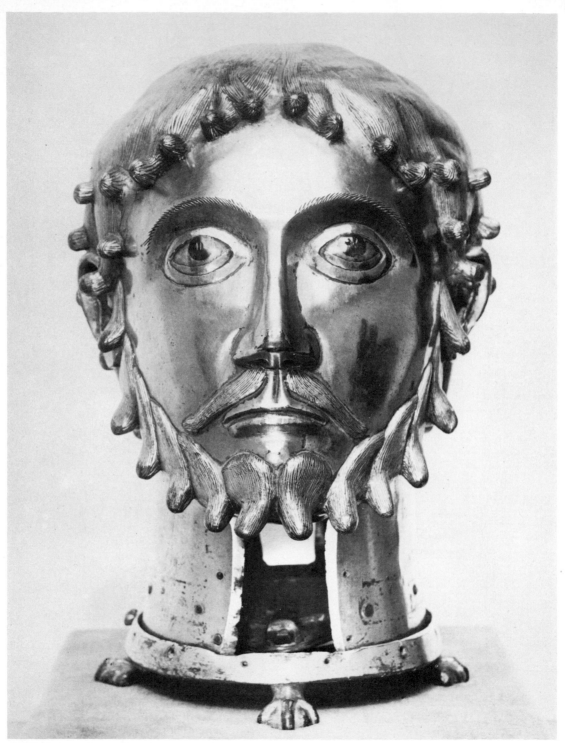

A gilded bronze reliquary from Fischbeck; 12th century. Now in the collection of the Kestner-Museum, Hannover, which was formed from bequests of the Kestner family to the town. (Courtesy Landeshauptstadt Hannover)

Chapter One

Materials and Methods

Pigments

WHITES

Bismuth white: (also termed: Bougival and pearl white) Bismuth nitrate. It had a brief use in the early 19th century, but it is fugitive, and darkens when coming into contact with other pigments that contain sulphur. **Now obsolete, being replaced by zinc white.**

Bone white: Obsolete; it was made by burning bones to a white ash. Cennino Cennini in his *Il Libro dell' Arte* says 'the best bones are from the second joints and wings of fowls and capons; the older they are, the better; put them into the fire just as you find them under the table.' It was used for a ground for panels.

Ceruse: (an obsolete name for White lead, Flake white, also Nottingham white and other local names) Basic lead carbonate. In use since the prehistoric Greek period, **the second oldest artificially produced pigment**. It was the only white oil-colour available to artists until the middle of the 19th century.

Chalk: Calcium carbonate. **One of the whitest substances known.** It has no use as a pigment with oil, although when mixed with an aqueous glue medium it makes an excellent ground for tempera or oil and retains its brilliant whiteness. It is the basis for most pastels and coloured chalks.

Chinese white: Zinc white specially prepared for water-colour. It was introduced by Winsor and Newton in 1834.

Cremnitz white: High-quality corroded white lead made by a 19th-century method that uses litharge instead of metallic lead that is employed with the Dutch process to make flake white.

Flake white: (see Ceruse)

Foundation white: A cheap pigment prepared from barium and zinc salts. Main uses being for grounding and with some low-grade paints.

Tin white: Tin oxide. First mention is in a 16th-century manuscript where a recipe for its manufacture is given. It was used for illuminating with little satisfaction. Van Dyck experimented with it ground in oil, also Mytens who found that it blackened in sunlight.

Titanium white: Titanium dioxide. Its properties as a pigment were known from 1870 or earlier, but it was not marketed for artists until 1920. Advantage over flake white is that it is non-poisonous and not so liable to be affected by atmospheric pollution.

Zinc white: Zinc oxide. First made and sold in France towards the end of the 18th century. It has similar advantages to titanium white.

YELLOWS

Aureolin: Cobalt-potassium nitrite. A strong transparent yellow discovered by N W Fischer, Breslau in 1830; first introduced as a paint pigment by Saint-Evre, Paris, 1852; introduced to England about 1860. Supersedes gamboge for water-colour, is suitable for glazing with oils.

Aurora yellow: Sulphide of cadmium. A variety of cadmium yellow introduced by Winsor and Newton in 1889.

Cadmium yellow: Sulphide of cadmium. Discovered in 1817 and introduced commercially in 1846. There is a range from pale yellow through to orange. For many artists cadmium colours have superseded the chrome colours.

Chrome yellow: Lead chromate. Originally discovered when chemists were examining a natural lead chromate, chrocoite, in the Beresof gold-mine in Siberia in 1770. Introduced as a colour in 1797. As with cadmium there is a range of yellows through to orange; the chrome, however, can react with other pigments.

Fustic: A yellow dye that is obtained from the plant *Chlorophona tinctoria*, native to the Americas, introduced to Europe in the 16th century. It had a limited use with water-colour. Older name was *ffusticke yealowe*.

Gallstone: Prepared from the gallstone of an ox and gives a reasonably dark yellow. Nicholas Hilliard found it useful for shading with miniature work. John Payne in the 18th century found that dishonest colourmen were selling an inferior substitute. He suggested in his book on miniature-painting that artists should approach slaughter-houses and that the men there should be on the watch for gallstones. **In 1801 it was one of the top four most expensive colours**, Ackerman's showing a charge of five shillings a cake.

Gamboge: A native yellow gum from Thailand. A bright transparent golden yellow for glazing or water-colour, it is not a true pigment. It has been in use since medieval times. J Smith in *The Art of Painting in Oyl*, published in 1701, describes a method for preparing the colour, which usually comes in rough cylinders about $2\frac{1}{2}$ in (6 cm) in diameter. 'For a Yellow Gumboge is the best, it is sold at Druggist in Lumps, and the way to make it fit for Use, is to make a little hole with a Knife in the Lump, and put into the hole some Water, stir it well with a Pencil till the Water be either a faint or a deeper Yellow, as your occasion requires, then pour it into a Gally-Pot, and temper up more, till you have enough for your purpose.' (Pencil here would mean a small, soft, hair brush.)

Gold: Fifteenth- and 16th-century writers mention the use of gold for water-colour painting also for illuminating letters. **The grinding of gold to a powder presents difficulties as the metal is so soft.** There is a mention of this being done in honey, and then the honey being washed away.

Indian yellow: An obsolete lake (colour) that was produced by heating the urine of cows that had been fed on mango leaves. It came to England in 1786 although its method of manufacture was a mystery until the late 19th century. Owing to the cows being wasted by their diet, production was stopped in 1908, and a substitute synthetic colour introduced.

Massicot: An obsolete pigment prepared from lead oxide with possibly tin oxide. In use from the 14th to the 18th century in Europe. Hilliard found it helpful and told that it should be used with sugar candy, which could have made for problems as massicot is very poisonous. It tended to discolour and turn grey with exposure to the air.

Naples yellow: Lead antimoniate. Traces said to have been identified in Babylonian tiles dating back to the 5th century BC. Cennini in *Il Libro dell' Arte* supposed that it was a native earth from Vesuvius. It has been made artificially from at least the 15th century. Works well ground in oil.

Orpiment: (also termed: King's yellow, auripigmentum, arsenic yellow, Chinese yellow) It occurs in early Egyptian, Syrian and Persian work, also in Chinese cave-paintings. It is a yellow sulphide of arsenic and a number of natural deposits occur in Asia and Europe. Although quite widely used artists disliked it because of its highly poisonous nature and unpleasant smell.

Pink: The word pink was used for yellow when referring to a yellow pigment certainly up to the end of the 17th century and it is likely well into the 18th. The pink (yellow) was made by a skill in cooking. Several ingredients were used including: unripe buckthorn berries, weld, broom. Norgate in his treatise mentions 'callsind eg shels and whitt Roses makes rare pinck that never starves'.

Quercitron yellow: Obsolete yellow obtained from the bark of the black quercitron oak from America. It was introduced to Europe by Edward Bancroft, a Doctor of Medicine and Fellow of the Royal Society, in 1775. It appeared in Ackermann's treatise in 1801 masquerading as: 'Ackermann's Yellow, another new Colour, lately discovered, is a beautiful warm rich Yellow, almost the tint of Gallstone, works very pleasant, and is very useful in Landscapes, Flowers, Shells, &c.'

Saffron: A fugitive colour prepared from the dried stigmas of the *Crocus sativus*. **Its use dates from Roman times.** Used by illuminators although frowned on by Hilliard as of dubious value.

Turner's yellow: (also termed: Kassler yellow, Verona yellow, Cassel yellow) Lead oxychloride. Patented in 1781 by the English colour-maker James Turner. His recipe was plagiarized by his competitors and he nearly ruined himself with lawsuits against them. Obsolete today, but gathering from the mention in writings it must have had considerable popularity in the late 18th century and the 19th.

Turpeth mineral: Mercuric sulphate. Highly poisonous. Valued at one time for the fine greens it produced when mixed with Prussian blue. Discarded because it decomposed and turned black in some mixtures.

Yellow Ochre: An opaque pigment, a clay

coloured by iron oxide. **One of the artist's basic colours from prehistoric times through to the present day.** Ochres are mined all over the world; some of the best coming from the south of France.

REDS

Alizarin crimson: Dihydroxyanthraquinone. A derivative of anthracene, a coal-tar product. The alizarins also include red, scarlet, lake, violet and yellow. All colours are transparent, useful for glazes; but tend with oils to be slow driers. They were discovered in 1868 by two German chemists, C Graebe and C Liebermann.

Brazil wood lake: A blood-red natural dyestuff produced from the wood. It was introduced to Europe in the 16th century. The derivation of brazil, also earlier spelt brasil, is from the Old French *braise*, live coals. The South American country was named after this product not the colour from the country.

Cadmium red: Cadmium sulphide and cadmium selenide. Introduced by de Haen in Germany in 1907, general use in England after 1919, although Robersons were supplying it in 1912. **It replaces vermilion.** Both light and dark cadmium reds are clean, strong colours with considerable tinting strength.

Carmine: A warm rich red dyestuff extracted from cochineal insects found in Central America. It held an important place on the artist's palette from the 16th century until the late 19th and early 20th when it was passed over by the more reliable alizarins. An illustration of the pains some of the earlier painters were prepared to go to in preparing their colours comes in John Smith's *The Art of Painting in Oyl* published in 1701: 'Buy at the Drugists some good Cochinele, about halfe an ounce will go a great way. Take Thirty or Forty Grains, bruise them in a Gally-Pot to fine pouder, then put to them as many Drops of the Tartar Lye as will just wet it, and make it give forth its Colour; and immediately add to it half a spoonful of Water, or more if the Colour be yet too deep, and you will have a delicate Purple Liquor or Tincture. Then take a bit of Allum, and with a Knife scrape very finely a very little of it into the Tincture, and this will take away the Purple Colour, and make it a delicate Crimson. Strain this through a fine Cloath into a clean Gally-Pot, and use it as soon as you can, for this is a Colour that always looks most Noble when soon made use of, for it will decay if it stands long.'

Cinnibar: A native vermilion, inferior to the manufactured product. Traces of it have been found in Assyrian relics and other early cultures. Theophrastus in his *History of Stones* records an instance of an outcropping of cinnibar ore on high difficult Spanish cliffs, and how it was worked by shooting arrows to dislodge the ore.

Dragon's Blood: A warm ruby-red resinous exudation of *Calamus draco* found in eastern Asia. Its use in Europe in painting dates back to the 1st century. **Medieval illuminators employed it.** Pliny the Elder expounded his fanciful idea that the substance was actually the mixed blood of those legendary enemies, the dragon and the elephant, which was spilt during their mortal combat.

Folium: A red-purple colour from vegetable origins in use in medieval times. **Now superseded and replaced by alizarin.**

Indian red: Ferric oxide. Originally a native earth imported from the East. First manufactured in the early 18th century. Supposedly introduced to England by the American painter Benjamin West, the term was used by early American artists to describe a colour typical of an earth used by the Indians.

Kermes: A red dye that comes from the insect *Coccus ilicis* which inhabits an evergreen oak growing in southern Europe and North Africa. **Used in Roman and medieval times.** It was rendered more or less obsolete by cochineal and Brazil wood. The name kermes is Persian, the Old English name for it is grain.

Light red: (also termed: English red, Prussian red, colcothar, and Persian red) Ferric oxide. Calcined yellow ochre. **A pigment with considerable opacity and tinting power.**

Madder: One of the lake or dry pigments made from the root of the madder plant or garance, *Rubia tinctorum*. It is thought to be the rubia mentioned by Pliny the Elder. Traces have been found in paintings from Egyptian and Graeco-Roman times. Probably brought to Italy by the Crusaders. By the 13th century the plant was being cultivated in many places in Europe, notably Holland. The colour does not appear to have been used in medieval or Renaissance painting. It had its greatest vogue in the 18th and 19th centuries.

Minium: Red lead. The name in early times was applied by the Romans to their native vermilion and cinnibar; it was not applied to red lead until the Middle Ages. **The word miniature is derived from miniate, meaning to paint with minium. Obsolete as a pigment for the artist today**, its chief use is as a primer for steel work.

Realgar: Native arsenic disulphide. Small deposits have been found in all parts of the world, and traces of its use for colouring have come to light on relics from the most primitive civilizations.

Safflower: A red lake made from the dried petals of the safflower plant, *Carthamus tinctorius*. **In use from ancient times for painting and dyeing textiles.**

Sinopia: An ancient name for native red iron oxides, it takes its name from the town of Sinope in Asia Minor. Cennini says in *Il Libro dell' Arte* of its unsuitability for fresco and tempera. **Well watered down it was much employed by artists for laying in the under-drawing for fresco work on the** *arriccio.*

Terra Pozzuoli: A red earth originally produced at Pozzuoli (Puteoli), Italy. **A popular red used by the Renaissance fresco-painters.** It has a characteristic of setting very hard like cement.

Vermilion: (also zinnober) Mercuric sulphide. A very heavy, powerful and poisonous pigment. Used in China from an early date and introduced to Europe in the 8th century. John Smith in *The Art of Painting in Oyl* gives a description for making it that smacks of alchemy: 'Take six Ounces of Brimstone and melt it in an Iron-Ladle, then put two Pound of Quicksilver into a Shammy Leather, or double Linnen-Cloth, squeeze it from thence into the melted Brimstone, stirring them in the mean time with a wooden Spatula, till they are well united, and when cold, beat the mass into a Powder, and sublime it in a glass Vessel, with a strong Fire, and it will arise into that red substance which we call artificial Cinaber, or Vermillion.' **Now superseded by cadmium red.**

BLUES

Azurite: Native basic copper carbonate. **Dates from at least as early as the Roman times.** Had a long run as an important blue for tempera and water-colour until being **superseded by smalt in the 17th century.**

Cerulean: Cobalt stannate. It is made by roasting cobalt and tin oxides. Late-18th-century colour-makers tried to find a successful method, but the process was eventually perfected by Höpfner, Germany, in 1805. It was introduced as an artist's colour by George Rowney, England, in 1870.

Cobalt blue: (also Thénards blue)
A compound of cobalt oxide, aluminium oxide and phosphoric acid. **It was discovered by**

Baron Thénard, France, in 1802. Introduced for artists about 20 years later and **replaced unsatisfactory colours, such as smalt.**

Cornflower blue: A blue dye made from the petals of the flower, and which was used by some water-colourists in the 18th century.

Egyptian blue: (also termed: Alexandrian blue, Vestorian blue) **The oldest manufactured colour**, a mixture of copper silicates. It was used in Egypt from about 3000 BC. It was imported into Mesopotamia, Crete and other Mediterranean lands. Vitruvius records that the process was brought from Alexandria to Pozzuoli in the 1st century BC.

French ultramarine: An artificial ultramarine, a complex combination of alumina, silica, soda and sulphur. The process was made viable by a Frenchman, Guimet, in 1828. **Economically it has replaced true ultramarine.**

Frit: A vitreous blue with a low tinting power **known from early Egyptian times.**

Indigo: It is produced by extraction and precipitation, using the leaves of the plant *Indigofera* found in India. It was an important trading item with the East India Company, which had to keep a strict eye open for contamination. **Today the natural colour has been replaced by a synthetic made from coal tar.** Pliny the Elder writes about: 'the slime of India's rivers [indigo] and the blood of her dragons and elephants'. The Greeks and Romans would have had it in use from the 1st century AD.

Phthalocyanine blue: A very powerful blue lake, produced from copper phthalocyanine. In its prime state it is so strong that there is no sign of blue, almost black with a coppery sheen. **Introduced into England in 1935, replacing Prussian blue for many artists.** Trade names include Monastral, Winsor, Thalo and Bocour blue.

Prussian blue: (also termed: Berlin blue, bronze blue, Paris blue and paste blue) Ferric ferrocyanide. It was discovered by Diesbach, Berlin, in 1704, and it is the **first synthetic pigment with an established date.** The process was kept secret until 1724, when it was published in England by John Woodward. Varieties of the colour are called: Antwerp blue, Brunswick blue, celestial blue. Monthier blue and soluble blue.

Smalt: A type of cobalt blue glass or frit. Manufactured in the Netherlands in the 16th century. It survived until the introduction of French ultramarine and then lost popularity owing to its weak tinting power and rather unpleasant gritty feel in working.

Ultramarine: The most expensive pigment, worth more than twice its weight in gold. It is produced by grinding the semi-precious stone lapis lazuli, and then by a complicated process separating the blue from the grey gangue rock associated with it in nature. One of the best deposits is in the Kokcha Valley in Afghanistan. **Its use as a pigment in Europe dates from the 12th century.** It has a distinctive very slight warm blush which makes it very difficult to match with synthetic blues. Hans van Meegeren used it in his Vermeer forgeries, and the fact that Winsor and Newton had supplied him with fairly large quantities was evidence at his trial.

Ultramarine ash: A delicate blue-grey pigment with low tinting power. It is made from lapis lazuli and the grey gangue rock with which it is found in nature.

Woad: Produced from the leaves of *Isatis tinctoria* which was extensively cultivated in Britain from very early times. **It was replaced by the arrival of indigo.** The two colours are so similar in chemical composition, that it is almost impossible to differentiate them in old paintings.

GREENS

Emerald: (also termed: Schweinfurt green, Scheele's green) Copper aceto-arsenite. Discovered by Scheele, Sweden, in 1788, and first produced commercially as a pigment by Russ and Sattler in Schweinfurt in 1814 and marketed in 1816.

Hooker's green: A dark green produced from a mixture of gamboge and Prussian blue.

Malachite: Native basic carbonate of copper. **Used as a pigment by the earliest civilizations.**

Oxide of Chromium: Known since 1809, although there is a possibility it was being used by the Sèvres factory and at Limoges at an earlier date. A strong, safe tinting pigment with a pleasant cool willow-green hue.

Sap green: (also termed: Bladder green, Iris green and verd vessie) **An obsolete lake produced from unripe buckthorn berries.** It faded rapidly and is today replaced by a blend of chlorinated copper phthalocyanine.

Terre Verte: A native clay coloured by small amounts of iron and manganese. It was popular in Italy from earliest recorded times, especially for fresco and tempera. It is quite transparent and has a low tinting power.

Verdigris: Hydrated copper acetate. **One of the earliest manufactured pigments; it was produced by the Greeks and Romans and lasted in fairly general use until the 19th century.** Montpellier in France was an important centre for the making of the pigment, where according to travellers strange processes were involved. A quantity of sour red wine would be poured into an earthenware jar to a depth of about 3 in (80 mm); grape stalks previously soaked in wine, and small copper plates were placed in alternate layers on a grille above the level of the sour wine; this allowed the acid fumes from the sour wine to penetrate. After a few days the copper plates were turned. Another lapse of several days and then the plates were removed and placed in small piles then to be soaked with the same sour wine. Lastly each pile was subjected to pressure for about a week, and then the plates were separated and the accumulated verdigris was scraped off, moulded into balls with a little wine and sold.

Viridian: (also termed: Casali's green, Mittler's green, Guignet green) Hydrated chromium hydroxide. It was first made in Paris in 1838 by Pannetier and Binet by a secret process. In 1859 the method was published by Guignet and the colours made available to artists. It was introduced to England in 1862 by Winsor and Newton.

BROWNS

Asphaltum: (also termed: Bitumen) A dark brown mixture of asphalt and oil or turpentine. Sources for the asphalt include Trinidad and the Dead Sea area. It is not a true pigment. Sadly during the second half of the 18th century and the 19th many painters in oil fell a victim to the lure of this warm brown that could provide a glow to areas of the canvas. **Whether asphaltum is applied as a glaze or thickly, it is always very detrimental to an oil-painting, cracking severely, coalescing into hard lumps and often bleeding into neighbouring colours.**

Bistre: A yellowish-brown soot produced by charring beech wood. **It was widely used for water-colour wash-drawings and monochrome work from the 14th century until the 19th.** This in spite of the fact that it fades appreciably.

Mummy: (also termed: Egyptian brown) **The most macabre of artist's colours.** In the 16th century mummified bodies were imported to

Effects from using asphaltum, both as paint and with varnish. Part of the varnish has been removed from the face to show excess cracking in paint layers. (Photo by the Author)

England from Egypt, generally being taken from the mass graves near the Pyramids. At first they were used for making internal medicines and then tried out as a pigment. The dry powdered mummy is a warm dark brown in colour and has a faint odour, rather pleasant, of spices and embalming materials. It was safer than asphaltum in an oil-glaze, and many artists in the 19th century liked it for water-colour. In general it should be obsolete today as the export of the mummies is forbidden, but examples of the colour can still be found.

Sepia: A semi-transparent warm brown pigment obtained from the ink sac of the cuttlefish. **It was used by the Romans and its greatest popularity was between 1780 and the end of the 19th century.** It was used not only for wash work but also as an ink for sketching.

Sienna: A native clay that contains iron and manganese. In the raw state it has the appearance of dark and rich yellow ochre. Burnt sienna is made by calcining or roasting the raw sienna in a furnace. **The two, raw and burnt siennas are amongst the most stable pigments on the painter's palette.**

Umber: A native earth similar to sienna but with a greater proportion of manganese. Burnt umber is made in a similar manner to burnt sienna. The burnt is considerably warmer than the raw which is of cool green-grey-brown tint.

Van Dyck brown: (also termed: Cassel earth, Cologne earth) A native earth composed of clay, decomposed vegetable matter, iron oxide and bitumen. In colour it has a black-brown appearance and is a bad drier in oils and it will fade in water-colour.

PURPLES

Archil: A dye obtained from various lichens that was then treated with an alkali to develop a violet tint.

Tyrian purple: The famous imperial purple dye of the Romans; it was also used by the Greeks. The dye was prepared from the shellfish *Murex trunculis* and *Murex brandaris.* In 1908 Friedländer found the colouring matter of this ancient dye was the same chemically as purple coal tar that had been introduced in 1904.

GREYS

Davy's grey: A weak pigment made from powdered slate, that had some popularity in the late 19th century and the beginning of the 20th.

Neutral tint: A prepared artist's colour made up from lampblack, Winsor blue and a little alizarin crimson. Popular for monochrome work or rendered drawings.

Payne's grey: Another prepared colour similar to the above. This time it is mixed from alizarin crimson, lampblack, Prussian blue and French ultramarine for water-colour. For oil it is mixed from Davy's grey, lampblack, mars red and French ultramarine. Main uses are as for neutral tint.

BLACKS

Black lead:
An obsolete name for graphite, which was also known as plumbago, before the composition of graphite was known. The names continued up till the early part of the 19th century.

Graphite itself is an allotropic form of pure carbon that is greyish black, crystalline and greasy. These characteristics limit its use as a pigment but make it ideal for pencils or drawing-sticks. Graphite was first mined in England at Borrowdale in 1664. It is mixed with clay for use in pencils, a discovery that seems to have been made simultaneously in 1795 by Nicolas Jacques Conté in France and Joseph Hardmuth in Austria.

Ivory black: The vast majority of this pigment is made by charring bones. A very fine quality is still made by charring ivory chips. Equally satisfactory in oil- or water-colour. **Mixed with cadmium**

yellow it will produce a range of luscious, strong, bright greens.

Lampblack: Pure carbon pigment, a fluffy light powder collected from burning oils and fats. The most widely used pigment in the black group. It is not so intense or velvety as ivory black. **It has been used since the earliest periods.**

Vine black: (also termed: grape black, mare black and yeast black) A somewhat impure carbon pigment with a slight bluish undertone that is made by burning selected vegetable materials, such as grape vines and other such substances.

Supports
(including 'grounds')

ACADEMY BOARD
An economic board for oil-painting. It is made from several sheets of paper sized together. The face is then primed with a ground of white lead, chalk and oil. The back was often painted grey. **It had a considerable vogue in the late 19th century.**

ALTARPIECE
A decorated screen, panel or series of panels, fixed or movable, placed on or behind the altar. Normally it would carry paintings or reliefs. **Two hinged panels comprise a diptych, three a triptych, five or more being a polyptych.** A fine example of a polyptych is the 'Adoration of the Lamb' in the Cathedral of St Bavon, Ghent, Belgium. It is by the two brothers Jan and Hubert van Eyck. At Chatsworth there is the 'Donne Triptych' by Hans Memlinc. The term diptych was originally applied to the Roman codex or book of two leaves or tablets fastened together.

BRISTOL BOARD
A stiff durable ply-produced cardboard suitable for pen and ink work or water-colour and gouache.

CANVAS BOARD
A heavy cardboard with a cotton or linen canvas glued to one side, with the edges folded over to the back. The face is primed in the same manner as an Academy board.

CASSONE
An Italian word for the marriage coffer. In the Renaissance period it was the fashion to have painted *cassoni*. Florence led with this vogue, and artists who decorated them included Botticelli and Uccello. The craze ceased towards the end of the 15th century when it was replaced by carved oaken chests.

CERAMICS
Small ceramic platters, round, oval, square and rectangular have been used by some painters, either working on miniatures or pictures up to a maximum over all of about 12 in (300 mm). The top measurement is because of the frailty of the material. The painting could be carried out in oils, water-colour, gouache, tempera, or with more modern media such as casein, acrylics and alkyds.

COQUILLE BOARD
An illustration board intended for the commercial artist. The working face has a shallow dotted, stippled or textured embossing. When this is drawn upon with crayon or pencil a type of half-tone is produced.

FASHION BOARD
A heavy laminated card with a white quality paper face that may be finished rough, 'not' or hot pressed.

GESSO
In the broad sense it is a mixture of a plaster or like substance and a glue. Its purpose was to present the painter with a smooth, hard, white ground on which to paint. Owing to its hard brittle nature it could not be applied successfully to canvas or metal sheets. It was from the start intended for application to wood panels.

    The method was to first either size the wood panel or to size down coarse muslin or linen. When this was dry the first coat of the gesso would be put on, the coarse *gesso grosso*. Two or three days later would be put on the fine *gesso sottile*, and nearly always there would be a number of coats of the latter. When the gesso had hardened it could be smoothed flat, if there were any imperfections, with a block of pumice. The resultant surface would have an ivory smoothness and hardness. The earliest type of gesso made in medieval times used parchment glue and well-slaked plaster of Paris. The curds from long-soured milk were also used in place of the glue. Later recipes included rabbit-skin glue and precipitated chalk and whiting. The gesso could also receive the imprint of tools with decorative gilding, and be coated over mouldings or other decorations included with the panel.

GLASS
The support for back painting. It is important that it should be reasonably stout; plate glass is best, and it should be of a good quality so that there will be no distortions in the glass to falsify the image created.

GOLD GROUND
Many of the painters of the 15th and 16th centuries used grounds either covered or partially covered

with gold-leaf. The underlying ground would be gesso. On top of this would be brushed a thin coating of a red earth, a bole, often the one termed Armenian. Next an adhesive was put down, this was often glair, which was egg white with a little water. Then the fragile thin gold-leaf would be picked up on a wide, soft, hair brush called a tip, and laid in position. Each sheet was normally given a lap of about $\frac{1}{4}$ in (0·5 cm) over its neighbours. Before it was totally hardened, decorative work with patterned iron or steel dies could be done. When completely hardened an agate burnisher could bring up the desired sheen.

The most suitable medium for working on the gold was tempera; oil could be used but with this there could arise adhesion problems. A number of the German school were particularly adept with the handling of this method. Stefan Lochner left some exquisite religious scenes, with his strong blues and deep crimsons reacting against the gold in a splendid manner.

Another method of exploiting the richness of gold in a painting was worked through by Botticelli, notably with his 'Birth of Venus' when he brushed powdered gold into the hair of the figures.

The extreme fragility of the gold-leaf is exemplified by Cennini when in *Il Libro dell' Arte* **he records that 145 leaves could be beaten from one ducat, enough to cover an area of about 70 sq ft (6 m²).**

GROUND
The name that is applied to the coating of the surface on which the painting is to be carried out. Thus gesso is the ground for a wooden panel. A canvas is given a ground by sizing and then priming. Painting surfaces such as water-colour papers, boards and parchment act as ground and support at the same time.

HARDBOARD
(also termed: Beaverboard, Masonite, Upson board) These boards are made from wood-pulp and/or waste paper. The front presents a smooth hard surface, the back having a textured tooth resembling a reverse canvas texture. Suitable for oils if sized and primed, also for acrylics if grounded with acrylic primer. Sizes over 20 × 24 in (508 × 600 mm) need some kind of battening for support.

IVORY
Sheets of ivory about $\frac{1}{16}$ in (1·5 mm) thick or less are considered the standard support for the miniaturist. Other substances that have been used include: ceramic platters, various cards, parchment, and at times such as stretched and treated chicken skin.

LEATHER
Not a happy support for oils as it is a substance that is open to deterioration from a number of sources. It is difficult to control the movement of the material. There have been some instances of it being used by French painters for small-size pictures. There is a ceremonial parade shield of leather in the National Gallery of Art, Washington, DC, which carries a painting of 'The Young David' by Andrea del Castagno; it is 45 in (1143 mm) high and tapers being 32 in (812 mm) wide at the top and 17 in (431 mm) at the bottom.

METAL
Copper sheets have been used primarily, although works have been painted on aluminium, iron, steel and zinc. Media suitable are acrylics, alkyds and oils. The metal sheets should be de-greased and then given some kind of grounding. Paintings on metal are susceptible to damage by temperature change and if the sheets are thin, by careless handling.

PAPER
A substance produced from wood-pulp, rags or other material with fibres. **It is thought that the art of making paper started in China with Tsai-lun about AD 105.** It is likely that the invention was carried from the Far East by the Turks during the Dark Ages. It first appears in Europe in Spain, and Italy during the 12th century.

The varying methods used by artists have called for a large number of different papers. Broadly they can be divided into two categories: cartridge and rag. The former is made from wood-pulp and is generally used for schools, or work in a draughtsman's office. Rag papers are produced from good-quality cotton or linen rags and are finished to three surfaces. 'Hot pressed', with a smooth slightly shiny surface; 'Not' with a matt and 'Rough' with a quite coarse-grained, rough appearance. The papers beside being white can have tints ranging from black right across the palette to pale creams and greys. Pastel-workers often choose coloured papers. There are also heavy textured papers such as Cox, de Wint, Ingres, Turner and specialized makes including Montgolfier, Carson and Hodgkinson. Papers for printing on can be made from mulberry, and with some of the Eastern types, grasses and reeds are used. Papers will take all media except oils and alkyds without further treatment, for the two latter some form of isolation and priming should be used. Hans Holbein was a painter who used treated paper for preliminary oil-sketches.

PAPYRUS
A form of paper made by the early Egyptians. It was made from the reed *Cyperus papyrus*; strips of the reed were laid over each other, then they were soaked with water and pounded, lastly being dried in the Sun.

PARCHMENT
Animal skins that have been treated by scraping, use of lime to remove hair, and rubbing. Skins of sheep, pigs, goats have been used. For vellum those of young calves or still-born lambs are favoured. **Pliny the Elder claims that parchment was discovered by Eumenes II (197–159 BC) of Pergamum. It remained the principal support for writing on until the advent of paper in the 12th century.**

SCRAPERBOARD
A cardboard that is covered with plaster, white clay or chalk mixed with glue. It may be left white or black. It was introduced during the 19th century. The artist can work on it with nib-like scraperboard tools to simulate engraving; pen and ink may be also added. It is largely used for commercial work, as from it excellent line-blocks can be made.

SPIDER'S WEB
Some painters have chosen strange materials to paint on, often regardless of permanency and suitability. One of the strangest examples is in Chester Cathedral, England, where there is a small religious picture painted on a woven material made from spider's web.

STRETCHER
The wooden frame that is used to strain a canvas when preparing it for painting on. The four corners are mitred in such a way that wedges can be driven into them to increase the tension on the canvas.

TAPA
A Polynesian word meaning bark-cloth. It is made by taking the barks of various trees, including breadfruit, fig, mulberry, etc. and pounding them together. The masters of making tapa are the natives of Oceania, although it is prepared in South America, Indonesia and tropical Africa. It is painted on but more often decorated by block-printing.

TEXTILES
Woven fabrics that have been and are used for painting on include: linen and cotton canvas, cambric, silk, hessian, sailcloth, sacking and synthetics. They all call for some form of grounding. Generally this will consist of isolation of the fabric by sizing and then priming with an oil, glue or acrylic-bound primer.

It is not certain why the painters of the Renaissance started to forsake panels for canvas. The trend may have started with the painting of banners and so-called Flemish cloths, substitute hangings for tapestries.

Some of the Egyptian mummy portraits were done on a canvas and then glued to a board, a method that is called marouflage. Traces of early canvas-painting have been found in Pre-Columbian art of South America. Theophilus records the use of canvas and the authoritative Pliny the Elder writes that Nero in a moment of excessive aggrandizement commissioned a portrait that was to be on canvas and to be 120 ft (36·57 m) high.

TONDO
A circular panel, plaque, relief or stretched canvas (from Latin *rotundus*: round).

WALLS
A painting made directly on a wall or a ceiling is termed a mural. The surface of the wall has to be carefully prepared so that the paints will have the maximum chance of adhering. For fresco-painting this implies special plastering. (See in detail under Fresco in Methods section.) For oil-painting the wall will need to be thoroughly isolated and then given an adequate lead priming. For acrylics the surface should be treated with acrylic medium and water to consolidate the plaster. Large oil-paintings on canvas are sometimes marouflaged (attached) on to walls using a variety of adhesives; the traditional one being white lead and oil, with sometimes a little dammar varnish to increase initial tackiness. Several of the modern synthetics have been used, although they may have the drawback that they are not reversible if it is necessary to move the painting.

WOODEN PANELS
Up till the 15th century and the coming of canvas nearly all the portable paintings in Europe were executed on wooden panels. The Flemish and the French painters preferred oak, the Italians white poplar, the Germans pine. But all used the woods of other trees including: beech, cedar, chestnut, fir, larch, linden, mahogany, olive and walnut.

Large panels would be made with several planks supported not only by gluing together but also with battens at the back or some system of cradling. All panels would have some form of grounding (see gesso) and isolation if they were to be used for oils. Tempera can be painted straight on to the gesso.

As a side-look at economies for the early artists, it is interesting to note that during the restoration of the flood-damaged paintings in Florence, direct evidence came to light that some of the panels must have been made up with timbers that were worm-eaten at the time the pictures were painted.

Tools

AIRBRUSH

An implement that resembles a thick fountain-pen and which has a small container near the nozzle. By air pressure supplied from a container or a mechanical compressor, varnish, fixative or colours can be applied. **It can produce effects from fine lines up to wide sweeps.**

BRUSHES

The first known examples are probably those used in Egypt which were simple bundles of thin reeds bound to a handle; the British Museum has one of these and its date is put at about 1900 BC. Since that time many strange hairs and bristles have been used. Apart from attempts to use human hair; at least the following animals have been tried: **horse, cow, ox, black sable, kolinsky, weasel, squirrel, ring-cat, skunk, civet, fitch, badger, pony, goat, bear, hog bristle from China, India, Poland, France and the Balkans**; and from the sea the Blue, Fin, Sei and Humpback whales have contributed baleen. Plant fibres from Agave, Yucca, Sisal, Bahia, Gumati, Palmetto and Hickory splits have also been used.

Brush and ink drawing of a stag beetle by Albrecht Dürer (1505). (Courtesy Sotheby Parke Bernet)

Broadly stated, hair brushes are for water-colour, gouache, miniature work, inks, tempera while the hog bristle is for oils and acrylics.

Brush shapes that can apply to both hog and hair are: round, bright, flat, filbert, sword, rigger, fan or sweetener, mop.

In the 18th century small sable or other hair brushes generally set in quills were termed pencils.

BURNISHER

An instrument to polish either a metal surface or other substance that will take it. It is either shaped from hardened steel or the semi-precious stone, agate. In the 15th century Cennini in *Il Libro dell' Arte* mentions using a piece of hematite.

CAMERA LUCIDA

An optical device which, by the use of a prism, makes it possible to copy an object. The rays of light from the model are reflected by the prism and produce an image on the paper. By adjusting the prism and inserting magnifying lenses the size of reproduction can be made smaller or larger. It was invented by Richard Hooke about 1674.

CAMERA OBSCURA

Another optical copying device which is much larger than the lucida. It relies on the principle that rays of light will pass through a small lens in the side of a darkened cabinet and then be projected either straight or by reflection from a mirror on to a piece of paper. The principle was first noted by Aristotle. Early astronomers found it helpful to use an obscura to observe the stars; it is described in this connection by the Arabian Alhazen at the beginning of the 11th century. The artist's obscura was probably first constructed by Leone Battista Alberti (1404–72).

Giovanni Battista della Porta is likely to have been the first to make a written account of its use for drawing, in his *Magia Naturalis* (1558).

CLAUDE GLASS

A small convex mirror that instead of being silvered was blackened at the back. The idea was that being convex it would reduce the scene and by being blackened it would only reflect the main masses of the subject. The artist would sit with his back to the view and hold the glass in front of him so that he could look over his shoulder. It is said to have been devised by Claude Lorrain (1600–82). It was popular in the 17th and 18th centuries and may still be seen in action today.

DRAWING-FRAME

A rectangular frame crossed with wires or threads to form squares, which the artist sets up between himself and his sitter at such a distance that his view is the same as the drawing he intends to

Brush shapes – left to right – round, flat, filbert, sword or cut, rigger and sweetener or fan

make. The frame isolates his subject and if his paper is squared to correspond to the squares on the frame, he can quickly place the main outlines and details. Leonardo describes such a frame; Albrecht Dürer made a woodcut of one being used; Van Gogh wrote to Theo his brother enclosing a self-sketch he had done using one.

EASEL
A wooden or metal stand for holding a canvas, a panel or a drawing-board. It may range from a small, light, tripod sketching-easel up to a large studio easel which will take canvases up to 12 ft (3·65 m) high and which can be raised, lowered and canted by worm-gears and winding handles or wheels. There are also small easels for resting on a table that will allow a drawing-board to be almost vertical or gently sloping as for water-colour wash work.

ÉCHOPPE
A needle that has had its point bevelled to an oval facet that can be used in etching and engraving. It will make lines of varying thicknesses, and with

Burin for metal engraving

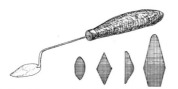

Painting knife with examples of further blade shapes

engraving can be used to rework and expand certain lines. A favourite tool of the 17th-century graphic artists such as Jacques Callot (1592/3–1635).

GRAVER
(also termed: Burin) A hard steel instrument for metal- or wood-engraving. The section of the cutting tool can be lozenge, diamond or rect-angular. The tool is set in a small wooden handle designed to fit into the somewhat unusual grip the artist adopts, and also the handle is flattened underneath to allow the graver to be lifted from a line.

LAY FIGURE
A jointed wooden figure, either quite small or life-size, that may be used as a substitute for the sitter. The figure is so made that the limbs can only be moved in the same way as an actual human figure. **Popular 18th-century portrait-painters used them dressed in the clothes the sitter demanded and thus saved the clients many arduous hours sitting still.**

MAHLSTICK
A long wooden rod with a pad at one end that is used by the painter to steady his hand when working on fine details. He holds the mahlstick in his left hand and lays the pad on the canvas and then rests his right with the brush on the stick.

MULTIPLE TINT TOOL
A tool used particularly in wood-engraving with a thick rectangular rod which is so made that it will cut up to five or six lines at a time. It can be used for cross-hatching or pecking textures.

PAINTING- AND PALETTE-KNIVES
Both of these are made of fine tempered steel that is flexible. The palette-knife has a straight handle and is intended for mixing colours on the palette or for cleaning it. The painting-knives have cranked handles to keep the fingers clear from the painted surface; they also have a wide variety of shape ranging from small trowels to long spatulas.

PALETTE
(1) The instrument the artist mixes his colours on. This may be a traditional mahogany, or other wood, as a rectangular shape or 'hook' or balanced studio. Artists also use metal and ceramic palettes, glass-topped tables, and for outside work with oils there are disposable greaseproof-paper blocks available, which allow a sheet to be torn off and discarded with the colour remnants.

(2) The selection of colours that the artist uses. In general the early masters used fewer colours than the painters of this century. Partly this can be explained by the fact that the chemist

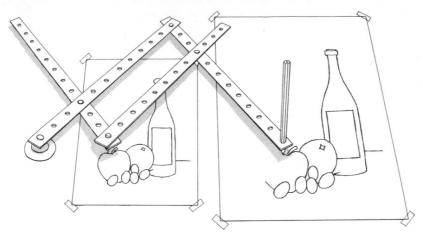

A pantograph that can be used for enlarging or shrinking a drawing

has provided a far greater selection for today's painter; but also the Renaissance masters and those around them normally employed a well-thought-out scheme of underpainting that gave greater scope to the pigments applied on top.

PANTOGRAPH
An instrument for reducing or enlarging designs or sketches, that uses a simple system of levers; known since the 17th century.

PEN
The English word, the French equivalent, *plume*, and the German, *Feder*, originally meant a wing-feather. St Isidore of Seville in the 7th century writes about a quill-pen. The hand-cut quill, from birds such as geese, swans and turkeys, was the principal drawing instrument for use with ink until the 19th century. In 1809 Joseph Bramah patented a machine for cutting up a quill into separate nibs. In 1818 Charles Watt patented a process for gilding quills, which could be regarded as the forerunner of the gold nib. In 1822 J I Hawkins and S Mordan patented a method for making nibs from horn and tortoise-shell, the points being made long lasting by attaching small pieces of diamond or ruby. Steel nibs of various trial types, successful and not, started to appear late in the 18th century and by the mid 19th had taken over.

The earliest example of a metallic pen was found at Pompeii and is now in Naples Museum. Many Oriental artists have in the past and through till today used pens cut from thin bamboos.

PENCIL
(see Brushes)
Graphite wooden jacketed pencils as are known

today date from the end of the 17th century when there was a prosperous British business in the north country and when, as Sir John Pettus commented, 'Black Lead . . . is curiously formed into cases of Deal or Cedar, and so sold in Cases as dry Pencils.' On the Continent they were nicknamed *crayons d'Angleterre*.

It was Nicolas Jacques Conté (1755–1805), the French inventor, who worked out the process of mixing a clay with the graphite to give a selective range of hard and soft pencils. An account of his inventions is given in Jomard's *Conté, sa vie et ses travaux* (1852).

ROCKER
A broad-ended tool with a sharp-toothed curved base which is used to ground a mezzotint plate. It is so called because it is applied with a rocking motion.

ROULETTE
A small-toothed wheel set in a handle that can be used for working on a metal plate. A second use is for helping to transfer a large cartoon to a wall, a canvas or a panel. The roulette is run along the main lines on the cartoon, laying a series of tiny holes. When the cartoon is placed in position the artist dabs along the lines with some powdered graphite or charcoal wrapped up in muslin.

SCORPER
A solid metal tool with a square or rounded end used for clearing out non-printing areas on a wood-block.

SCRAPER
An etching and engraving tool, triangular in section, which is used to remove burrs or unwanted roughnesses, also to slightly lower areas for subtle grading.

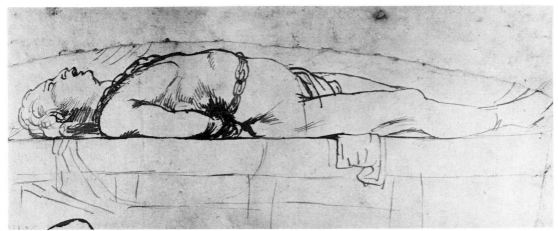

Pen and ink drawing of St Lawrence chained to the gridiron by Van Dyck. (Courtesy Christie's)

STRETCHING PLIERS
Heavy pliers with a wide mouth for gripping the canvas when it is being stretched, and so assisting in giving an equal tension as the holding tacks are being driven into the stretcher frame.

TORTILLON
A paper stump made of rolled-up blotting-paper or soft thick paper, which is used with pastel, chalk or charcoal for blending. It may also be used for mopping small areas with water-colour.

Additional Materials

ALKYD RESINS
The vehicle for the latest colours to be prepared for the painter. In 1927 a substance was made from acids and alcohols and it went under the name of Alcid, a title coined by Kienle. From this word comes the name for a range of paints introduced by Winsor and Newton in 1976.

ARMENIAN BOLE
A rich, fine, red clay used as a ground on a gesso panel for gold-leaf. The strong colour serving to enrich the optical effect of the very thin metal.

BEESWAX
Has many uses in art, including: mixed with turpentine to make a wax polish for finishing oils, tempera and alkyds; mixed with varnish and turpentine to prepare a painting medium for oils; as a stiff paste with a small amount of turpentine to assist impasto; mixed with Venice turpentine and resin as an adhesive for relining a painting.

BINDER
The cementing ingredient of a paint vehicle, its purpose being to hold the pigment particles in a cohesive coating. It can also describe the gum that holds pastels, water-colour and inks.

BLADDER
From the mid 17th century artists' pigments when mixed with oil were stored in small bladders. To use them the painter made a small hole with a tack, squeezed out some colour then pushed the tack back into the hole. Towards the evolution of the tube, the bladder was followed by a form of syringe. **In 1840 the collapsible tube came into being.**

CASEIN
A milk protein used as a binder for casein colours. It is prepared by drying the curd from sour milk, then grinding it into a yellowish powder. Casein is only water-soluble in the presence of an alkali such as ammonia, thus casein paints once dry are waterproof. **A type of milk curd glue was used**

'The Mocking of Christ' by Rembrandt. A pen and brown ink drawing. (Courtesy Sotheby Parke Bernet)

by the Egyptians, Greeks and Romans. It has also served as an adhesive for joining the planks of a panel.

DILUENT
Any liquid that will dilute or thin a substance, as opposed to dissolving it.

DRIERS
Substances that are added to oil-paints to hasten the drying. The idea is, if possible, to make all the colours dry at an even speed. Quick-drying pigments include: umbers, siennas, ochres and flake white; slow-drying are such as alizarin crimson, ivory black and vermilion.

EAR-WAX
An occasional additive to some lake colours to improve their flow, an idea of the the late 17th century.

FIXATIVE
A liquid, that may be shellac in methylated spirits or synthetic cellulose solution, that is intended to be sprayed as a fine mist on to charcoal, soft pencil, chalk or pastel to consolidate the drawings. This spraying must be done with care because too heavy an application can flood and float the drawing away. Tests should always be made with pastel as it is likely the fixative can alter tone and tint values.

GUMS
The principal binder for water-colour is gum arabic, it comes from certain acacia trees growing in Africa, Asia and Australia.

Gum tragacanth is used as a binder for chalks and pastels; it is procured from a shrub *Astragalus*, a native of Asia Minor.

MEDIUM
The method in which an artist works; oil-painting, gouache, pastel, pen and ink, etching, collage, sculpture, etc., are all media for his expression. **In another sense medium may be used to describe an additive to the colours when painting**, linseed to oil-paints, egg yolk to tempera, gum to water-colour.

MEGILP
(also termed: McGuilp, magilp) An 18th-century oil-painting medium, a mixture of linseed oil, mastic varnish and lead driers. It is a jelly-like substance slightly cloudy and yellow. It does impart an ease of working to the colours, but it is liable to make the paint film brittle and cause heavy cracking.

OILS
Painters have used an extraordinary variety of oils in their efforts to attain the perfect personal paint

'Jason Sovereign Bratby' by John Bratby; an acrylic painting using heavy impasto. (Courtesy George Rowney)

consistency and working quality. **The chief oil for oil-paints today is linseed**, although there might be additions of poppy oil if it was desired to slow the rate of drying. In history such as these below have been experimented with, sometimes with injurious effects to the finished painting: walnut, sunflower, hempseed, safflower, rosemary, cloves, pine, poppy, spike and tung.

SICCATIVE

A substance added to oil-colours to considerably hasten their speed of drying. Faster than driers, it is intended as an accelerating agent only, not as an equalizer across the whole palette.

VARNISHES

Protective coatings for oil-paintings, tempera, acrylic, alkyd, gouache and water-colour. Varnishes may be made from natural or synthetic resins, with additions of natural or synthetic waxes to lower the gloss or induce a desired sheen. Desirable properties include, fairly quick drying, some plasticity, resistance to cracking, blooming and yellowing.

The natural resin varnishes that have been used for centuries include: mastic, copal and dammar. In general the modern synthetic resin varnishes behave far better than the natural resins, not producing unpleasant optical appearances or impermanent features. The notorious 'gallery varnish', beloved by dealers and at times the Academy, was a brown deep copal which would impart to a picture a feeling of 'age and respectability' but at the same time make it almost unrecognizable. **John Constable was one who suffered at the hands of those who liked to brush this disfiguring 'treacle' over paintings.**

WETTING AGENT

A liquid to be added in small amounts to water-colour to reduce the surface tension and thus increase the flow of the colours. Ox-gall has been the traditional agent, but now synthetic preparations akin to detergents have been introduced.

Painting, Drawing and various Picture-Making Methods

ACRYLICS

Pigments dispersed with acrylic resin (synthetic resins made by polymerization of acrylic acid esters). **A medium for painting introduced during the early 1960s.** It offers considerable freedom to the painter. Almost any support can be used, and only needs a single coat of acrylic primer. The colours can be put on with an impasto of upwards of $\frac{1}{2}$ in (12·5 mm) without danger of flaking or cracking. The acrylics can be diluted with water to simulate wash work. They dry out quickly and may be varnished or not as desired.

ALKYDS

(see previous section)
These recently introduced colours act as an extension to oil-painting. They have a uniform speed of drying. They may be used for under-painting, and are excellent with glazing over dried-out oil films. As a painting medium by themselves they do not retain brush marks and impasto to quite the extent of oils; but these characteristics may well suit some manners.

ALLA PRIMA

To paint a picture in one sitting, particularly applicable to oil-painting. The French use the term *au premier coup*. **It is the wisest method where heavy impasto is to be used.** The paintings often have a virile life and freshness of colour and effect, not always attained by more precisely planned methods.

AQUARELLE

(see Water-colour)

BACK GLASS PAINTING

Painting pictures on the back of sheets of glass. With this manner the artist has to work his picture backwards, starting with what would be the finishing strokes with the conventional method. Such a painting has high permanence and brilliance, as the colours are sealed behind the glass. A sheet of tinfoil is generally applied to the back.

BARK PAINTING

(see Tapa under supports)

BLEEDING

Describes the action of one colour running into another. Most applicable to water-colour, where a second or third colour can be dropped on to an already applied wash while it is wet. To a certain extent the result is uncontrollable, but a wise hand will be able to judge approximately. Bleeding in oil-colours is associated with pigments such as asphaltum that can mix with other colours after application and drastically affect the optical and physical qualities of a painting.

The First Sketch made with Moist Water Colours, carried out in 1835 by Mr Henry Charles Newton. (Courtesy Winsor and Newton)

'Self-portrait 1937' by Henri Matisse. A charcoal drawing. (Courtesy The Baltimore Museum of Art, Cone Collection.)

BLENDING

A term concerned mostly with oils, acrylics or alkyds. It implies the softening of hard edges between colours, and the artist would be likely to use a fan brush or the tip of a finger.

BLOOM

A phenomenon that occurs with varnish on paintings, and occasionally on polished furniture. Causes can include damp conditions during varnishing, picture hung in a chilly, draughty position, or exposed to gross humidity such as can be generated by some gas heating devices. The condition appears rather like the bloom on a black grape. If it is on the surface of the varnish it can normally be removed by gentle wiping with a piece of cotton wool. If underneath the varnish, which is rare, the only cure is to remove the varnish and to revarnish.

BLOT WORK

A manner worked on by Alexander Cozens, which is elaborated on in *A New Method for assisting the invention in drawing original compositions*

of *Landscape* (1786). The idea is that an accidental blot or brush mark on the paper can act as a trigger for an imaginative composition. Leonardo makes mention of a similar approach to marks on walls that could be worked in with a painting. Twentieth-century Surrealists have experimented with the child ploy of folding paper over a blot of colour to produce a fantastic shape from which some idea could grow.

BODY COLOUR
Descriptive of opaque colours as opposed to transparent.

A charcoal drawing heightened with white chalk, by Louis Le Nain. (Courtesy Christie's)

A modern water-colour grinding mill for bulk production. (Courtesy Winsor and Newton)

A stage in the making of a top-grade sable brush. (Courtesy Winsor and Newton)

Page from Psalter, all in Latin, illuminated on vellum. Paris c 1220–40 (7 × 5 in (183 × 132 mm)).
(Courtesy Sotheby Parke Bernet)

CABINET PICTURES
An old-fashioned name for small easel paintings.

CARNATION
An obsolete term which described the rosy pink, flesh colour of a female portrait.

CASEIN COLOURS
Paints produced by mixing the pigments with curd, a casein milk protein. They may be used on paper, cards, hardboard and walls, but not on a flexible support such as canvas; for the reason that the colours dry brittle they should not be applied too thickly. During the 1920s and 1930s American painters built up a considerable tradition with the manner. The dry finish resembles true tempera. It is best left unvarnished or waxed as such treatments can alter the tone values.

CHALKS
Sticks of prepared calcium carbonate left white and either used as a drawing material on a dark-tinted paper, or for heightening a wash or pen and ink drawing.

CHARCOAL
One of the oldest drawing materials, charred sticks were used with the early cave-paintings. The Romans used them and throughout the history of art the material crops up again and again. It was often the medium for preliminary drawings. Various types of wood produce different characteristics; willow and beech tend to produce brittle sticks, vine twigs the softest and blackest. The charcoal can be applied to the paper

Graphite, crayon and heightened drawing by François Boucher. (Courtesy Christie's)

in a direct manner, and then manipulated with a tortillon, a hog brush, a fingertip, a piece of rag or a plastic rubber. The drawings need to be fixed when finished.

CLEAVAGE
It implies that the adhesion of paint layers in a picture has failed.

COLLAGE
A method of picture-making which incorporates a wide variety of materials and often a certain degree of relief besides actual painting. Papers, cards, textiles, wood fragments, fur, small stones, metal foil, etc., can be used. The manner evolved from *papiers collés*, a 19th-century leisure pastime. The Surrealist Kurt Schwitters did considerable work in this manner early in this century. Max Ernst was another attracted to the freedom of the technique.

CONTÉ CRAYON
Introduced by Nicholas Jacques Conté, they are sticks of compressed compound of binder and pigments; the colours being sanguine, sepia and black. They are grease-free and can produce very sensitive work.

CROSS-HATCHING
A technique for making depths of tone in pen and ink and pencil drawings, also in etching and

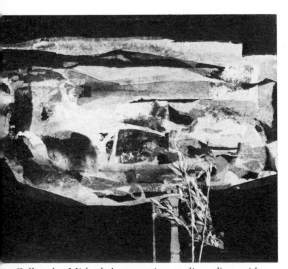

Collage by Michael Ayrton, using acrylic medium with many layers of tissue paper, smoking from tapers, and oatheads. (Courtesy George Rowney)

engraving. Regular lines are drawn in series, first one way and then across each other. The manner can also be used with wood-engraving to obtain light tones by the use of the graver in a similar manner.

DRY BRUSH
The brush should be loaded with the minimum of colour and then lightly dragged over the surface of the canvas or paper. A bright or flat brush will give the best results.

ÉBAUCHE
In oil-painting it signifies **the first underpainting**. It should be low in oil content to enable subsequent layers of colour to adhere properly. In oil-paintings early layers of colour should always be lean; if they are too rich and thick paint is put on top, varying speeds of drying will almost certainly cause cracking.

ENCAUSTIC
One of the oldest methods of painting, being practised from at least 3000 BC. Some of the finest existing examples are the mummy portraits from Fayum executed about the 3rd century AD. The colours are applied to the support with hot beeswax, either with spatulas or brushes, finally being driven in with a heavy hot iron. The method was more or less abandoned in the 9th century. Count Caylus, a French archaeologist

Mummy painting from the Hawara Roman period, c AD 100/250. (Courtesy National Museum of Ireland)

and engraver, sought to revive interest in the 17th century but failed. A number of scenes were painted in the Residenz at Munich in 1831 by Julius Schnorr von Carolsfeld. The invention of electrically heated spatulas has brought a slight revival of interest, but it is a laboursome and awkward technique at the best.

FRESCO
(see Walls)
An art started by Minoan and other early civilizations. In antiquity they had the idea of painting fairly small portable frescoes. Some of these have been found in Crete and date from about 1000 BC.

Frescoes are painted with pigments which have been ground in water and which are then applied directly on to a freshly plastered wall, while **still moist, this method is known as *buon fresco*. When it is painted on to dried-out plaster it is termed** *fresco secco*. This preliminary drawing is done on the under plaster, known as the *arriccio*. It is usually brushed in with a mixture of reddish brown clay and water and is termed the *sinopia*. In some cases, if it is an elaborate design, a cartoon may be prepared and then transferred (see Roulette). It is then worked out as to how large an area the artist can paint in a day and a top layer of plaster, the *intonaco*, is applied. Into this the artist has to work directly and without mistakes.

Fresco secco can be carried out with tempera, glue or casein colours. Before making a start the wall should be well soaked with lime-water.

The Renaissance produced a host of the world's greatest fresco-painters. It started with Giotto with such as his Arena Chapel in Padua and went on with the likes of Masaccio, for instance his work in the Brancacci Chapel, Florence, as well as Raphael's 'The School of Athens' at the Vatican and reached its peak with the stupendous ceiling in the Sistine Chapel by Michelangelo.

FROTTAGE
The process of making rubbings through paper of objects or textures underneath. Brass-rubbing is frottage. Max Ernst was one Surrealist who explored the idea, he was seeking to find some visual stimulus for his subconscious. In his 'The origin of the pendulum' it can be noted how he has rubbed rough boards for parts of the design.

GLAZE
Applied to painting media, the term glazing means the **laying of a transparent colour over previously laid and dried-out pigments, that may be opaque or transparent**. With water-glazing only water need be added to the colours, with acrylics just the acrylic medium and water. For glazing with oil-paints, the diluent can be such

A sinopia, under-drawing for buon fresco. (Photo by the Author)

'Flatford Mill from the lock' by John Constable. Oil sketch, pochade (10 × 12 in (254 × 305 mm)). (Courtesy Christie's)

Egyptian Middle Kingdom faience hippopotamus, c 1900 BC (9 in (228 mm)) long. Sold in 1977 for £26 000. (Courtesy Christie's)

The god's father and priest of Anum in Karnak. Set of Ii. Grey-green granite. Sold for £110 000 in 1977.
(Courtesy Christie's)

'Study of a Miner' by Josef Herman. The painting is built up by numerous superimposed applications of glazing. (Courtesy George Rowney)

'Self-portrait No. 2' by Leon Kossof, born in 1926 near St Paul's, London. The painting shows very heavy oil impasto sometimes up to a thickness of $\frac{1}{4}$ in (6 mm). (Courtesy Fischer Fine Arts Ltd)

as: linseed oil, poppy oil, turpentine or white spirit.

Glazed colours appear to advance while opaque recede. Very rich translucent effects can be gained; for example, note the extreme richness of crimson in some of Titian's paintings, obtained by glazing over these areas with lake.

GOUACHE

In a broad sense it is **a water-colour carried out with opaque or body colours instead of just transparent**. **The earliest signs of the method are traced back to the Egyptians**, when they bound their pigments with either gum tragacanth and/or honey. Dürer used the colours in his closely observed nature studies, and it was popular with the French, Italian and Swiss water-colourists who saw the possibilities of the attractive chalk-like finish that comes up when gouache dries. François Boucher became extremely adept at handling the medium and judging how it would dry out.

GRADING

The handling of a water-colour wash to give it a lightening or darkening effect as the colour flows down the paper. This is done by adding water to the bowl of colour or more colour.

GRANULATION

An effect that can be achieved with wash work when using colours with heavy pigment particles. French ultramarine, ivory black, umbers, siennas and ochres will leave a granulated broken effect if the washes are put on with the board lying nearly flat.

GRISAILLE

A type of monochrome painting executed in greys. The results often resemble sculpture. Excellent examples are St John the Baptist and St John the Evangelist which are on the backs of the two folding sections of the 'Adoration of the Lamb' by the Van Eych brothers in St Bavon Cathedral, Ghent.

HARD EDGE

A painting manner in which the component parts of the composition are crisply defined presenting a somewhat austere effect. A number of painters since the war have chosen this technique. It is more often geometric than representational.

HERALDRY

An art that dates back to the ancient custom of distinguishing nations, such as the Greeks and the

Romans, military leaders and important officers by a badge on their shields. Later came examples like the white horse of the Saxons and the leopards of William the Conqueror. These were not hereditary but just the owner's choice. Heraldry proper starts at the beginning of the 13th century. From the artist's point of view he has to work to a limited palette of special colours which are: Or – gold (yellow); Argent – silver (white); Azure – blue; Purpure – purple; Gules – red; Verte – green; Sable – black.

ILLUMINATION

The decoration of manuscripts, which may have started from the simple addition of minium to the script, the general part being written in black. From this grew quite extraordinary elaboration, fantastic interwoven strap patterns, decorative motifes, zoomorphic imagery, plant forms. miniature portraits of religious figures. **It was one of the most important arts of the Middle Ages.** Wherever there were monasteries the art seems to have been practised. The monastic scribe worked about six hours a day. After he had finished the work was proof-read. Then the sheets went to a rubricator who put in titles and headlines, then to the illuminator. The last worked miracles of miniature presentation with the materials at his command.

The oldest known illumination is an Egyptian papyrus, the 'Book of the Dead'. The Greeks and Romans produced some work, but very little survives. The Byzantine manuscripts contain some perfect examples. Fourteenth-century Persian editions of the Koran, exquisite delicate designs. Among the famed European manuscripts are the 'Book of Hours' of the Duc de Berry produced by the Limbourg brothers (1410–13), and 'The Book of Kells', 8th century, now in Trinity College Library, Dublin.

The manuscripts were worked on vellum, using not only colours, but also gold-leaf and other metals, tiny fragments of precious and semi-precious stones and raising paste.

IMPASTO

Colours applied heavily so that they show distinctly the brush- or knife-strokes.

IMPRIMATURA

A coat of colour that is applied over the priming.

Many painters dislike working directly on white so this thin, lean veil of a tint such as red ochre, burnt umber, cool green is brushed on.

INDIA INK

A very intense black ink containing lampblack with shellac or glue as a binder.

LIMNING

An obsolete term for drawing or painting.

MINIATURE

A small picture not normally larger than 6 in (150 mm) in any one direction. **The greatest schools of miniature-painting flourished in England during the 16th and 17th centuries.** The leaders were such as Nicholas Hilliard, the Olivers and John Hoskins. Portraits were nearly always mounted in elaborately worked gold lockets. Miniatures can be painted in oil, water-colour, gouache and tempera and the smallest brushes No. 000 are known as triple goose, and are made from fine sable hairs. Although Hilliard painted small heads that would fit in rings, **the smallest ever are by a Canadian, Gerard Legare of British Columbia, who manages to work on pin-heads with diameters from $\frac{1}{32}$ to $\frac{1}{4}$ in (0·8–6·3 mm).**

Brush first made by Winsor and Newton after the last war. It is in Series 7 and the size of the brush is 000 – sometimes called by artists 'a triple goose'. It is made from the very finest Kolinsky sable hairs, and the brushmaker estimates that there are between 150 and 200 individual hairs in each brush, depending upon the thickness from the animal concerned. The weight of the hair in one brush is 15 milligrams. The origin of the Series No. 7 is that Queen Victoria asked this firm to make a size 7 sable brush. This was produced with an ebony handle and a silver ferrule, and to commemorate the presentation they christened their top range of sable brushes Series 7.

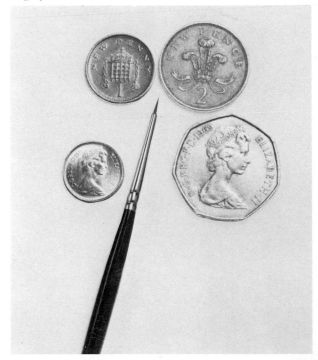

MIXED MEDIA

One or more medium used in the same picture. Thus pastel and ink, pastel and water-colour, tempera and water-colour, etc.

MONOCHROME

A picture painted in different tones of any one colour.

MONTAGE

A picture made up of portions of existing pictures, such as prints, photographs, so arranged that where they join they match or blend in with each other.

MOSAIC

A method of decorating floors, walls and ceilings with tiny fragments (tesserae) set into mastic plaster or cement. **It has a beginning in Crete and with the early Greeks.** The largest mosaic is on the walls of the Library of the Universidad Nacional Autónomao de Mexico. There are four walls, the two largest measuring 12 949 sq ft (1203 m²) which depict the Pre-Hispanic past. The largest in Britain is Roman and is the Wood-chester pavement, Gloucestershire of about AD 325. It was excavated in 1793 and measures 48 ft 10 in square (14·88 m²), and is made of 1½ million tesserae. It is kept covered with protective earth. The next showing, the eighth, will be in 1983.

MURALS

Paintings that are executed directly on to a wall. Media can include fresco (buon and secco), oils, tempera, casein and acrylics. In all cases the painter

Centre portion of a mosaic from Hinton St Mary, Dorset. (Courtesy Royal Commission on Historical Monuments (England))

must take great care to see that the wall is stable, the surface firm and that it has been prepared correctly for the chosen medium. Jacopo Robusti, nicknamed Tintoretto, painted the largest mural during the Renaissance. With the help of his son Domenico he produced 'Il Paradiso' on Wall 'E' of the Sala del Maggior Consiglio in the Palazzo Ducale (Doge's Palace) in Venice. It is 72 ft 2 in (22 m) long and 22 ft 11½ in (7 m) high and contains more than 100 figures. The largest painting in Britain is the great oval 'Triumph of Peace and Liberty' by Sir James Thornhill (1676–1734), on the ceiling of the Painted Hall in the Royal Naval College, Greenwich, Greater London. It measures 106 ft (32·3 m) by 51 ft (15·4 m) and took Thornhill 20 years (1707–27) to complete.

OIL-PAINTING

This technique was not suddenly invented; the story that accredits its invention to the Van Eyck brothers is incorrect, although they did much to help the evolution of the new medium. Previous to the 15th century the painter had to rely on fresco and tempera, both of which media, as beautiful as they are, lack the power to give the full richness and glow to the pigments. The exploratory steps of adding oil and varnish to egg tempera to raise a brighter, stronger palette were taken by such as Piero della Francesca (c 1410/20–92), Filippo Lippi (c 1406–69) and particularly Antonello da Messina (1430–79).

Today the colours are principally ground in linseed oil. Supports can be canvas, hardboard, wooden panels or prepared paper. Brushes are largely hog bristle as they have the strength to control the thick colours; painting-knives are also used for application. The technical procedure is always to start with a lean underpainting and then finish with richer thicker paint if desired. Heavy impasto and glazing can be employed for special passages. When completed and thoroughly dried through, a process which can take up to and more than twelve months, then a resin or wax varnish can be applied.

PASTELS

A method of painting or drawing with sticks of dry colour which have the minimum of binder; a reason why pastel pictures keep their bright fresh look almost indefinitely. The main danger for them is concussion, a sharp knock can cause particles of colour to fall off the paper. They are composed of pure pigment mixed with an inert filler such as kaolin with a minimal amount of gum tragacanth, casein or skimmed milk and are then formed into sticks by extrusion or a pressure mould.

The technique was an evolution from early

Part of the dining-room murals by Rex Whistler (1905–44) at 'Plas Newydd', Wales. (Courtesy The National Trust)

chalk-drawing. In the 18th century a number of French artists worked in the medium, including: Maurice Quentin de La Tour (1704–88), Jean-Baptiste Chardin (1699–1779), Jean Etienne Liotard (1702–89), and Edgar Degas (1834–1917) who exploited pastels to the full with his charming scenes from the ballet.

Pastels should be used on a paper that has some 'tooth' to grip and hold the pigment particles. It can be manipulated with a brush, a fingertip or a tortillon. Fixing a pastel is debatable by some artists as the fixative is liable to change the tones and tints.

PENTIMENTO

A reappearance of a design, a drawing or a picture that has been painted over; **It is a phenomenon particularly associated with oils**. It is caused by the medium or vehicle with the overpainting acquiring a higher refractive index and thus becoming more transparent. Some of the paintings by the 17th century Dutchman De Hooch are prone to this condition. He overpainted somewhat thinly, and black and white tiles can be seen ghosting through women's dresses and furniture and misty figures appear.

POCHADE

A rapid rough sketch of a landscape executed out-of-doors from nature; generally it is the intention that it should act as a guide for working up a larger, more finished picture.

SAND-PAINTING

(also termed: sand mosaic, sand altar, earth picture, ground-painting) A rather odd method of pictorial expression first practised by the North American Indians, especially the Navaho. The pictures are often up to 20 ft (6 m) across. They are prepared on the ground, sand being spread to a depth of about $\frac{1}{2}$ in (13 mm), smoothed out and then various coloured sands are sprinkled on this surface to form symbolic patterns which are often used in magic and religious rites.

Sand-painting does turn up in Tibet, Japan and with the Australian Aborigines.

SCUMBLE

The applying of an opaque or semi-opaque colour over an area of an oil-painting without completely obscuring the underpainting.

SFUMATO

Derived from the Italian word for smoked. It is a

'Brewery in Ireland'. A water-colour drawing by James Malton (c 1766–1803). (Courtesy Fine Arts Society, London)

Study of a German castle by Turner, showing broad treatment accented with detail and in the centre the use of abrasion. (Courtesy Christie's)

well-controlled and subtle method for graduation of tone; it leaves a soft hazy effect. **Leonardo used the manner most effectively with the 'Mona Lisa'.**

SGRAFFITO

Scratching or cutting through a layer of colour to expose the ground or support, or to bring up a second layer of colour underneath the last laid. The term was introduced by Otto von Falke in 1907.

SILHOUETTE

A small picture, often a profile of a head, a whole figure or some simple scene. The name was in memory of Monsieur Etienne de Silhouette (1709–67), an unpopular French Minister of Finance, whose extreme parsimony suggested his name for this somewhat economic and rigid manner for expression.

SILVER-POINT

A method of drawing that was popular during the 15th and 16th centuries. A silver-tipped stylus is used on a paper that has been given a slightly abrasive surface. One way to achieve this is to grind up calcined bones and mix them with a gum and water, and brush over the sheet of paper. When the drawing is actually being made the silver point leaves only a very faint grey line; later with exposure to the air the minute particles of deposited silver tarnish leaving an attractive warm sepia tone. Albrecht Dürer used silver point for a self-portrait in 1484 when he was 13.

STEREOCHROMY

(also termed: water-glass painting and mineral painting) A method introduced by Von Fuchs in 1825. Water-glass is used as a medium for the paints and is afterwards applied as a protective coating.

STIPPLE

Applying small dots of colour with the point of a brush, which is often held at right angles to the support. William Holman Hunt was one who experimented with the manner, as was also Georges Seurat when working in the Pointillist manner.

TEMPERA

Broadly put this term implies using pigments which are mixed with substances such as: egg white, the whole egg, egg yolk, casein, glue and gelatine. **In the specialized sense it means the true egg tempera where only the egg yolk is used.** This is one of the most permanent media available to the artist.

UNDERWATER PAINTING

A technique has been evolved by Antonio de Havo in 1977 for working submerged. He mixes his colours with oil, glycerine and wax, and paints on panels that have been waterproofed. Wearing a frogman's outfit he has worked at depths of up to 100 ft (30 m) in spells of 20 minutes. The sketches are then finished on the surface.

WASH

The application of dilute water-colour to a support. The paper on the board should be at a slope of about ten degrees. Plenty of colour should be mixed up in a bowl and a large mop brush used. Start at the top and continue with horizontal strokes only just touching the paper. While the colour is still wet it may be bled into or wiped or mopped out with a dry brush, blotting-paper or a piece of rag.

WATER-COLOUR

In the purist sense this implies working only with transparent colours on white paper; attaining many of the colour mixes and tones by overpainting again and again; for example, yellow over blue, produces green. As with the term tempera, water-colour broadly includes such as: gouache, poster colours, show-card colours and designers' colours. If paper is the support for water-colour or allied techniques it should always be stretched to prevent buckling during the working of very wet passages.

Print-making

ANASTATIC

A relief-etching method, the opposite to the normal which is intaglio. The picture that is to be printed is painted or drawn with a pen directly on to the plate using an asphalt varnish; the result being that when the plate is put in an acid or mordant bath all the areas to be white or unprinted are etched away. **William Blake used the manner for text and illustrations with many of his books.**

AQUATINT

An etching method that uses areas of tone rather than lines and cross-hatching. The plate is grounded with either powdered asphaltum or resin. The plate is then heated; this causes the powder to melt and separate into thousands of tiny specks. The control of tonal areas in between dips in the acid bath is done by brushing on stopping-out varnish.

ARTIST'S PROOF

One of the proofs (or prints) in a limited edition of original prints. These would all be signed and bear a number such as, 7/32; this would mean it was the seventh pull of an edition of 32.

'The Giant' by Goya. An aquatint demonstrating the broad power possible with this method. It fetched £20 500 in the auction room. (Courtesy Sotheby Parke Bernet)

BAREN
A smooth, flat pad with a handle that is used for hand-proofing wood-blocks. Closely associated with the Japanese print-makers.

BAXTER PRINT
A method of printing using oil-colours developed by George Baxter (1804–67). Among his best works are, a copy of 'The Descent from the Cross' by Rubens and 'The Opening of the First Parliament of Queen Victoria', for the latter he was awarded the Austrian gold medal.

BLIND PRINTING
Placing damp paper over an uninked plate or block to achieve an embossed image.

BRAYER
A roller used to work up the ink and apply it to the block or plate. Made of hard rubber or a gelatine compound.

CHIAROSCURO WOODCUT
A monochrome relief-printing manner, that is built up by using a number of blocks with varying depths of tone with the same colour ink. Developed largely by Ugi di Carpi (1450–1525) and experimented with by the Germans Lucas Cranach and Hans Baldung; it was Cranach's wood-cutter Jost de Negker who did much to perfect the method.

CLAY-BLOCK
A simple process which uses stiff clay that has been pressed into a shallow rectangular box. Line work is then scratched into the clay with a knife-point or similar instrument.

COUNTER-PROOF
An impression of an engraving or etching printed from a wet proof. This is done by placing a piece of damp printing-paper over the wet proof and passing both through the printing-press. It is a help for the artist to be able to see in the counter-proof what the plate looks like, and assists in spotting mistakes.

CURRIER AND IVES PRINTS
Hand-coloured lithographs published by Nathaniel Currier (1803–87) and James M. Ives (1824–95). Their subjects ranged over the contemporary American scene; sporting, sentimental, political, disasters, city life, railways and steamboats.

DABBER
An instrument somewhat similar to a muller used for grinding pigments, only the bottom is a thick pad of wool covered with leather; the purpose of the dabber is to ground an etching plate. A second type is covered with a heavy woollen material and is used to force the ink into the intaglio lines during printing.

Dabber for grounding or inking etching plates

DRY-POINT
An intaglio-printing method related to engraving. It is worked on copper and zinc plates with the

'Black Lion Wharf' by Whistler. An etching carried out almost in the manner of a pen and ink drawing with little shading or hatched areas. (Courtesy Sotheby Parke Bernet)

design being cut by a hard steel tool, called a dry-point, or a diamond-tipped stylus. The main characteristic is the slightly softer lines than those with an engraving. The reason for these is that the steel dry-point or diamond raises a slight burr, which retains some ink during the wiping of the plate (see Etching).

DUST BOX
A box with a fine gauze bottom that is partly filled with powdered asphaltum or resin, and then shaken to ground a plate to be used for an aquatint.

DUTCH MORDANT
An alternative to nitric acid for biting a plate when etching. It is a solution of hydrochloric acid and potassium chlorate. Smillie's bath, that some prefer for aquatint, is a more concentrated version of the above.

ETCHING
One of the favourite print-making methods for the artist. The word is derived from the Dutch *etsen*. The plate is generally copper or zinc; iron has been tried but is erratic and will only produce rather unsuccessful prints. The plate has to be first meticulously prepared; the surface must be without blemish. This is achieved by grinding and smoothing with fine abrasives such as emery, tripoli and crocus powders. Then the plate is heated and grounded with asphaltum or resin with the aid of a dabber.

The artist now has to work his design, as with all printing methods except serigraphy, backwards. Some use light guide-lines of weak

Bark-strippers by Frank Brangwyn. An etching using a very heavy line and dark veiling with selected areas wiped clean. (Courtesy Royal Academy of Arts)

'The Artist Drawing from a Model'. A rare unfinished plate by Rembrandt. It shows not only the build-up of the etched areas, but also work done with a dry-point and burin. (Courtesy Frederick Mulder Esq)

Chinese white, others sit with their back to the subject looking into a mirror; but most go straight in with the etching needle cutting through the ground to expose the metal. When the needling is finished the back of the plate is brushed over with acid-resistant stopping-out varnish. The plate is now cautiously lowered into the acid bath. A careful watch has to be kept that too great an accumulation of bubbles does not cause the acid biting to be erratic; to stop this the bubble groups are dispersed with the tip of a feather. After the bath the ground is removed with white spirit and for the first time the artist can see exactly what he has done.

The plate is inked with a dabber, then the surface is wiped, first with retroussage, stiff canvas, next with muslin or cotton rags and lastly with a *coup de main*, the palm of the hand; the idea being to leave a subtle veil of ink on the surface. The printing is done with a strong press, the inked plate being laid on to a firm bed, damped paper is laid over the surface, backed with blotting-paper and thick wool blankets. It is then drawn through the

Etching needle set in a wooden handle

rollers of the press; and the blankets and blotting-paper are removed and the print is carefully lifted. With intaglio prints the inked lines are always slightly raised, a fact that can be picked up with a magnifying glass and a raking light. **There are oustanding masters of the method; Rembrandt, Goya and Whistler.** Rembrandt is perhaps the supereme genius who could bring to this difficult medium an intense feeling with superbly controlled light and shade and great variety of velvet tones.

FLAT BED

A simple term in print-making to identify methods such as serigraphy and lithography that rely on neither relief nor on intaglio for image production.

FLOCK PRINT

The surface of a wood-block is brushed with glue and then finely chopped textile material is dusted on to the glue. The resultant print is soft and tends towards indistinct outlines. Some artists experimented with the manner in south Germany during the latter part of the 15th century. The Ashmolean Museum in Oxford has a rare example, 'Christ on the Cross with the Virgin and St John'.

GLYPHOGRAPHY

An electrotype process by which it is possible to take a copy of an engraved plate which can be used for letterpress printing.

'Wet Afternoon' by E L Spowers. Colour lino-cut (9·4 × 8·3 in (240 × 213 mm)). (Courtesy Frederick Mulder Esq)

'Le Parc' by Gaston de Latenay. Colour lithograph (9·6 × 12·9 in (245 × 330 mm)). (Courtesy Frederick Mulder Esq)

Above: 'Woman Reading' by Franciso José de Goya y Lucientes. A lithograph, one of only eight or nine proofs, which was picked up for only £1 in a sale and then resold for £16000. (Courtesy Sotheby Parke Bernet)

Right: 'Umbrellas in the Rain' by C R W Nevinson. A lithograph which shows the crayon-like texture possible. (Courtesy The Leicester Galleries)

INTAGLIO
Lines and areas that are sunk into the plate to take the ink as opposed to relief where the printing areas are left upstanding.

JAPANESE PRINTS
A broad term for the period from about 1650 to 1868. Early examples were black outline blocks that were coloured by hand, *tan-e:* vermilion and *beni-e:* a gentle rose-red. Full colour prints often using up to 30 blocks came into being in the middle of the 18th century, early examples being called *benizuri-e* and the later *nishiki-e.* The subtle tone and tint changes combined with transparent inks at times give an impression of water-colour. Masters of the later periods include Utamaro, Hokusai and Hiroshige. Works were not only single prints but often diptychs, triptychs and polyptychs.

LINOCUT
A relief method, the block being made from high-grade linoleum. Cutters are small gouges that are fitted into a handle in the same manner as nibs into a penholder.

'Knight, Death and the Devil' by Albrecht Dürer. A line engraving carried out in 1513, probably one of the greatest masterpieces in this method. Note his control of almost velvet-like tone areas. (Courtesy Christie's)

LITHOGRAPHY
A flat-bed method which uses a stone or specially prepared zinc plate. The principle is the mutual repulsion of grease and water. The drawing is made either with a wax crayon or wax ink on the stone or plate. To print, the stone or plate is damped with water, which adheres to all areas not treated with the wax ink or crayon, and repels the oil-bound lithographic ink which is then rolled on and the print made.

The method was developed by Aloysius Senefelder, a German (1771–1834) in 1798. It has found much favour with artists since then as it gives greater freedom with drawing and executing a design. An account of the method by Senefelder, published in 1818, made it universally known.

MANIÈRE CRIBLÉE
A relief method that is a combination of engraving on a copper plate combined with the use of dies and prickers to produce areas of white dots for tones.

METAL-ENGRAVING
Cutting the design into copper, zinc or steel plates with a burin or graver, an intaglio method. Sometimes, after printing, the metal plates were filled with a black composition of metallic alloys and were known as niello plates and framed as they were.

MEZZOTINT
A method of engraving said to have been invented in 1642 by Ludwig von Siegen (1609–c 1680), an artist born at Utrecht, Holland, of German parents. He is said to have communicated the process in 1654 to Prince Rupert, who introduced it into England after the Restoration. The British Museum has a number of Siegen's mezzotint portraits. Mezzotint relies on tones rather than lines. The plate can be copper, zinc or steel, and it is first roughened by a rocker and then highlights and various tones are worked into it with scrapers and burnishers plus any lines called for by the use of a burin.

MONOTYPE
A method that will only allow for the taking of one print of a design. The plate can be a sheet of glass, metal or formica. On this the design is painted in oils, acrylics, gouache, inks or tempera and the print is taken by placing a sheet of paper over the wet colours and smoothing it down with the hand. Marc Chagwall is one who has experimented with the method.

OFFSET
A manner for printing lithographs which uses an intermediary roller that picks up the inked image

and in the transferring reverses it to the right way round. The advantage for the artist being that when he makes his drawing he does not have to reverse it.

PLASTER-BLOCK
A relief method in which a block of plaster of Paris is prepared on to which the artist works with various scrapers and cutters.

PLATE-MARK
The impression made on the damp paper by the plate as it passes through the press.

REGISTER
A mark made in colour-block or plate printing that acts as the key, to ensure perfect registration.

RELIEF ETCHING
(see Anastatic)

RELIEF PRINT
The opposite of intaglio, it is the upstanding parts of the block or plate that take the ink.

REMARQUE-PRINT
A print where the artist has made small sketches in the margin which relate to the state of the print.

SERIGRAPHY
(also more popularly known as silk-screen printing) The name is reputed to have been coined by Carl Zigrosser, Curator of prints at the Philadelphia Museum of Fine Arts. Basically it is a method of refined stencilling. A screen of silk, organdie, or fine wire mesh is stretched over a frame. On this screen is placed a stencil, which may be of thin card, shellac film, or *tusche* (a mixture that may include: wax, soap, lampblack, spermacetti, shellac or tallow). The ink which is of thick consistency is forced through the screen on to the paper by a squeegee. Multi-colour work is just a matter of preparing as many screens as the colours called for. Subtle water-colour effects can be gained by thickening water-based colours with isinglass.

SOFT GROUND ETCHING
The ground used has an excess of wax. The plate is then covered with a piece of thin hard paper and the design is drawn on this with either a hard pencil or a stylus; this causes the ground to adhere to the paper when it is lifted. The attraction for artists is that the print has the qualities of a pencil drawing. Thomas Gainsborough and Paul Sandby were exponents of the method.

STEEL FACING
A method of electroplating nickel or chromium steel on to an etching or engraved plate, which will allow a much larger edition to be printed.

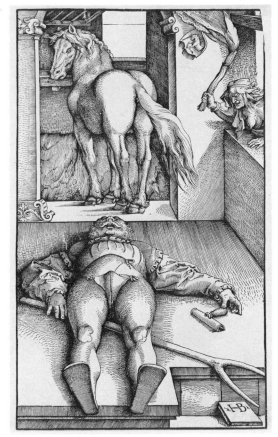

'The Bewitched Groom' by Hans Baldung. A woodcut. (Courtesy The Cleveland Museum of Art)

With etching on copper in particular, owing to the pressure applied in printing, an edition of over 50 is undesirable as the plate tends to squash and to lose crispness and detail.

STIPPLE-ENGRAVING
A copper or zinc plate is grounded and then worked on by a multi-etching needle and a roulette. The plate is then bitten and finished with further roulette work or a stipple-engraver.

SUGAR AQUATINT
A method of making an aquatint that has qualities akin to a brush drawing. The design is painted on the plate with black, gamboge and caster sugar mixed with a little water. The plate is then brushed over with stopping-out varnish, and left to soak in water overnight. During this time the water somehow penetrates the varnish, causes the sugar mixture to swell and lift the varnish exposing the metal. Picasso used this manner a number of times; it provides the artist with great freedom.

'The Lion' by Thomas Bewick. A fine wood-engraving by the great master. Note the different textures he employs. (Crown Copyright. Victoria and Albert Museum)

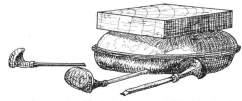

Boxwood engraving block sits on a sand-filled leather bag for working with engraving tools such as those shown

SURI-MONO
Japanese miniature colour-prints.

TEXTURE BLOCK
A relief method in which the upstanding relief areas are stuck on to the block. Materials used include pieces of cardboard, string, sand, coarse textiles, etc.

WAX-BLOCK
Molten wax is spread over a piece of hardboard or planking. The design is scratched into the wax and the print made in a similar way to an etching.

WOODCUT
Relief printing using the plank or long grain of the wood. **The oldest method for print-making;** it was introduced during the Middle Ages into Europe. **Hans Holbein the Younger and Albrecht Dürer were among the talented early users.** More recently the manner has been exploited by Paul Gauguin and the Norwegian Edvard Munch, also members of the leading German movements, notably Emil Nolde.

WOOD-ENGRAVING
Carried out on the end grain of hardwoods such as box, holly and cherry. **One of the finest exponents has been the Englishman Thomas Bewick (1753–1828)** with his exquisite studies of animals and birds. Both with engraving and cutting, the wood-blocks are held on a circular, flat leather bag filled with sand. In general the cutting tool is held steady and the block is moved on the bag to effect the cut.

Rules

ABAT-JOUR
A device for controlling the quantity and quality of light for a studio; it can use louvres and tinted shades.

ABOZZO
The first lay-in lines for an oil-painting, also called dead colouring as it is generally in monochrome.

ACHROMATIC COLOURS
White, black and greys.

ADVANCING COLOURS
Warm strong colours such as red and orange. These tend to give the impression of coming forward.

ANALOGOUS COLOURS
Those that are very close to each other in the colour circle and may be mistaken for each other; such as grass and yellow greens.

ANAMORPHOSIS
A distorted image produced by some form of optical device or by practice. Looked at straight on it can be incomprehensible, but looked at from a selected angle the image appears correct. A well-known example is **the skull in the foreground of Hans Holbein the Younger's 'The Ambassadors'**. A form of anamorphosis has to be used when painting on curved ceilings or other awkward shapes to produce a satisfactory image. **The principle was understood by the early Greek painters.**

ANATOMY
From the artist's point of view it is the **study of the whole human figure, bones and muscles**. Leonardo was one of those who made numerous studies by dissection of corpses.

ARABESQUE
A motif featuring scroll work, plant forms and flowing lines which is typical of Islamic ornament. It is also a motif which is found in Hellenistic art, notably in Asia Minor. Profuse use was made of it at Pompeii.

ASYMMETRY
The opposite to the perfect equal balance of symmetry; in painting asymmetry generally produces a happier composition.

ATELIER
The studio or workshop of an artist or craftsman, the Italian equivalent is *bottega*.

BAROQUE
The dominant style in Europe particularly on the Continent during the 17th century. It used overdecoration right across the visual arts. Artists associated with the High Baroque manner include: Giovanni Lanfranco (1572–1647) and Guercino (1591–1666).

BROKEN COLOUR
A phrase introduced by the French Impressionists

Decorated baroque sculpture. 'St Jerome' by Zimmerman, in Wies Church in Bavaria. He shows the saint with a cardinal's hat, which of course did not exist at the time of the saint's life. (Photo by the Author)

in the 19th century. It means the putting down of different colours with separate strokes. This can give a vital look to a picture, and to a degree the eye of the viewer makes many of the colour mixes optically.

CAPRICCIO

A fantasy in painting or sculpture that overthrows the order of normal composition, often depicting collections of grotesque, imaginative, bizarre subjects and invented landscapes and townscapes. Goya's 'Los Caprichos', a series of etchings and aquatints, are examples. In painting there was a considerable vogue for this style in the 18th century.

CHIAROSCURO

The handling of light and shade. Many masters of the past have been fascinated by the effects of light, its direction, reflected lights and what goes on in the shadows. Caravaggio, La Tour, Joseph Wright and the incomparable Rembrandt are among the peaks in this moving way of presenting a subject.

CHROMATIC COLOURS

All colours except black, white and grey.

COLLECTOR'S MARK

A special stamp impressed on the margin of a print or a drawing, or a seal pressed into sealing-wax on the back of a panel or the stretcher of a canvas.

COLOUR CIRCLE

For hundreds of years artists have worked on and discussed numerous colour theories, mainly concerned with the interplay of one tint next to or in relation to another. The whole subject is complicated by individual reaction, no eye and mind seeming to respond in exactly the same way; furthermore, light conditions altering only minimally can drastically affect a carefully balanced scheme.

The easiest and proven the most effective theory is the complementary which uses a simple colour circle; it was the one that the French Impressionists used in general. The circle has the primary and secondary colours arranged in this order: yellow, orange, red, violet, blue and green. The painter working from this circle obtains the maximum effect by using opposites beside each other; thus: green next to red, violet next yellow. A more elaborate circle was constructed by Dr Wilhelm Ostwald, whose book *Letters to a painter on the theory and practice of painting*, was published in New York in 1907. He broke up the main colours for the circle thus: yellow, orange yellow, orange, red orange, red, red violet, violet, blue violet, blue, green blue, green, yellow green.

Using the Ostwald circle gives greater subtlety with the complementary manner.

COMPLEMENTARY COLOURS
(see above)

COMPOSITION

The putting together and organizing the arrangement of the different components of a picture so that they register with the eye as a harmonious whole. There are no definite direct rules that can in truth control this operation. Theories have proliferated only to be disproved by a painter working directly against them. Meindert Hobbema (1636–1709) with his 'The Avenue, Middleharnis', showed that symmetry could be used, although many scoffed saying it made for dullness.

In the main, composition is something the sensitivity of the artist guides. A few basics need care, including: not slicing a landscape in two with the horizon, not having the main group so small that it appears lost in the picture or too large so that it appears to be bursting out of the frame or that it has been cut down. Composition is attaining a total harmony of shape and form arrangement, and integral with this, colour happiness and a realized interpretation of light and shade.

CONTRAPPOSTO

The posing of a figure in painting or sculpture with parts of the body in opposition such as: the left leg forward with the right arm, the hips turned in a different direction to the shoulders.

COOL COLOURS

Blues and blue-greens, the opposite side of the colour circle to red and orange.

DEGRADED COLOUR

One that has been mixed with greying tints and so has had its normal brilliance or truth brought down. Also over-mixing colours particularly with oils, alkyds and acrylics reduces them to a muddy dull appearance.

ECLECTIC

Can be applied to artists who in **a gentle way plagiarize the work of others**, leaning on them for a composition idea, details, tricks of lighting, etc.

ÉCORCHÉ FIGURE

An anatomical figure of a man or an animal that is shown without its skin so that muscle arrangements can be studied. Andrea Vesalius (1514–64), the Flemish anatomist, was one who left some remarkable drawings in this vein. His chief work was *De Corporis Humani Fabrica*, 1543.

FLOATING SIGNATURE

One that is put on top of the varnish. If found it is very often the signal that either the signature is fraudulent and/or the painting. It is very unlikely a painter of a particular picture would sign on top of the varnish, it would almost always be on the actual paint.

FORESHORTENING

The technique, particularly with figures, of giving the impression that a particular part is turned towards the viewer.

GOLDEN SECTION

(also Golden Mean) A rule of proportion which lays down a relationship of line or figure to the whole composition. It was worked out by Vitruvius geometrically. Basically this establishes that the division of a straight line into two unequal parts shall be so that the proportion of the smaller part to the greater is the same as the greater part to the whole.

GROTESQUE

A painted or carved ornament of fantasy, using figures, animals and foliage. Found in the Greek and Roman houses and revived during the Renaissance.

HALF-TONE

One that lies between white and black, bright and dark.

HIGHLIGHTS

The brightest touches in a painting.

HUE

The actual colour of anything.

KEY

(1) General effect of colour and light. Predominance of light colours would be high-key, of dark colours low-key.
(2) Roughness or texture in a support or ground.
(3) A small wedge driven into the corner of a stretcher to increase the tension.

'LIBER STUDIORUM'

A collection of etched or mezzotint prints either made or designed by Joseph Mallord William Turner (1775–1851). It was brought out by him between 1807 and 1819 with the intention that it should act as a guide to landscape. He took the idea from Claude Lorrain who made up a similar folio as a tally of paintings he had already sold, which was called 'Liber Veritatis'.

LOCAL COLOUR

The true colour of an object in normal daylight as opposed to its apparent colour when affected by unusual lighting.

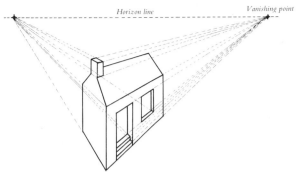

Bird's-eye perspective; the horizon line is raised well clear of main objects

Worm's-eye perspective; the horizon line is dropped right down

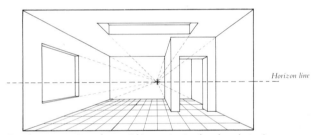

Interior perspective; the vanishing point is placed as desired on the horizon line within the picture area

Two vanishing point perspective; by moving the points along the horizon line the aspect can be altered

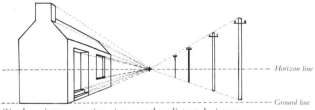

Single-point perspective gives a rather distorted picture

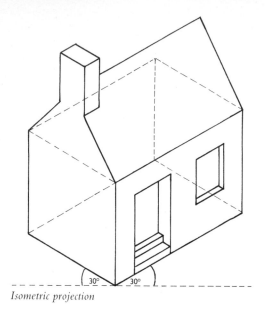

Isometric projection

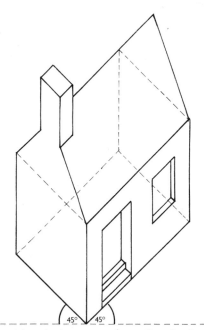

Axonometric projection operates from 45° angles, but retains an accurate undistorted plan

MEANDER
A pattern that appears endless with no great strength to it.

MONOGRAMS
An arrangement of initials in a decorative manner often used to sign pictures.

MOTIF
A predominant feature which recurs or holds together a composition.

OBJET TROUVÉ
An object that is found or selected by an artist and then exhibited as a piece of art without being worked on in any way. It is a slight manner of the 20th century played with by the Dadaists and Surrealists. Marcel Duchamp was one who flung many such objects into the public's eye. The objects could be natural: a piece of wood, a bone; or manufactured: a beer bottle, a can, etc.

PASTICHE
A painting made up of portions from other painters' works or in a direct imitation of another's style.

PERSPECTIVE
A geometric solution to producing the impression of a third dimension in a picture. For the artist's work externally one or two vanishing points may be used; for interior work normally only one. Other forms of perspective include worm's-eye view, which gives the feeling of looking up; and bird's-eye view, looking down.

PLANE
In perspective the picture plane is an invisible plane through which pass all the lines of sight to the viewer, and which corresponds to the image that the artist finally depicts on his paper or canvas.

PLEIN AIR
An expression in vogue during the latter part of the 19th century to signify a painting executed out-of-doors. Principal exponents of this manner were the Barbizon school and the Impressionists (see Chapter 5).

PRIMARY COLOURS
In pigmentation they are red, blue and yellow.

PROJECTIONS
Geometric methods of producing a non-natural third-dimensional appearance. They are axonometric, which uses a 45 degree angle from the horizontal and always produces and works from an accurate squared plan, and isometric, which uses a 30 degree angle from the horizontal and produces a distorted plan.

QUADRATURA
A method of painting on ceilings and walls to give the impression of great space with illusory treatment of figures and details.

RETREATING COLOURS
Cool, light tints that suggest distance.

ROCOCO
An escapist decoration for decoration's sake style that flourished in the mid 18th century. Lavish use of motifs and embellishment characterize the style, which often gives the viewer a feeling of claustrophobia so overbearing can the ornament be.

SECONDARY AND TERTIARY COLOURS
Two primary colours in combination create a secondary colour. Two secondary colours in combination create a tertiary colour.

SIGNATURES
These may often be accompanied by one of the following descriptions, which can mean as indicated.

Pinxit (pinx, pictor) – Painted, painter
Delineavit (delin, del) – Drew

'Mr Lytton Strachey' by Max Beerbohm, a caricature. (Courtesy Christie's)

Fecit (fec, f) – Made (painting or print)
Invenit (inv) – Designed
Sculpit (sculp, sc) – Sculptor (engraved, engraver)
Incidit (incid, inc) – Engraved, etched
Excudit (excud, ex) }
Impressit (Imp) } – Printed
Lith – Lithographed (either artist or lithographic printer)

SOTTO IN SÙ
Extreme foreshortening to give an effect of floating figures in ceiling work as with some of the work of Giovanni Battista Tiepolo (1696–1770). The over-all appearance is one of imaginative fantasy.

TONE
Tones are depths of the same colour, tints are different colours.

VIGNETTE
A picture which has the edges softened and graded away.

Subjects

ABSTRACT
A picture or sculpture which is non-representational in purpose.

ALLEGORICAL
A pictorial narrative that has its meaning veiled, a story-telling picture using implication.

CARICATURE
A manner of accentuating features and characteristics when drawing or painting a portrait or group of figures.

CARTOON
A full-size drawing prepared as a guide for a mural or easel painting (also mosaic and stained glass).

CONVERSATION PIECE
A painting which incorporates two or more portraits of people, sitting or standing naturally in a room as if they were at home; or outside in a garden or rural setting. Popular during the 18th century, they were often painted by Arthur Devis (1711–87) and Johann Zoffany (1734/5–1810).

DIORAMA
A method of display, generally reserved for museums, which starts with third dimensions in the foreground and as it goes back this is gradually lost until it ends in a painted two-dimensional background.

FÊTE-CHAMPÊTRE
(also Fête galante) A style of painting that was in

'Still Life' by Georges Braque. (Courtesy Tate Gallery)

revolt against the Academic manner; it flourished in France during the early part of the 18th century. Escapist in subject and theme. The foremost protagonist was Jean Antoine Watteau (1684–1721).

FIGURATIVE
The opposite of Abstract, a name for Realism or Naturalism.

GENRE
Representation of everyday life without imaginative or romantic treatment. This manner more or less started with the Dutch painters of the 16th and 17th centuries.

HISTORY PAINTING
The reasonably free showing of historical happenings. The largest known painting now in existence is 'The Battle of Gettysburg' by Paul Philippoteaux and 16 assistants; it took them two and a half years and is 410 ft (124·96 m) long and 70 ft (21·34 m) high.

ICON
A formalized religious painting popular with the Eastern Church. Generally on a wooden panel, and showing the Virgin and Child, single figures of Jesus Christ or different saints or groups of religious figures.

MAESTÀ
Virgin and Child enthroned.

MYTHOLOGICAL
Main themes, subjects and characters drawn from mythology.

PANORAMA
A landscape that could be painted along a wall or walls or on a long roll of canvas or paper. **The largest picture ever painted** was a panorama painted by John Banvard (1815–91) in 1846. It showed the Mississippi for 1200 miles (1930 km) and was itself probably 5000 ft (1525 m) long and 12 ft (3·65 m) high, an area of 1·3 acres (0·52 ha). It is thought to have been destroyed by fire.

PIETÀ
A painting or sculpture of the Virgin Mary holding the body of the dead Christ. One of the most moving of these is the one in marble by Michelangelo Buonarroti (1475–1564) in St Peter's, Rome.

SACRA CONVERSAZIONE
The Madonna and Child grouped with various saints.

SINGERIES
Paintings which showed monkeys dressed and behaving as people. Watteau was an artist who produced some examples.

STILL-LIFE
An art form that has its roots with the Greeks and Romans. It applies to any subject on which the painter works that consists of inanimate objects. In the narrow sense it does not include fruit and flowers.

Example of Trompe l'œil *'The Old Cupboard Door' by W M Harnett (1889). (Courtesy Sheffield City Art Galleries)*

TROMPE L'ŒIL

A style of painting that sets out to be a complete illusion. The medium most suited to the exact style is tempera.

Sculpture
Materials, tools and techniques

AGALMATOLITE

A softish stone, ranging from grey to green and ochre. It was used by the Chinese for images and decorative architectural details.

ALABASTER

A type of gypsum with fine texture which has been and is used for carving ornaments and small figures. Colour generally white, but this may be tinged with yellow if impurities are present; it also has a pleasant translucent appearance.

ARMATURE

A metal or wooden or a mixture of both framework used by a sculptor when modelling in clay or other pliable materials.

BARDIGLIO

Italian marble recognizable with its distinctive dark veining on a grey-blue mass.

Male head from Bin' Aqil the Necropolis of Timna, capital of ancient Qataban. It is worked in alabaster. (Courtesy Gimpel Fils)

BASALT

Volcanic rock, very dark and even black; it possesses a brittle splintery nature. It was used a great deal by the early Egyptians.

BAS-RELIEF

Sculpture in which the figures and forms are kept in very low relief.

BATH STONE

A collective term for limestones that occur in a belt diagonally across England. The texture is open, and they vary in colour from warm ochre to cool yellow-grey. Henry Moore has used them for larger works.

BENIN BRONZES

Bronzes that were produced by the natives of Benin in southern Nigeria; they were cast by a type of *cire-perdue* process.

BRASS

An alloy of about two parts copper to one part zinc. Little used for casting, although there are some heads of kings and religious leaders existing.

BRAZING

A method of welding that uses non-ferrous metals.

BRONZE

An alloy of copper and tin which is harder and more durable than brass. It is easier to cast than copper as it has a lower melting-point and is more liquid when melted. Originally bronze was cast solid, but this limited the size of the piece. The invention of hollow casting, sand casting and *cire perdue* (lost wax) gave the sculptors considerably more freedom. The *cire perdue* was and is the method most favoured as it will give the greatest truth in the cast from the model. The method consists in making a model with a wax surface of a suitable thickness, then the outside mould is formed over this and is heated so that the wax melts and runs away, is lost, through vents and thereupon liquid bronze is poured in to replace the wax.

BRONZE DISEASE

A surface blemish on a bronze which consists of greyish-green spots. More prevalent in older pieces; the cause is likely to be the presence of damp and chloride, also damp could react with impurities in the alloy.

BRONZING

Painting plaster casts with bronze powders and a suitable medium to imitate bronze.

BUSH HAMMER

(also boucharde) A fairly heavy double-ended

'The War' by Eugene Dodeigne. An example of free expressionist carving in stone. (Courtesy Henie-Onstad Art Centre, Hovikodden, near Oslo)

hammer that has the striking faces covered with regular teeth; used in shaping and clearing.

CAEN STONE
French limestone, soft and cream coloured.

CARRARA
A milk-white marble that comes from the famous quarries north of Pisa, Italy. Favoured by sculptors since the Renaissance.

CARVING
To cut into wood, stones, marbles, bone, ivory, etc. It is one of the oldest art techniques and dates back thousands of years.

CASTING
(see Bronze)

CIMENT FONDU
An aluminous cement used in preparing forms for casting.

CIRE PERDUE
(see Bronze)

CONSTRUCTION
Building up a metal sculpture using welding and riveting, often from scrap material. Applicable also to wood and glass.

Bronze fibreglass main doors to the Metropolitan Cathedral, Liverpool. (Courtesy Crispin Eurich)

FIBRE-GLASS

A material combined with polyester resins that can be used for casting. It may be mixed with bronze powder for special or imitative effects.

GANOSIS

A process developed by the Greeks for toning down the shine on polished marble or protecting painted marble sculpture. It involved warming the marble and applying a little wax and then buffing to a soft sheen. The method could also be applied to painted plaster walls.

GLYPTIC ART

Derived from the Greek, it means the carving of precious and semi-precious stones. Engraved stones are called *intagli* and those in relief are cameos. The art was practised in Mesopotamia as far back as 400 BC.

IVORY

A material that comes from a variety of mammal sources including elephant and walrus, that has been used for carving since prehistoric times. Examples from Assyria and Egypt go back to the period of Moses.

KAMAKURA BORI

Wood-carving in relief that is treated with black and red lacquer, so that the black ghosts through. It has been made in Japan since the end of the 12th century.

LEAD

The Greeks used it for casting small figures. Since then it has had but little use except for garden figures.

MAQUETTE

A sketch figure prepared by a sculptor as a guide for a larger work; usually modelled in clay or wax.

MOBILE

An innovation by the American Alexander Calder in 1932. The constructions are so made that parts of them can be moved by the action of wind or water. The largest recorded mobile is 'Quest' by Jerome Kirk. It was installed at Redondo Beach, California in 1968; 32 ft (9·75 m) long a wind-driven pivotal mobile weighing 5·35 tons (5442 kg). The term mobile was coined to contrast with stabile sculpture by Marcel Duchamp in *c* 1832.

NETSUKE

Very small carving in wood or ivory. It is a form of button to attach the *inro* (a small box for medicines) to its silken cord and belt. Examples from the 18th and 19th centuries often display brilliant and imaginative carving, even if a little grotesque and macabre at times. One of the favourite categories for collectors of Japanese art.

Ivory figure of Christ Jesus, from the Cistercian Monastery of Carrizo. Originally the cross was decorated with gold and enamel. (Courtesy Spanish National Tourist Office)

POINTING MACHINE

An instrument to assist a sculptor in transferring the proportions and form of a model or maquette to a block of stone or wood for producing the full-sized piece. It is comparable to a pantograph which works for two dimensions in drawing.

POLE SCULPTURE

Descriptive term for works carved from naturally cylindrical objects such as a tree-trunk or an elephant's tusk.

POLYCHROME

A painted sculpture in several colours.

RASPS

File-like hard steel tools of varying shape, which are used for abrading and forming stone.

RELIEF

(also relievo) Collective term for sculpture which stands out from but which belongs to the wall on which it is carved. There are three categories:

Basso – slight projection.

Mezzo – standing about half-way out.

Alto – almost completely clear, only held by slight attachments.

A massive example of relievo and also **the world's biggest sculpture** is the group of mounted figures of Jefferson Davis (1808–89), General Robert Edward Lee (1807–70) and General Thomas Jackson (Stonewall) (1824–63) which covers 1·33 acres (0·5 ha) on the face of

Stone Mountain, near Atlanta, Georgia. They are 90 ft (27·4 m) high and took from 1958 to May 1970 to work by Walker Kirtland Hancock. A still vaster work will be a statue of Tashunca-Uitco (*c* 1849–77) the Indian Chief, known as Crazy Horse, of the Oglala tribe of the Dakota or Nado-wessioux (Sioux) group. It will be 561 ft (170 m) high, 641 ft (195 m) long and require the removal of 5 890 000 tonnes of stone. It was begun in 1948 and is expected to take until at least sometime in 1978, being the life-work of Korczak Ziolkowski.

REPOUSSÉ
A method for producing a design in relief with metal sheet by hammering from the back.

RIFFLER
A type of rasp.

SANDSTONE
A material used since the early Egyptian times for carving. It is often textured and comes in a variety of colours, grey, brown, yellow and red.

SOAPSTONE
(also called steatite) It carves easily and is slightly greasy to the touch, hence its name.

WAX
Used for making maquettes or small models either to stand in their own right or to be cast. During the Renaissance life-sized portrait busts were produced. From these probably sprang the complete Naturalism of today's waxworks.

'*The Madonna and Child*' *in the Lady Chapel, Liverpool Metropolitan Cathedral, by Robert Brumby. This is in terra cotta, that was first fired to biscuit, then oxide of iron was washed in and firing was then taken to 1300 °C (2372 °F). (Photo by the Author)*

Chapter Two

Roots of Western Art

Art in any recognizable form seems to have evolved some 25000 to 30000 years ago, although there is evidence that art activity may pre-date even this. The world of the archaeologist is at times a patient plodding on, following clues, and suddenly a dramatic discovery is made and there is more material for the art historian.

How and why did man first begin to draw, paint or scratch, model or carve visual images? We know pretty well how, and also what materials he used; but the answer to why is still largely in the realms of thoughtful guesswork. Did those first artists draw and colour the animals from some superstitious basis to help them in the hunt for the vital food for survival, or was there some deeper ritual magic behind the action, or could it be possible after all that they just did it from a similar creative urge to that which has fired many painters and sculptors during the intervening periods?

The art forms of long-ago peoples can tell vividly so much about them; how they looked, dressed and often some of the things that they did. It is also from these roots that so many of the basics of the visual arts since have sprung.

ACHILLES PAINTER

An unknown Greek artist, who has been given this nickname, he was among the leading *lekythos*-painters. A *lekythos* was a flask for oil, that might be used for anointing bodies.

ACROLITH

A type of statue especially associated with the ancient Greeks. The head, arms and legs were generally made from white marble with the body constructed of wood covered with a metal that could be gilded.

AEGICRANES

The heads or skulls of rams adopted in classic sculpture as a decoration for altars and temples.

AEGINETAN ART

For some 50 years before the conquest of Aegina by Athens about 457 BC, the island was a centre for the arts which culminated in the Temple of Aphaia. It was characterized by the skilled

Nike, Niki or Nice. Greek goddess of Victory. The statue was found in Athens in 1891. (Courtesy Archaeological Museum, Athens. Photo Voula Papatoanno)

grouping of human figures in decoration and the lively sensitive sculptured forms.

AGALMA
From the Greek, it is a word meaning a statue or image generally of a god, a primitive image.

AGASIAS
The name of two Greek sculptors who were followers of the Ephesian school dating about 100 BC. One of them who was the son of Dositheus worked the **'Borghese Warrior'** in the Louvre, the other, the son of Menophilus, made the impressive statue of a gladiator in the Athens Art Museum.

AGATHARCHUS
Athenian painter, 5th century BC, concerned with the theatre.

AGELADAS
Greek Sculptor of the Argive school who was active at the end of the 6th and beginning of the 5th centuries BC. Known as a carver of statues of Olympic winners, and it is thought that **he may have had Phidias and Myron as pupils.**

AGESANDER
First-century BC Greek sculptor said by Pliny to have worked on the **'Laocoön'** with Athenodorus and Polydorus.

AGORACRITUS
Greek sculptor of the 5th century BC, a pupil of Phidias and he worked with him on the **Parthenon sculptures.**

ALCAMENES
Another pupil of Phidias; he used gold, ivory and bronze as well as stone. He worked a 'Hecate' which stood near the Temple of Athena Nike at Athens.

ALDOBRANDINI WEDDING
First-century BC wall-painting found in Rome, which was taken from a 4th-century BC Greek original (the name Aldobrandini came from its previous owners). It depicts the wedding of the god Dionysus to the wife of the Athenian High Priest with Aphrodite present. It is remarkably lifelike with advanced modelling of the anatomy and well-handled drapes.

ALEXANDER MOSAIC
An amazing pavement dating from about 50 BC found in Pompeii, Italy; it shows with considerable detail and feeling for movement the battle between Alexander the Great and Darius. It has been lifted and is now in the National Museum in Naples.

ALTAMIRA
The site of Palaeolithic cave-paintings in northern Spain; it lies about 19 miles (30·5 km) from Santander. As with a number of finds, the entrance to the cave was discovered by accident in 1869. During clearing work in 1879 a little girl was the first to sight the paintings of horses, stags, boars and other animals that covered the limestone roof in the first part of the cave. The paintings have been dated back to the Upper Magdalenian period about 12000 BC. The materials the long-forgotten artists might have used could have included mineral-coloured clays and charcoal, with either fat or possibly blood as a binder; chewed sticks or pieces of fur as well as fingers could have served as brushes.

ANDROSPHINXES
Allegorical figures found in Egyptian art: they are lions with human heads. **The largest is the one at Giza which, except for the fore-paws, is hewn out of the solid rock in front of the Cheops Pyramid. It is 60 ft (18·28 m) high and 189 ft (57·61 m) long.** There was originally a small temple between the fore-paws. There is a winged sphinx at Delphi which is on top of an Ionic pillar, and there is another on the Temple of Apollo at Aegina, also there are a number at Thebes. In Minoan art they are carved in bone and ivory and appear on glass and gold plates. The Egyptian sphinx has no wings, while the Greek one is winged.

ANTENOR
Sixth-century BC Greek sculptor, son of the painter Eumares. His most important work was a bronze group called 'The Tyrannicides' which

'Jockey'. A small bronze early Greek figure with an exquisite sense of movement. (Courtesy Tourist Organisation of Greece, Athens)

Tibetan Thangka. (Courtesy Christie's)

'A mahout riding an elephant decked with full regalia in a procession up a hillside.' Mughal c 1580 (11 × 7½ in (286 × 192 mm)). (Courtesy Sotheby Parke Bernet)

Tlingit Indian head-piece for ceremonial dancing. The crown would be filled with down that would fall out with movement and sprinkle as falling snow. (Courtesy Christie's)

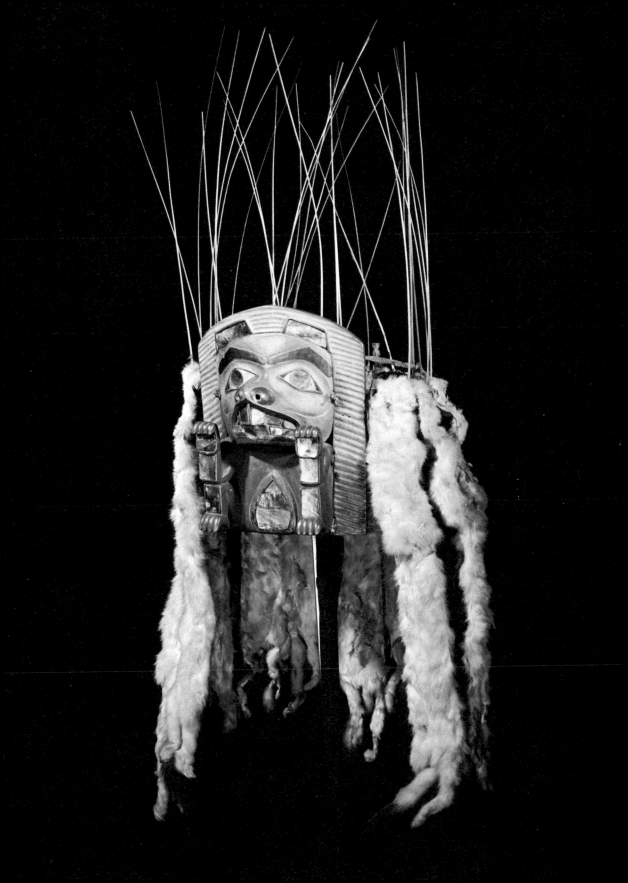

Xerxes took from Athens during the Persian Wars, but it was returned by Alexander the Great; it has, however, now vanished.

ANTIPHILUS
Fourth-century BC Greek painter, a contemporary of Apelles, he worked for Philip of Macedon and Ptolemy I. He it was who thought up a strange hybrid, part man, animal and bird called a gryli.

ANTIQUE
Used here it implies the arts, painting, sculpture, gems and medals, of the ancient Greek and Romans.

APELLES
The celebrated Court Painter to Alexander the Great, he was active in the middle of the 4th century BC and he was probably a pupil of Ephorus. Unfortunately, there are no works of his extant, but he is supposed to have painted a striking portrait of Alexander holding a thunderbolt. In the 'House of the Vetii' at Pompeii there is a likeness of Alexander as Zeus which could be a copy of a picture by Apelles.

APOLLO
In Greek mythology he represents youth, manly beauty, and is also identified with Helios, god of light. He had seven festivals in his honour. Statues of him include the **'Apollo Belvedere'** in the Vatican and the **'Apollo Sauroctonus'** in the Villa Albani, Rome.

APOLLODORUS
Greek painter of the Athenian school active about 600 BC. He worked on perspective and the handling of light and shade.

ARA PACIS AUGUSTAE
A temple put up by the Roman Senate in the Field of Mars close to Rome about 10 BC in honour of Peace brought by Augustus. The altar was decorated by beautiful relief-carved slabs, some of these are now in the Louvre and the National Museum of Rome.

ARISTIDES
Fourth-century Greek painter, born at Thebes, he may have been related to Nicomachus with whom he studied.

ASSYRIANS
(see Mesopotamian Art)

BABYLONIANS
(see Mesopotamian Art)

BERLIN PAINTER
A highly important Greek vase-painter, real name not known, who was active early in the 5th century BC.

'The god Bes'. (Courtesy Pilkington Brothers)

BES
An ancient Egyptian godlet of children, dance, mirth and music. He was worn sometimes as a talisman against serpents and evil spirits, as well he was a god of war and on the domestic scene a guard for the home. There is a fierce-looking one of blue-glass paste, just over an inch (25 mm) high, in the Pilkington Museum of Glass, St Helens, Lancashire; it dates from 600 BC, from the Sais dynasty.

BRYAXIS
Fourth-century BC Greek sculptor who worked with Scopas, Leochares and Timotheus on the Mausoleum at Halicarnassus; also he is reputed to have made a great statue of Serapis the Graeco-Egyptian god.

CALLIMACHUS
Fifth-century BC Greek sculptor who by the writings of scholars was possessed of considerable elegance and skill. It is probable he studied with Calamis, who worked a statue of Aphrodite at the entrance to the Acropolis, and who also made a huge figure of Apollo which was taken to Rome. Callimachus is suggested as being the **maker of the first Corinthian column.**

CATACOMB-PAINTING

The catacombs were subterranean cemeteries, generally made with passages and galleries with recesses on the sides for tombs; they were also used by the early Christians for secret places to worship and as refuges. The best known are in or near Rome; on the Appian Way, and under St Sebastian Church. The interiors had at times a form of rough fresco from which can be noted some of the roots of the evolution of Italian painting. In the assembly room of the Catacomb of Petrus and Marallinus, Rome, 3rd century, the walls and ceiling are painted with well-contrived architectural features and decorative devices, to alleviate the depressed underground place and to suggest greater room. **It is possible the oldest Roman examples of such painting are in the Ampliatus Crypt in the Domitilla Catacomb on the Via Ardeatina.** In the Coemeterium Maius of the Trasone Catacomb there are the great praying figures which stylistically would appear to be about mid 4th century.

CELTIC ART

A form of creative expression apart from the classically inspired arts, it was in many ways like that of the Scythians. The Celts, a race of warriors and chieftains, were principally nomadic and their particular art form seems to come to light in the latter part of the 5th century BC, based on an area covering parts of southern Germany, Austria. Switzerland and eastern France, and at two places in particular, Hallstatt, in the Salzkammergut, Upper Austria, and Lake Neuchâtel in Switzerland, near the village of Marin, some 4 miles (7 km) from the town of Neuchâtel. **La Tène culture** spread out over large areas of Europe and crossed the sea to Britain and Ireland partially being overrun by the Romans but continuing in areas until about the 8th or 9th centuries.

Largely it was an art coming from the metalsmiths, who worked in gold, bronze and iron, with advanced techniques, and who used decorative motifs and features which may have had some influence from the zoomorphic of the Scythians, but which were in the majority fresh and new; forms which had sinuous movement and continuity that found an echo in the Art Nouveau of the late 19th century. The heads of chieftains in stone were symbolized, with formal hair arrangement and swirling moustaches. Their designers were masters at space-filling, as with a bronze helmet from Amfreville, Eure, in the Louvre and a roundel from a bronze trumpet from Lough na Shade, Co. Armagh, Ireland, in the National Museum, Dublin; there is no cramping or padding, the pattern is harmonious and perfect. There are also examples of the intricate champlevé enamel process, as with a medallion from the early 8th-century belt shrine or bronze found at Moylough, Co. Sligo, Ireland, and now in the National Museum, Dublin (with the champlevé method the areas to receive the enamel are lowered, by gouging or etching).

CEPHISODOTUS

The name of two Greek 4th-century BC sculptors, the elder is thought to have been the father of Praxiteles, the younger is supposed to have been the son of Praxiteles.

CHALK-CARVING

In England are two very big examples of this, sgraffito on a grand scale, the cutting away of turf and subsoil to show underlying white chalk. **The largest is the 'Long Man' at Wilmington**, on the north side of Windover Hill, Sussex; he is 240 ft (73 m) tall and may have been cut out as early as the mid- Saxon period. The other figure is **the giant of Cerne Abbas**, on a hillside close to the Dorchester–Sherborne road, Dorset. The giant is shown naked and is some 180 ft (55 m) high. There are references to 'Helith' which was apparently his early name, and he is likely to be a representation of the Roman Hercules, and related to spring fertility rituals, probable dating is Romano-British.

CHORAGIC MONUMENTS

These were set up by the ancient Greeks to honour the winners in musical competitions. The best-preserved example is the monument to Lysicrates, Athens, 334 BC.

CHRYSELEPHANTINE SCULPTURE

Cult religious statues in Greek temples of the 5th century BC, they were made of ivory and gold, a technique probably imported from the Orient. There was the Athena by Phidias and the 'Dionysus' by Alcamenes. Small figures with the flesh in ivory and clothes and hair of gold were found in the Tomb of Tutankhamun. But the manner of the Greeks veered towards size. **Phidias made a 40 ft (12 m) high figure of Athena**, the underlying material was well-seasoned wood on to which were fixed plates of ivory for flesh and sheets of gold for drapery.

COLOSSUS

Both the Egyptians and the Greeks produced huge statues. There is the great figure of **Rameses II** in front of the Temple of Abu Simbel, and the **Memnon colossus.** Probably the best known by word alone is the **'Colossus of Rhodes'**, one of the Seven Wonders of the World; this has now disappeared, apparently destroyed by an earth-

'La France' by Henri Matisse. (Courtesy Christie's)

'Sail Boats' by Roy Lichenstein. Born in 1923, the leading American exponent of Pop Art. (Courtesy Christie's)

quake about 200 BC. It was a bronze statue of
Helios, the sun god, by Chares of Lindus, over 100
ft (30 m) high, and set up in 285 BC. In Rome at a
later date Nero saw himself as the sun god and had
made another hundred footer, which gave the
name to the Colosseum.

COPTIC ART
The visual expression of the early Christians in
Egypt from the 4th to the 7th century. Coptic
wall-paintings were stylized and related to the
formality of the Byzantine. In Abyssinia this form
of primitive Christian Art has worked through to
the present day.

CRESILAS
Fifth-century Greek sculptor who worked the
portrait of Pericles.

CYCLADIC ART
The form of visual expression of the Bronze Age
civilization of the Cyclades, Greek islands in the
centre of the Aegean. White marble was shaped
with bronze tools into mainly symbolic simple
figures of small size; pottery was decorated with
primitive bird and plant motifs.

DAEDALIC FIGURES
Images that were preserved in sanctuaries in

*Calf of Man Crucifixion carving, dating from the 8th
century. It was discovered in 1771. The Eastern
Mediterranean influence shows clearly. (Courtesy
The Manx Museum, Douglas, Isle of Man)*

*Carved figure from Ballintober Abbey, Co. Mayo,
Ireland. (Courtesy Commissioners of Public Works in
Ireland)*

*'Harpist'. Marble carving, Greek, Cycladic 3000–2500 BC.
(Courtesy Metropolitan Museum of Art, New York)*

memory of Daedalus. Most applicable to statues of wood, decorated with bright colours and gilding with real drapery; **they were among the earliest known images of the gods.**

EARLY CHRISTIAN ART
That of the period from about 350 to 800. The emergence of Biblical themes in painting; for example, Jonah, and Christ, who at this time was always shown without a beard, and the portrayals of the Madonna and Child (see also Catacomb Painting). Most sculpture was that of talented stonemasons, principally confined to the production of sarcophagi; other materials included ivory. There was also the emergence of the illuminated manuscript. The period to a degree runs parallel for a time with the Byzantine, but the latter ran on until the 15th century. Specifically the Byzantine was at its height during the rule of Constantine and the later Emperors in the Eastern Roman Empire. Much use was made of mosaic, not only on floors but also on walls and ceilings. The paintings had a stiff, formal, linear style, but the colours were beautiful.

EGYPTIAN ART
It is characterized by a strange treatment of the human figure; the head being shown in profile, the torso towards the viewer and the legs again in profile. **Painting media included: forms of fresco and tempera and encaustic.** Figures were generally graded in size according to their importance in life. There was much relief-carving, also the making of large statues, and using different available stones and rocks. Largely the visual arts were attendant on the dead, being associated with his or her prowess in life, showing often interesting domestic and other details.

Etruscan carving in volcanic stone of a leopard, c. 550 BC. (Courtesy Museum of Fine Arts, Boston)

ELGIN MARBLES
Sculptures and decorative architectural fragments from the Parthenon and other Athenian buildings brought to London by the then Lord Elgin, who was Ambassador to Turkey in 1799 when Greece was part of the Turkish Empire. The marbles are now in the British Museum.

ETRUSCAN ART
Centred in Tuscany, Italy, the period lasts from about 700 to 100 BC. A seafaring people, the Etruscans were influenced largely by the Greeks in their arts. In their tomb-paintings there are some remarkably vivid and lively presentations of people, and decorative use of plant forms. Sites include Tarquinii, Orvieto and Vulci. Sculpture with some examples shows an affinity to the Greek Tanagra figures. Materials used were: terracotta, bronze and stone but not marble.

EUPHRANOR
Fourth-century BC Greek painter and sculptor; Pliny the Elder claims various works by him, included such as statues of Paris and Alexander the Great and numerous paintings and copies, all of which have disappeared.

EUTYCHIDES
Fourth-century BC Greek sculptor who was possibly a pupil of Lysippus. Finest work was said to have been the goddess of fortune with the River Orontes at her feet, a copy of which is in the Vatican.

EXEKIAS
Highly talented Greek vase-painter of the 6th century BC. The Vatican has one of his best works, an amphora decorated with two figures playing draughts.

'FARNESE BULL'
A Roman copy of a lost original by Apollonius and Tauriscus. It is in the National Museum in Naples and shows Dirce being tied to a bull by her stepsons Amphion and Zetus.

'FARNESE HERCULES'
A copy of an original by Lysippus in Naples, probably made by Glycon, an oversize statue of the great hero.

GRAVETTIAN
An Upper Palaeolithic culture, likely to have been a descendant of the Châtelperronian in central France. They made cave-paintings and were in all probability the painters of much of the work in the Lascaux Caves; also they produced small female figurines in mammoth ivory.

GUNDESTRUP BOWL
The vessel was found in a dismembered state

Animals in the caves at Lascaux. (Courtesy Direction Générale du Tourisme-France)

Animals at Lascaux. (Courtesy Direction Générale du Tourisme-France)

Relief carving of an antelope, Hittite, c. 1000 BC. (Courtesy Detroit Institute of Arts)

in a peatbog at Gundestrup in Himmerland, Denmark. The individual plates from which it was made were profusely decorated with repoussé work showing mounted warriors, horn-blowers and foot-soldiers with the long Celtic shields. It is likely to be of east Celtic origin of about the 1st century BC.

HITTITE ART

The Old Empire dates from about 1750 BC to 1450 BC, and the New from the latter date to the collapse about 1200 BC. Principal area was Anatolia, Asia Minor; works included zoomorphic pottery, small bronze figurines related to Cycladic, bronze animals and gold geometrically patterned ewers, pots, fibulae and bracelets, also stone figures and rock relief carvings.

KORE
A kind of Greek statue of a girl in long robes, usually dedicated to a particular sanctuary.

KOUROS
Greek statue of a standing, nude boy with one foot slightly forward. The Glyptothek Museum in Munich has an excellent example called the 'Apollo of Tenea'.

LASCAUX
The most famous painted cave in France, it was discovered in 1940, situated near Montignac, Dordogne. It displays fine paintings and engravings from the Upper Paleolithic period. The first evidence of Paleolithic art in France was discovered in 1834 at Chaffaud, Vienne, by Brouillet, when he came across a flat bone bearing an engraving of two deer dating from about 20000 BC. Older still are blocks of stone engraved with animal figures and symbols found at La Ferrassie, near Les Eyzies in the Périgord.

LOW HAM VILLA
Situated in Somerset, it is the site of a 4th-century Roman house which has a particularly fine mosaic pavement. This shows scenes from the Romance of Dido and Aeneas in Virgil's *Aeneid*. **One of the first examples which tells a story in a series of chronological scenes.**

LYSIPPUS
Greek sculptor working about 450 BC. Noted for his modifications with the treatment of the human figure, making the body more slender, muscles more prominent and heads slightly smaller.

MESOPOTAMIAN ART
The visual expression of the Assyrians, Babylonians and the Sumerians. The work of the Assyrians was to a degree an offshoot of the Babylonians and dates from about 1500 BC until the destruction of their capital city Nineveh in 612 BC. The produced striking relief work on limestone such as those in the Palace of Ashurbanipal, at Nineveh, hunting and battle scenes, which are now in the British Museum; the reliefs were originally polychromed. **A splendid example of sculpture is the ferocious 'Lion of Nimrud'.**

The Babylonians main medium was with bas-relief illustrated by **the magnificent Red Dragon on a blue ground from the Ishtar Gate at Babylon**; it is now in the Detroit Institute of Arts. Wall-paintings have been found at Mari, and these people were also expert in the cutting of green schist into small cylinder-seals.

The Sumerians settled in southern Mesopotamia in about 4000 BC. They also made cylinder-seals, intended to guarantee authenticity of messages; writing having been invented in the Uruk period. **The introduction of the potter's wheel resulted in an improvement in the making of ceramics.** Building interiors were decorated with cone mosaics in colours. They made bas-reliefs and in-the-round sculpture.

Standing male figure, Sumerian, c. 2700 BC. (Courtesy Detroit Institute of Arts)

MUMMY PORTRAITS

Mainly produced at Fayum in central Egypt, the method was to use colour mixed with wax on thin wooden panels or canvas on panels bound to the face by bandages. They are often extremely lifelike. Dating is from the 1st to the 3rd century.

MYRON

Fifth-century BC Greek sculptor, best remembered for '**Discus-thrower**', which is known by the numerous Roman copies.

NICHOMACHUS

Greek painter active about 350 BC who was a pupil of Aristides and praised by Cicero, Plutarch and Pliny.

OSEBERG TREASURE

Queen Åse, who was the grandmother of Harald the Fair-haired, the first King of Norway, was buried in AD 850 in a large longship under a huge mound at Oslo Fjord. Fortunately the circumstances preserved the materials, and as well as the ship, many objects including a bed, chariot and sledge have been recovered, pointing to the excellence of the Viking wood-carving.

PARRHASIUS

Fifth-century BC Greek painter; early writers record that he was **the first artist to give an individual character look to his portraits.**

PARTHIANS

These people came to Iran in about 250 BC from the steppe between the Caspian and the Aral. Their art was strongly influenced by the Hellenic; Greek motifs being used on coins. The best example of Parthian sculpture is the Tomb of King Antiochus I of Commagene at Nimrud-Dagh in North Syria. The most impressive object of the time is a bronze male figure from Shami.

PECH-MERLE CAVERN

In this place, situated in France, is some evidence that man's first attempts at visual expression goes back further into time than has been thought. On the ceiling of one of the chambers is a patch of natural clay which still holds simple drawings made with the fingertips of some Stone Age man, who experimented possibly about 30000 years ago; close by he or others left footprints.

From East Africa has come another sign that man may have indulged in picture-making long before the date of the earliest so far discovered paintings and engravings. At Olduvai Gorge there has come to light a living floor of Abbevillian Man, Lower Palaeolithic, that dates back to 400000 BC. Strangely on this floor were found lumps of red ochre, the nearest deposit of such

Rock engravings from Norway, probably Bronze Age. (Courtesy Historical Museum, Bergen)

material being about 40 miles (70 km) away. The man must have had a liking for such colour to have troubled to carry it that far. It could have been for an intended drawing somewhere, still not found, or perhaps just for body-painting in the way the ancient Britons used woad.

PHIDIAS

The greatest of Greek sculptors, he was active during the 5th century BC. Information about him and his works has come down with the writings of such as Plutarch. There were the chryselephantine statues of Athena Parthenos in the Parthenon and Zeus at Olympia. He advised Pericles on the construction of the Parthenon and may have actually made the design for it.

POLYCLEITUS

Fifth-century BC Greek sculptor, **said to have had more copies made of his works than any other sculptor.** His best-known work was 'Doryphorus', the spear-carrier, the finest copy of this being in the National Museum, Naples.

PRAXITELES

Greek sculptor, active in the middle of the 4th century BC. There is a copy in the Vatican of his 'Aphrodite of Cnidus'. He had particular success with the undraped female nude.

ROCK-ENGRAVINGS

These have been found in many countries including: North Africa, Russia, Siberia, Italy, France, Norway and Sweden. In most cases they are of hunting subjects, although with Scandinavian examples in particular they include boats, pin-head figures and sometimes abstract symbols. In the Vingen area of Norway there are at least 1500 known examples, and in northern Sweden at Nämforsen in Ångermanland there are 2000 figures. Strangely in next-door Finland such work is almost unknown. Engravings at the Lake of Onega and the White Sea area in Russia show signs of being carried out by the same culture as those in Norway.

ROMAN ART

At the start, in about the 3rd century BC, it was heavily influenced by Greek Art, not only by examples and copies being imported but also by the employment of Greek craftsmen and artisans. To purists Roman Art is a continuation or a vulgarization of the Greek; certainly with their portrait busts the Roman sculptors or those working under them often achieved a most convincing realism. Influences apart from Greek were various: Etruscan, Asiatic, Syro-Mesopotamiam and Palmyrene. Historical reliefs were popular, **the greatest here being 'Trajan's Column' covered completely with a corkscrew relief 650 ft (198 m) long** which provides a continuous narrative of the Dacian wars of the Emperor. Trajan had this erected in Rome in 114 with his statue on top, and later his ashes were deposited inside the pedestal. In 190 Marcus Aurelius put up his own column in Rome, with the relief showing his campaigns against the Marcomanni.

Painting was mostly used as an adjunct to interior decoration, adding three-dimensional

Head and shoulders of Flavius Sabinus Vespasianus Titus (AD 40–81). Typical of the realist Roman portrayal of the period. It is in the Museum of the Hunting Lodge at Erbach, West Germany. (Photo by the Author)

effects, to make the wall appear to dissolve into a series of receding vistas; this had its peak in the Flavian period about AD 80. The use of mosaic pavements was widespread and the craftsmen employed developed a high degree of realism. Nearly everywhere the Romans went signs of their art remain as evidence, whether in the middle of the desert, or at towns still being excavated like Conimbriga, Portugal, or at Colchester, where in the Museum is a fine gravestone commemorating one Marcus Favonius, a Centurion of the 20th Legion. Made of Bath stone, it may have originally been polychromed; it dates from AD 60. In London sculpture recovered from the Roman Temple of Mithras includes a life-size head of Mithras. An excellent stone relief of the god stabbing the sacred bull, was found near the little stream, called the Walbrook in 1880.

SAXON ART

The excellence of the metalsmiths and craftsmen-designers can best be judged by the Sutton Hoo Treasure, recovered in 1939 from a ship-burial in a barrow on rising ground close to the River Deben, to the east of Woodbridge, Suffolk. The objects recovered included: a gold buckle 5 in (127 mm) across with interlacing animal pattern, a gold purse still containing 40 Merovingian gold coins, and many pieces of jewellery decorated with garnets, mosaic glass and filigree; a sword pommel in gold and jewelled with sheath decorated with jewelled scabbard bosses. There was a fine iron helmet with a silver-covered crest, silvered bronze eyebrows and an iron visor with gilt nose and moustache.

Another striking example of the period are the jewels of King Alfred which were found at Newton Park, between Athelrey and Bridgwater, Somerset, in 1693. The most outstanding of these is the so-called 'Alfred Jewel', the gold mount being decorated with cloisonné enamel (with this method, small fences of gold or silver are soldered to the base first and then the interstices are filled with the vitrified enamel pastes) and bearing the inscription 'Alfred had me made'; at the base is a finely wrought gold animal head with a socket, the supposition being that it probably originally had a thin rod attached and may have been used as a pointer reading aid with manuscripts, and that the figure in the jewel could represent Sight.

SCOPAS

Fourth-century BC Greek sculptor who had a considerable reputation for the emotion, notably of sadness, he could instil into his faces.

SCYTHIAN ART

These wild nomads from the steppes of southern Russia were widely known for the skill and originality of their treatment of designs with gold, silver and bronze. Most of the motifs were zoomorphic, and the stag appears many times. In the burial-mounds of the chieftains, magnificent objects have come to light. Techniques included repoussé, solid modelling and beating in sheets for manipulation.

TANAGRA
A centre in Boeotia, Greece, where many small terracotta figures have been found. The making of these had its peak about the 4th century BC.

TIMOTHEUS
Fourth-century BC Greek sculptor who made the models for the Temple of Aesculapius at Epidaurus; he also worked on the Mausoleum of Halicarnassus.

USHABTI FIGURES
These were made of faience, wood or stone and were generally between 4 and 9 in (100 and 230 mm) in height and have been found in large numbers in Egyptian tombs. The word *ushabti*, in ancient Egyptian means 'answerer'; the idea of placing these figurines in the tomb was that they would 'answer' when the deceased was called to work. They would do this for him, so that, included in the modelling or carving, would be tools and agricultural implements.

VENUS FIGURINES
Among the earliest known examples of sculpture are these small objects from Aurignacian sites, dating about 25 000 to 22 000 BC. One of the most celebrated is the 'Venus of Willendorf' from Austria; it is a limestone carving only $4\frac{3}{8}$ in (111 mm) high.

'VENUS OF MILO'
Probably the best-known Greek statue of Aphrodite. It was found on the island of Melos in the Aegean Sea in 1820. It was worked in the 2nd century and appears to show the influence of Scopas. It is now in the Louvre.

ZIWIYÈ
A site, connected with ancient Zibia, a fortified town of the Mannians, is close to Sakiz in the Azerbaijan about 75 miles (120 km) from Lake Urmia. A treasure hoard was found there accidentally in 1946 by a young shepherd boy. The items that came to light included: ivory, gold and bronze objects of talented workmanship, originating dates being somewhere between 800 and 600 BC. Influences evident are Assyrian, Scythian and Mannian. Motifs included anthropomorphic symbols, such as winged genii, man-bulls, also in evidence are gryphons and sphinxes. **The finest piece is the Lion Armlet in gold.**

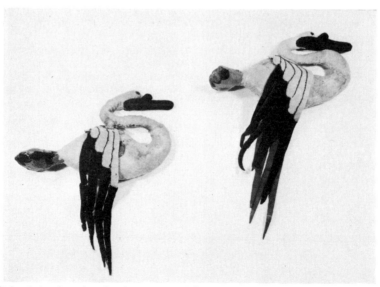

Swans in flight, dating from the 5th–4th century BC. They were made by the Altaians and were found in the frozen tombs of Pazyryk, and are from the Hermitage Museum in Leningrad. The Alti mountains border on Outer Mongolia. During the excavations in the Pazyryk tombs a number of treasures were found in a barrow at Bashadar in the Central Alti, including a wooden coffin with a procession of tigers carved on the lid. (Courtesy Hermitage Museum, Leningrad)

Chapter Three

Roots of Eastern Art

The Eastern civilizations have long and varied histories. In their beginnings there are Palaeolithic and Neolithic creative art forms. Then later Buddhism stands as a common factor for the three main peoples, the Indians, the Chinese and the Japanese, although each of these developed and progressed with a series of styles of their own. Indian sculpture has been almost from the start an evocative and living representation of the human body, combined with a masterly use of decorative motifs, rich in imagination and detail; her paintings reflecting the same mannerisms whether as large murals or miniatures. The Chinese shone in the early periods with the use of bronze, both technically and with their designs; later dynasties produced exquisite painting styles, in some cases long before their Western counterparts. Japan, the individual, set apart from the continental mass, may have suffered repetitive foreign influences, mostly from China, but she returned each time to her own traditions, a use of subtle decorative design, a particular way with painting and in the later periods the exquisite coloured wood-block prints.

AJANTA
The site of a Buddhist cave complex in the Aurangābād district of Bombay. Here is one of the finest collections of this religion's art, and a moving memorial to Buddhism. Ajanta's rock-cut temples had been undisturbed for centuries. There are 29 main cave chambers, the earliest dating from about 300 or 200 BC.

The fresco paintings discovered in 1817, are one of the most interesting features of Ajanta. The technique differs from traditional fresco, in that the rock walls were first prepared with plaster bound with cow-dung containing rice-husks and then surfaced with a fine white gypsum, and on this the paintings were made, the earliest being put at about 100 BC. Some, however, were made in the reign of the Vākātakas. (Harishena the last known ruler of the Vatsagulua branch of the Vākātakas has been assigned the period between 475–500.) The subjects range round the *Jakata* stories (popular tales of the former incarnations of the Buddha); on the walls are the representations of all human emotions: love, sacrifice, sorrow, all set down with a moving sincerity. The Bodhisattvas (potential Buddhas) and other figures, animals, plants and flowers are deployed to leave an unforgettable impression.

AMIDA BUDDHA
The work of an early Japanese sculptor Jōchō, who died in 1057. It differed from the earlier statues in that Jōchō made the figure in separate pieces and then put these together. **It is one of the traditional patterns for many statues to come.**

ANGKOR VAT
A huge terraced temple to Vishnu built by King Suryavarman in the 12th century, the greatest work of its kind by the Kmers in Cambodia. An inner gallery is covered with carved figures, **an area of 7000 sq ft (650 m²) with 1750 life-size dancers.**

ANURADIIAPURA
In Sri Lanka; the occupation of the site lasted from 200 BC until 780. One of the stupas has a semicircular threshold stone covered in carvings of geese, elephants, horses and lotuses.

BIHZAD, KAMAL UD-DIN (1468–1515)
The greatest Persian miniature-painter, orphaned while young, he was raised by Mirak Naqqash, another painter. Bihzad, nicknamed by some the Raphael of the East latterly ran the celebrated Herat Academy.

Ganesha, or Gana-pati, is the name of a Hindu god, the son of Siva. Images of Ganesha usually show him as a stout man with an elephant's head and four arms. One example shown here is of a volcanic stone figure on a lotus base. It is Javanese and dates from the 10th century. Its simplicity is interesting in contrast to the other illustration, which is carved from grey stone. Here there is a wealth of decoration, from the grotesque mask at the top to the elaborately worked head ornament. This one is from Hoysala and dates from the 11th century. Both of these figures illustrate clearly the extraordinary virility of the early arts from India and nearby countries. (Courtesy Spink & Son)

BOROBUDUR

In Central Java. Here is a collection of Buddhist buildings with **the largest stupa in Indian architecture, which includes 500 statues and about 1400 bas-reliefs.**

CH'ANG-SHA

In the Hunan Province, China; tombs from the period 300 to 100 BC have produced a number of interesting finds, including **the earliest known writing-brush**, also wooden figures of monsters with strange protruding eyes and tongues.

CHIA-HSIANG HSIEN (c 147–68)

Wu family burial-site in the Shantung Province, China. These Han dynasty tombs have produced much to add to the knowledge of this culture. There are notable murals of figures, animals and birds, which are drawn in black and coloured with red, blue and yellow.

CH'IEN HSÜAN (c 1235–90)

Chinese painter and calligrapher, he was the last of the four masters of flower- and bird-painting who set a fashion in the Five Dynasties and Northern Sung period. The Smithsonian Institution, Freer Gallery of Art, Washington DC, has a good example of his exquisite treatment of flowers.

Silver repoussé plaque of an ibex. T'ang Dynasty. It is 5 in (127 mm) in length. (Courtesy Barling of Mount Street)

CHINESE DYNASTIES
(to which works of art are classified)

Shang Yin   c 1766–1028 BC

Chou   c 1027–256 BC

The Warring States   481–221 BC

Ch'in   221–206 BC

Han   206 BC–AD 220
From this time can be traced the foundation of sculpture and painting in general.

The Six Dynasties   221–589

Northern Wei   386–535

Sui   581–618

T'ang   618–907
The Golden Age for all the Arts

The Five Dynasties   907–960

Sung   960–1278
One of the finest periods for landscape-painting.

Yüan   1260–1368

Ming   1368–1644
The great period of advance for ceramics.

Ch'ing or Manchu   1644–1912
High-quality jade- and ivory-carvings.

CHI PAI-SHIH (1863–1958)
Leading traditional Chinese painter of landscapes, animals and flowers, **acclaimed as the great 'master of all masters'.**

CHU TA (c 1625–1700)
A dumb Chinese painter, also known as Pa-ta-shan-jên. He was a landscape-painter, but is best known for his delightful witty brush drawings of comic birds, insects and fish. These he indicated with the minimum of strokes.

DONG-SON
A site in the Than-Hoa Province of Vietnam of the south-eastern Asian Bronze Age. It produced evidence of an Indo-Chinese culture. Art objects found included huge bronze drums, tomb furniture and ornaments.

DVARAVATI
A kingdom of the Lower Menam, Thailand, from sites such as Pra Pathom and Pong-T'uk have come

White marble figure of 'The World Watcher'. T'ang Dynasty. For all its appearance the figure is only 19½ in (495 mm) high. (Courtesy Staatliches Museum für Völkerkunde, Munich)

'Guardian Lion'. Chinese white marble carving, probably early T'ang Dynasty. (Courtesy The Cleveland Museum of Art)

Ridge tiles carried out in the T'ang style, dating from the Ming Dynasty. The figure on the right is Lokapala. Such figures were intended as guards against evil. (Courtesy Spink & Son)

figures of Vishnu with head-dresses in the shape of mitres.

ELEPHANTA
On an island close to Bombay there is the large rock temple at Elephanta with **the huge statue of Trimurti, the Brahmanist Trinity.**

ELLORA
A rock-cut cave site in the Bombay State it lies about 30 miles (50 km) to the south-west of Ajanta on a tributary of the Godavari River. There are 35 caves, with Buddhist and Hindu about equal in numbers and a few devoted to Jain (of or pertaining to a saint). The largest is the Mahawara monastery cave, 58 × 117 ft (17·67 × 35·66 m). There is also the 'assembly' cave 86 ft (26·21 m) deep with a huge statue of Buddha in the teaching posture. To be noted is the 'Siva's Paradise' temple, cut out from rock to imitate the Kailasanatha temple of Rajasinha Pallava at Kanchipuram.

GANDHARA SCULPTURE
Unique sculptures in which can be seen a marriage of Hellenistic with the Buddhist sincerity. Dating is from about 100 BC to about 500. Here **Buddha is shown for the first time as a human and not as some stylized symbol.** The area is now in Afghanistan.

GUPTA DYNASTY (4th to 7th centuries)
The Golden Age of Indian Art which is named after the Northern Indian Gupta rulers.

HAN KAN (*c* 720–*c* 780)
The famous Chinese painter of horses. He has been the strongest influence on those who have followed him in this subject. His brushwork is virile and the images evoke strength and action.

HARUNOBU, SUZUKI (*c* 1725–70)
The great Japanese master of the Ukiyo-E or colour print. His subjects included romantic idylls, night scenes and snowscapes, the last beloved by the Japanese print-makers. As with others he has been the victim of considerable forgery.

HINDU IMAGES
These included a large number of divinities and allied figures; Shiva and his wife Parvati, Vishnu, Devi, Brahma, Surya, Narasimha (the man-lion), Krishna, Rama and Ganesha.

HIROSHIGE, ANDO (1797–1858)
The last of the great masters of the Japanese colour-print, he studied with Toyohiro. He was especially skilled in creating atmospheric effects. His subjects included: 'Eight Views of Omi' and the famous series 'The Journey to Tokaido'.

Standing Buddha made from brass, 38⅝ in (981 mm) high, from North India or Kashmir, early 8th century AD. (Courtesy The Cleveland Museum of Art)

Standing figure of Abalokitesvara with arms held forward as a frontal likeness of the Buddha. It dates from the 7th century. (Courtesy Spink & Son)

Sculpture in Hindu style from Khajuraho in Central India. (Courtesy Government of India Tourist Office)

'Guardian of Shiva'. Stone sculpture from Mysore, India, 13th century. (Courtesy The Cleveland Museum of Art)

'Waterfall at Amida' by Hokusai. (Courtesy of the Trustees of the British Museum)

Some of his pupils labelled their work Hiroshige II, III, IV, etc.

HOKUSAI, KATSUSHIKA (1706–1849)
Renowned master of the Japanese colour-print, he was prolific not only with print-making, but also when working as an illustrator and a painter. Well-known series prints of his include: 'Thirty-six views of Mount Fuji' and 'Amusements of the Eastern Capital'. The British Museum has among others 'Waterfall at Amida' which gives a graphic example of his power and feeling for design.

HSIA KUEI (active 1180–1230)
Chinese painter who used an individual sharp brushstroke, and produced a great luminosity from just ink alone. He probably worked fast by the look of the strokes, and this produced lively atmospheric effects.

HUANG CH'ÜAN (active late 10th century)
He was the great flower-painter of the Early Sung period.

HUANG KUNG-WANG (1296–1354)
Chinese painter who was one of those to introduce the subtle technique of the dry brushstroke, also the eldest of the Four Great Landscape Masters of the Yüan dynasty.

JAPANESE ART PERIODS
Archaic   Prior to 552
Totem figurines back to 4500 BC

Asuka   552–710
Wood sculpture 603, Chūgūji tapestry 622, Tori the sculptor casts bronze Shake Trinity 623. Gigaku masks and **earliest oil-paintings in the world on the subject of the life of Buddha in the Tamamushi Shrine.**

Nara   710–94
Kōfukuji statues of eight guardian devas 734; sculpture of Ganjin c 763

Heian   794–1185
Statues of Shintō goddess, Nakatsu-hime, painting of the death of Buddha 1086, Ban Dainagon picture scroll by Tokiwa Mitsunaga.

Kamakura   1185–1333
Jigoku hell scrolls c 1200, portraits of 36 poets by Fujiwara Nobuzane, Tengu goblin scrolls.

Muromachi   1333–1568
Zen priest Kangan portrait by Kaō Ninga, c 1345, portrait of St Francis Xavier c 1550.

Momoyama   1568–1600
Pictures featuring Western visitors to Japan.

Edo-Tokugawa   1600–1853
Utamaro, Hokusai, Hiroshige, print-makers, all active.

Transition   1853–68
Pictures of Perry's ships, picture of Shōgun giving an audience to Townsend Harris.

Meiji   1868–1912

KAKEMONO
Japanese wall-picture painted on a roll of silk or paper and mounted on rollers for storage. It is intended to be hung vertically.

KANŌ SCHOOL
Japanese school that started in the middle of the 15th century; it was founded with Kanō Masanobu (c 1434–1530) and Kanō Motonobu (1475–1550) and went through in an unbroken line until the 19th century.

KENZAN (1663–1743)
Japanese painter and potter who was to produce exquisite ware with a delicate harmony between the shape and ornamentation. His brother Kōrin,

Sculpture from Khajuraho. (Courtesy Government of India Tourist Office)

also a painter, used to decorate a number of his wares.

KORIN SCHOOL
Founded in the Edo-Tokugawa by Honnami Kōetsu. It was primarily concerned with decorative art.

A stone statue of Narasimha from the Orissa province of north-eastern India (dating from between the 14th and 16th centuries). In Northern and Eastern India the main schools of art from about 850 onwards seem to have been those of Bihar and Bengal (Pala Dynasty), Orissa and Bundelkhand and Rajputana. Sculpture on the whole in Orissa is typically Brahmanical and can be connected to a series of great temples at Bhuvanesvara, Konarak, Puri, etc. (Courtesy Ashmolean Museum, Oxford)

KOSE NO KANAOKA

Japanese landscape-painter active in the latter part of the 9th century, esteemed as the equal of China's Wu Tao-tzu.

KU K'AI-CHIH (*c* 345–405)

He is credited as being the initiator of Chinese Landscape-painting

LI LUNG-MIEN (active late 11th century)

He was Chinese and his real name was Li Po-shih. He excelled with landscapes and also painted horses, ranking next to Wu Tao-tzu.

MAJAPAHIT

In East Java, at Candi Jago, there are notable reliefs clearly influenced by the cut-out figures used in the *wayang*, the attractive shadow puppet theatre, where the figures are operated by rods.

MAKEMONO

A Japanese painting on a silk or paper scroll that is intended to be hung horizontally.

MAMALLAPURAM

Near Madras, it is the site of a remarkable relief cut into a huge boulder, which depicts the Descent of the Ganges, and is 90 ft (27·43 m) long and 30 ft (9·14 m) high, with life-size humans and animals, and mingles with stories from mythology.

MARUYAMA OKYO (1733–95)

Esteemed realistic painter of the Edo-Tokugawa period; he founded a School that professed to follow the old Chinese artists, although in fact it showed a distinct influence from Western Art that was beginning to seep through.

MASONOBU, OKUMURA (1691–1768)

He worked during the development of the Ukiyo-E, colour-print technique. **He has been reported as initiating the *urishi-e* in which clear lacquer was brushed over the print.**

MATHURA

The old city on the River Jamuna in Uttar Pradesh, India, famous for the connection with Krishna and the *Mahabharata* (a sacred book of the Hindus, and one of the two great epics of ancient India. It is probably the longest epic in the world, being about eight times as long as the *Illiad* and *Odyssey* together). The sculpture displays influence from the West. There is an interesting stone-relief panel, with a tiny figure of Jain Tirthamkara in the centre, round which are a variety of motifs including a swastika, fish, flowers and a repetitive plant pattern.

MA YÜAN (active 1190–1224)
Chinese painter, a member of the Southern Sung Academy, and coming from a long line of painters. He was nicknamed One corner Ma from his manner of putting the emphasis into one corner of a landscape. He had a talent for handling soft, misty effects.

MI FEI (1051–1107)
Chinese Sung dynasty calligrapher, also a writer on the theory of painting.

MU CH'I (active mid 13th century)
Chinese painter who spent part of his life as a Zen monk in a temple near Hangchou. He was one who started the free brush and drop of ink manner that gave an unrestricted effect.

MUGHAL ART
A school of Indian miniature-painters which flourished under the Mughal Emperors; the first of these was Babur, the Turk, who conquered India and started the dynasty in 1526, during this period there was a fashion for copies of Persian miniatures. A school was founded by his son Akbar, some teachers being drawn from the Herat Academy.

NANGA SCHOOL
School of painting in Japan; it was largely based on influences from China, and confusion can be caused with Japanese work from this source and some Chinese because they are so alike; perhaps the former, however, is more gentle at times, and yet in some examples the Chinese motifs become exaggerated by the Japanese. It started to flourish at the end of the 17th century and lasted until the latter part of the 19th century.

NARA PERIOD
Lasted from about 710 until 794, during which time those Japanese who travelled to China, often returned full of inspiration and ideas for emulation. They styled their towns with a Chinese look, and set out to adopt Chinese manners and methods. Many fine bronze casts of religious figures and painted clay figures of Bodhisattva and others were made, showing clearly Chinese and even Indian influence; but in contrast to the sublime harmony of such were the *lokapala* or ferocious guardian figures of the Buddhist faith.

NI TSAN (1301–74)
One of the Four Great Landscape Masters of the Yüan dynasty, the other three being, Huang Kung-Wang, Wu Chên and Wang Mêng. Ni Tsan employed a deceptively simple looking dry-brush manner, that has been considerably aped and copied since his time and causes great confusion with authentification.

PALEMBANG
A collective heading for finds in Sumatra which point to the flux of influences passing along trade routes. Burial-places have produced unusual carvings of men riding buffaloes, and reliefs of men with drums of the Dong-Son type. There is also a notable figure of Bhairava just over 14 ft (4 m) high.

POLONNARUWA
In Sri Lanka, the centre of building activity between 1164 and 1197 under King Parakrama Bahu. There were numbers of large statues, including a figure of a scholar reading a palm-leaf book, over 11 ft (3 m), a Buddha in the Parinirvana attitude (lying down) nearly 50 ft (15·24 m) long and one of Ananda 25 ft (7·62 m) high.

PONG-T'UK
A site in Thailand about 34 miles (55 km) from Bangkok which somewhat unexpectedly produced among other objects, a Herculaneum-style Roman lamp dating from the 2nd century AD, pointing to either an exchange of trade items or perhaps more thought-provoking the passing of some early Roman mission.

SANCHI
In the Madhya Pradesh Province of India, is this small sandstone hill, a cache of some of the most interesting sculpture in the country. Hidden by encroaching trees and bushes, it was found in 1818. The principal interest lies in the carvings on railings and gateways round the stupas. These show, mostly in relief, motifs coming from the previous life of Buddha and his previous births; the handling of the whole theme is free and virile.

SESSHŪ (1420–1506)
Japanese painter in inks and colour who worked to the Chinese style; but who at the same time rose above just producing pastiche-like emulations; his brushwork is lively and his control of tones gives considerable atmosphere to many of his scenes, as with 'Winter Landscape' in the National Museum, Tokyo, which with its contrast of prickly trees and angled mountains has a striking atmosphere. His style was followed by Sesson (1504–89).

SHEN CHOU (1427–1509)
One of the leading Chinese painters in ink and a calligrapher. His landscape style was in the manner of the Four Great Landscape Masters. He founded a school in the Kiangsu Province.

SHIH CHAI SHAN
Lying about 15 miles (24 km) south of Kunming in Yunnan, China, it is a burial-site from which

'Quand tu tiendras cet éventail' by Utamaro. (Courtesy Huguette Bères)

have come a number of notable finds, including some bronze drums that have groups of figures engaged in rituals and battles on their top surface; these are of a Bronze Age culture of a non-Chinese type. There are also versions of a type of Dong-Son drum and one interesting bronze medal that points to the possibility that a form of bull-fighting may have taken place in the area.

SHUNSŌ, KATSUKAWA (1726–92)
Japanese colour-print master, who was the master of Hokusai. He produced among other subjects realistic and natural portraits of actors.

STUPA
Originally it was just a simple Indian burial-mound. Then it progressed to a dome-like structure of unbaked bricks. As holders of relics and other religious objects concerned with Buddhism, stupas became more and more elaborate; stone was used for construction; railings and gates enclosed them to guard against evil spirits. Sizes ranged from small miniatures to the enormous examples in Northern India that soar up to some 400 ft (120 m) in height. **Outstanding examples are the Great Stupa at Sanchi and the Stupa at Borobudur.**

SURI-MONO
Small colour-prints intended to be sent on anniversaries or for some special commemoration. They started to gain in popularity in the late 18th century.

SU SHIH (1036–1101)
One of the leading painters and calligraphers of the Sung dynasty, also a poet-philosopher.

T'ANG YIN (1470–1523)
One of the finest painters of the Ming dynasty and also one of the best known in Europe, he had a talent for subtle treatment with figures and a careful balance with his compositions. As with others of the popular Oriental artists he has been forged many times.

THANGKAS
Painted or embroidered banners in Tibet, they first appeared in the 10th century; themes included

'Pheasants and Flowering Trees' by Wang Wu dated 1662. Ink and full colours on silk, 5 ft 6 in × 3 ft 3 in (1676 × 990 mm). (Courtesy of the Trustees of the British Museum) (It is in the Eumorfepoulos Collection)

The Wheel at Konarak Sun Temple which was originally built around AD 1250.
(Courtesy Government of India Tourist Office)

Granite horse, Konarak. (Courtesy Government of India Tourist Office)

representations of deities and magic circles for calling a god. They could be hung in temples or carried in processions. Another form of painting common to Tibet is the 'wheel of life' which shows the states of possible transmigration for the human soul, in the shape of a large wheel which is held supported by strange monsters.

TOBA SOJO (1053–1140)
A Buddhist priest, also called Kakuyu, who had considerable talent for caricature, and also religious subjects.

TOKUGAWA PERIOD
That which started with the struggle for power after the death of Hideyoshi. It was one of his generals, Tokugawa Ieyasu (1542–1616), who took over and started a *régime* which governed Japan for more than 250 years, bringing peace and considerable prosperity. At the same time there was a rejection of the outside world, and Christianity was suspect as an excuse for Western Imperialism. In this shut-off alone atmosphere the arts flourished. It was the **time above all for the greatest production of fine colour-prints (ukiyo-e).** Towards the end of the period the Americans sailed in under Commodore Matthew C. Perry.

TOSA SCHOOL
A school of painting mainly in the hands of the Tosa family, which set out to use traditional themes and ideas as subject-matter. It was started by Tosa Mitsunobo (1434–1525) and was a reaction against the rather formal manners of the Chinese-inspired Kanō school. Their work was an underlying influence for the colour-printing to come later, also it was an inspiration to the development of the Japanese arts and crafts in general.

TOYOKUNI, UTAGAWA (1769–1825)
Japanese master of the colour-print and a highly productive illustrator.

TRA-K'IEU
It lies to the south of Tourane in central Vietnam, and although there are no buildings left on the site, an amazing amount of sculpture and carved stonework was discovered there. The finds include heraldically treated lions, idealized but realistic elephants, dancers executed in a lively manner suggesting movement; interesting information is given by the figures of two polo-players.

TSOU I-KUEI (1686–1772)
Chinese court painter, with a fine talent for flowers.

UKIYO-E
A term somewhat parallel to genre in Western painting. In Japan it was applied to pictures and coloured prints that had subjects connected with everyday life that the people as a whole could understand.

UNKEI (1142–1212)
Japanese sculptor who produced works of great realism such as the coloured carved wood statue of the priest Mūchaku.

UTAMARO, KITAGAWA (1753–1806)
Master of colour-prints with exceptional skill with his portraits of women.

WANGS
Four Chinese traditional landscape-painters influenced by the Four Great Landscape Masters. The Wangs were: Wang Shih-Min (1592–1680), Wang Chien (1598–1680), Wang Hui (1632–1720) and Wang Yüan-Ch'i (1642–1715).

WU TAO-TZU (c 700–c 760)
An almost legendary figure in the history of Chinese painting. Although none of his originals still exist, there are some copies. He is credited with the start of figure-painting and also a technique for printing on silk.

YAMOTO-E
A style of Japanese narrative-painting on scrolls; sometimes they are up to 100 ft (30 m) long. Scroll-painting was brought to Japan from China in the 8th century.

YÜN SHOU-P'ING (1663–90)
Chinese flower- and landscape-painter. He had a pleasant lively brushwork, and his flower pieces are models of elegance and observation.

Chapter Four

Primitive and Magic Art

Move into the area of the true primitive arts, and the accepted rules, aesthetic theories and judgements are left behind. The peoples who have lived and live in primitive societies act on tribal conventions and instilled rules that are to a degree governed by superstition, magic and taboos. The roots of these things of the mind go back rather deep, to the soul as a Mannikin, as it is glimpsed within the man, a phenomenon that must be protected. From the different areas of the world have come varying beliefs as to this Mannikin; the Australian Aborigines have seen it as a little body that on death jumps out of the man and runs behind a bush; the Eskimos feel that it is in the same shape as the man's body but more subtle and ethereal; the Malays have seen it as a little man the size of a thumb and the Nootkas of British Columbia believed that the soul is tiny and is standing up inside the human head, and as long as it remains erect all is well, but if it falls over the man will go mad.

It is with such problems of the supernatural and good and evil magics that the craftsmen working on their figures, masks, fetishes and art forms have been largely concerned. These arts are not the result of intellectual reasoning or working out; they are primarily an answer to intuitive impulses, that very often are connected with protection. To the primitive man there is a mystical element in many things that he sees and very often in what he has made or another tribe close by has made.

Symbols of life, fertility and divinity have been painted, carved and woven into objects. One of the most common of these is the swastika, a symbol of the Sun's movements; it appears in widely scattered places round the world. All primitive art seems to avoid any attempt at realism, it is much more concerned with evoking the power for good, for protection or for evil by some fetish figure that can often assume a quite horrific image.

But the primitive is still a man; and although his mind may be working along quite different channels from our own, what he creates artistically can result in objects which satisfy his search for what he interprets as pleasing or even beauty.

ABORIGINAL
The visual expression of the Australian Aboriginal is primarily concerned with painting and engraving or rather quite deep incising into the earth or rocks; it has some affinity to the work of the African Bushman. In eastern Australia the Bora ceremony for tribal initiation would have been that the candidate would be led blindfold along sacred drawings cut into the ground, a series of abstract geometric lines that formed concentric squares, wave forms and complicated curves. In another place the Warramunga tribe would prepare complex ground drawings in association with the Wollunqua totem. This would be a complicated ritual as the artists worked as a group and each one would be first painted as a part of the whole, and he would then transfer the design on himself to the main drawing on the ground.

APOTROPAIC IMAGERY
An abstract or symbolic representation made specifically to protect from ghosts and evil spirits. With many peoples it might take the form of a head or mask with grossly staring eyes, the image thus being given hypnotic powers in the mind of the maker. Many of the users would feel that the stranger and more terror-inspiring the image was made, the more it would destroy the evil devils, as they would find that they were being confronted by their own likenesses.

Lifesize Ashanti goldweight. (Courtesy Christie's)

ASHANTI

These people of Ghana have had in one way or
another their art related to gold. Some of the more
fascinating images are the tiny brass weights, cast
by the lost-wax method; they have been made to
act as weights for gold-dust. They take the form of
very small figures occupied with various activities,
and if they were 6 ft (2 m) high instead of about 2
in (5 cm) at the most they would be very striking
examples of modern sculpture; animals of strange
and weird shapes are also used. The delicacy of
workmanship is of a very high standard.

AZTECS

A warlike people who came from the north and
settled in the Valley of Mexico in the early part of
the 14th century; their capital was Tenochtitlan,
which stood on the site of the present-day Mexico
City. The Spanish completely razed the original
city during the Conquest in 1521, and tried to
destroy any works of art they found. Influences on
Aztec art came from the tribes who have lived
there before them; to such influences the Aztecs
added ingredients of their own, a distinct
savagery; they used past motifs and forms to pile
on the terror, as with the massive figure
'Coatlicue' (Lady of the Skirt of Serpent), which is
8½ ft (2·5 m) high, and is in the Museo Nacional,
Mexico City. She has the head of a great long-
fanged snake, a skull at the waist, and the whole
treatment speaks of brutality. They had the
Hieroglyph books, with pages of deerskin or a
kind of thick paper made from agave fibre, the
contents of which were magical or religious, and
some had forms of a calendar.

BENIN

A city in west Nigeria associated with some of the
most remarkable bronze-castings; a number of
them are among the few examples in Primitive
Art that are presentations of a striking naturalism.

*Maize Goddess (Pre-Columbian) – Aztec Culture.
(Courtesy The Baltimore Museum of Art)*

*'Seated woman'. Pre-Columbian, c AD 900–1200. 15¼ in
(394 mm) high in terra cotta. (Courtesy Gimpel Fils)*

The method of casting was the *cire perdu*, lost wax, and the modellers and the casters achieved a fine quality in surface finish and realism. Production is likely to have started at about 1400, and the discovery of the bronzes was made by a British expedition to the city in 1897. Many of the motifs and ideas seem to have a link with Europe, although there is no definite proof of this. Grotesque and fierce qualities are found; for example, there is the imaginative head of the giant Enowe, who has snakes coming from his nostrils, and a frog coming out of his mouth.

CALABASH
The wooden shell of a gourd which is often decorated with colour, poker work or engraving.

CARVING AND DECORATING
Many of these peoples have shown considerable skill with pattern-making and decoration, both with domestic utensils and ritual instruments. Those of Admiralty Island (Pacific Ocean) carve wooden bowls into the shape of crocodiles, tortoises and birds and then apply geometric designs in white, possibly made from ground-up sea-shells. Warriors of Borneo would paint their long shields with lurid tortuous designs which incorporated two great staring eyes. In the Nicobar Isles (Bay of Bengal) ritual evil, scaring models were made in the forms of simple carved dolls, small ladder-like shapes to represent rows of dancing men and women. Malayans made woven bird-shaped baskets for betrothal gifts; the Balinese decorated the handles of the kris by carving small figures and insetting fragments of coloured glass, they also produced fine large-size pieces such as Krishna riding on a boldly carved eagle with widely spread wings, a frequent motif in Bali where Hinduism has survived. The scope of decoration points to some need, sensed or unconscious, to label things or just to make them more satisfying.

CHAVIN CULTURE
The oldest civilization in South America, it takes its name from Chavín de Huántar in the northern highlands of Peru, on a tributary of the Marañón River. It dates from about 1200 BC to AD 200. Carved motifs on the buildings include fantastic birds and writhing serpentine forms, on some slabs are low-relief savage chunky stylized human figures.

CHILKAT
A branch of the Tlingit Indian tribe of Canada, they are centred on the west coast. The people produced some exquisitely decorated blankets, which were woven from mountain goats' hair and cedar bark. Their unusual patterns were made up

Benin Bronze Male Head. (Courtesy Sotheby Parke Bernet)

Canoe Prow Figure – Solomon Islands. (Courtesy The Baltimore Museum of Art)

Club of carved wood with fibre, feathers, and coconut wrapping. Oceanic – Polynesia, Marquesas Islands. (Courtesy The Baltimore Museum of Art)

A stone-cut by Eliajah, an Eskimo artist. The design is first routed out of the soft stone and then printed in the normal manner. The bird shown is a type of long-eared owl that appears in the North. (Courtesy Gimpel Fils)

Stone figure erected by Eskimos to assist them to encircle game when hunting. (Courtesy Charles Gimpel)

'Bird' by Davidee. Eskimo art in stone, measuring $10\frac{1}{4}$ in (260 mm). (Courtesy Gimpel Fils)

of dismembered natural forms and laid down in abstract relationship to each other.

CHIMU EMPIRE

Dating from about the beginning of the 14th century, it lasted to the middle of the 15th century until it fell to the Incas. Their pottery was shaped by hand out of a reddish or black clay; the pots were then covered with a thin coat of white, and by the use of sgraffito the under colour was exposed to form the design. Symbols in the patterns latterly had political or religious significance, but with the early wares symbols for the Moon point to lunar rather than solar worship; the dog often represented the Moon while the serpent stood for the interior of the Earth. A goddess would be shown as a snail. After the Moon, water was worshipped as the bringer of food and fertility. Buildings were decorated with mud-plaster reliefs with Cubist-like dragons, birds and fish interspersed with geometric angular shapes

and whirling curves. The goldsmiths were skilled and produced many examples of quality.

EASTER ISLAND

A small, hilly, volcanic island in the Pacific Ocean, practically treeless, that lies about 2000 miles (3200 km) west of Peru and nearly as far from the nearest Polynesian islands. One of the most isolated islands in the world, what a shock it must have given the first intrepid explorers under the Dutch Admiral Roggeveen when he found it on Easter Day 1722. For they would have come face to face with **400 huge grey stone figures averaging 30 ft (9 m) high**, which in those days mostly had big red cylinders of tuff weighing several tons on long heads, with deep-cut eyes, low foreheads and long pendulous ears. After the Dutch came Gonzales in 1770, followed by Captain Cook four years later and then the Frenchman La Pérouse in 1786. There have been many expeditions that have attempted to solve the riddle of the great heads which are on equally large torsos that splay out at the bottom, but which have no feet; another point is that the heads have no back to them, all interest being concentrated in the features of the face. In quarries are partially completed other heads, and apparently in 1680, when there was a civil war, all work was stopped. Other puzzles on Easter Island are ideographic inscriptions on wooden tablets, and the contents of secret family caves which include carved wooden images and grotesque sculptures in lava; excavations have also brought to light mural paintings of large reed-boats with sails.

ESKIMOS

They are probably the last of the Mongoloid peoples to cross to North America from Asia; several thousand years ago they would have come to Alaska and then they spread out across the bleak treeless tundra and ice- and snow-bound wastes of the north Canadian coastline, some 6000 miles (9650 km) long, from the Bering Sea to Greenland; their pattern of life being to winter on the ice in igloos, hunting for seals, whales and walrus, and in the brief summer living in skin tents and chasing caribou and the like and fishing from kayaks. Creative work in the older culture was concerned with **carving and working ivory, from walrus and bone into items for use; scratched designs were generally based on circles, ellipses and dots;** other symbols followed, geometric and abstract, heads of fishes or animals with the eye as the important feature. Later Eskimo craftsmen have produced **small carvings in wood, bone, ivory and soapstone, a material that works easily, of human figures, bears, caribou and walruses that have convincing forms.** Other strange creature carvings are likely to be images of good or evil spirits. The Alaskan Eskimos have been noted for the fanciful ritual masks they have produced; there is one in the Museum of the American Indian, Heye Foundation, New York, that has strange dangling leaves and branches, rather like a form of mobile, and one glaring eye and many teeth. Eskimos in general often decorate their clothes by an appliqué of pieces of fur, either dyed or plain, from dogs, seal and reindeer.

FETISH FIGURES

(the word Fetish comes from the Portuguese *feitiço:* meaning amulet, charm or talisman) One of the principal motives for creating art forms with primitive peoples is to have something with which to placate the numerous spirits which they feel are around them and which they see in the elements and in nature; they are concerned also with the spirits of their ancestors. So a man makes

an object, a head or a figure, in which a spirit can live. This creation is a fetish, an inanimate thing that can be worshipped, and given magical properties.

In the Lower Congo the Nkindu fetish would have its own little house and its body stuck with nails in memory of past benefits conferred. A Tchumbiri man of the Upper Congo would look after the material needs of his fetish by chewing kola-nut and then squirting spittle into the mouth of the figure, in that way assuring that his fetish would look after his affairs.

The shape and form of the fetish can vary considerably, many are plain wood-carvings, others are constructed from wood, fur, feathers, bones and shells. In size they may be only a few inches or as big as a man.

FUNERARY

Much care is often lavished on shrines for the deceased. In the Solomon Islands (Western Pacific) a gabled erection might be made, the box being supported on two carved, headless figures of women. The box itself might be painted with birds, and the sloping roof carved with simple

Fetish image from the Congo. (Rautenstrauch-Joest Museum, Cologne)

Tapestry weave panel (c. AD 500) – wool and cotton, measuring 38 × 22 in (965 × 559 mm). Early Nazca, Peru. (Courtesy The Montreal Museum of Fine Arts)

repetitive geometric patterns; also in the Solomons the grave might be guarded by a wooden symbolic fetish, carved and poly-chromed. In north-west Australia the bones of a certain tribe's dead ones were placed in a cave, on the walls of which are some strange paintings of a row of human-like figures with large staring eyes, but no mouths.

INCAS

The Golden Age of their empire in Peru and parts of Bolivia, Chile, Ecuador and Argentina lasted from 1438 until 1532. In their buildings they showed themselves to be superb masons; the stones sitting almost with hair-line joints, achieved solely by abrasion. The top stone would be fitted into slings and then rubbed to and fro over the one below until the required accuracy had been achieved; a similar method was used with the Kmer monuments, such as Angkor Wat in Cambodia. The goldsmiths and silversmiths were equally skilled. In the American Museum of Natural History, New York, there is an impressive model of an alpaca made from sheet silver with ornament applied in repoussé. There have been found a number of gold ornaments with the form of some irate threatening god figure surrounded by complicated symbolic patterning. Unfortunately much was destroyed by the Spanish, who melted down objects from the temples and palaces and sent the bullion back as loot. The pottery appeared in many shapes but nearly always with **a distinguished form and rather a restrained decorative manner.**

MAORI

Some of the most elaborate wood-carvings in Polynesian art have been made by the Maoris; these appear on their canoes, communal houses and on their intimidating war clubs. From greenstone (nephrite) they would work neck ornaments for the head of the family in the form of a weird little figure called a Tiki.

MASKS

Wherever primitive ceremonies or rituals take place, the mask is nearly always a predominant feature. The wearer can adopt unto himself a force from the image his mask creates and instil this into the rest of his tribe. Thus a great part of the art creative energy is used on the construction and decorating of these greatly varied objects.

In the Bismarck Archipelago (north-west of the Solomon Islands) the members of a secret society, called the Duk Duk, would deck themselves out with conical headgear topped with bright yellow feathers, on the mask would be painted large white concentric circles, the eyes, and a wide

Left: A shrine figure in wood and copper, 19th century, from the Bukote tribe, Gabun. (Courtesy Gimpel Fils)

Right: Shrine figure. Fang tribe, Gabun, 19th century. Wood with black patina. (Courtesy Gimpel Fils)

Maori wood figure. It may have been part of a canoe ornament and is 17 in (432 mm) high. (Courtesy Christie's)

'Legendary Ancestor' – Maori. (Courtesy The Baltimore Museum of Art)

Dance mask in wood and fibre (19th century) from the Ivory Coast. (Courtesy Gimpel Fils)

Bella Coola wood frontlet, carved in high relief with a raven's head with a humanoid mask below the beak. Inlays of brass circles and abalone shell. (Courtesy Christie's)

white slash of a mouth. At initiation ceremonies in the Torres Strait (South Pacific Ocean), the youths to be brought in would see for the first time the sacred masks, in the shape of a very large head, outlined in white with the staring eyes and a gaping mouth. There have existed many of these secret societies, all with their own particular type of mask, by which they would be instantly recognized by each other, and in which, when garbed with grass skirts and other garments, they would appear to be able to take all law into their own hands and beat up non-members, terrifying the women and chasing the children. The simple natives would believe that such as the Duk Duk, when masked, had supernatural power and were no longer human.

Ritual and other dancers very often have their symbolic masks. There is an odd one connected with the natives of Nissau Island, rather like a clown's face with rectangular wings standing out each side. Again in the Bismarck Archipelago they have a very human-like face mask with dark eyes set in very bright whites; it also has strange butterfly-like wings; this would be worn by men going round the houses collecting contributions for a funeral feast.

Materials used are various: in the Augusta River area, Melanesia (the term Melanesia embraces all the islands from the Bismarck Archipelago in the north-west to the Fiji Islands in the south-east), great bird masks with long spiked beaks have been woven from dried grasses; at Warapu a boy initiate would wear a ring round one eye set with shells and seeds, and long thin feathers would be stuck in round the outside. The care of these exotic objects has been of first importance; in some places they have been kept in secret caves; in New Ireland (to the east of New Guinea) the secret society built its own mask-house.

At times the roots of various beliefs seem to have become confused, and even the members of a society may not know exactly what all the symbolism means. In the Gazelle Peninsula (Melanesia) the dancers would wear some very odd masks indeed, the white face part having features indicated with almost a European Look, then there would be, too, a wide-brimmed hat surmounted by intertwined carved wooden snakes. In Fiji there were crop-protecting masks, about 6 ft (2 m) high with violent triple faces in white on black or dark red-brown.

In Africa the craftsmanship increases and the uses vary more. In the Congo, hunters would turn out before the chase with huge carved wooden masks with grotesquely exaggerated features of the animal to be killed. Heavy wooden helmet masks, with staring eyes jumping out of their sockets, would be worn by the Isambo who live near Lusambo. Some of the most refined and expressive masks came from the Dan tribes of the Ivory Coast, and in contrast some of the most violent and shock-inducing examples come from the Ngere, many of these not being made from wood, but being basketry which would then be covered with clay and modelled and coloured. An important masquerade cult called *mmwo* (ghosts or spirits) is connected with the Ibo of the Onitsha-Awka district of Eastern Nigeria. They would wear tall horned masks representing male spirits, they would also use smaller more gentle white-faced female masks.

In North America there were numerous tribes who had their own speciality masks, notably the Hopi Indians, whose Katchina masks were a compilation of various feathers, wood parts, and evergreens, put together with saw-toothed gaping mouths and overlarge staring eyes. These would be worn in honour of their ancestors, the ceremony being held between January and July, after which the magic of the mask is supposed to have driven the spirits back to Shipapu.

In South America the Tekúna Indians had a wickerwork demon mask, with huge eyes and large flapping woven grass ears, and the Yahuna Indians had one made of red bast material daubed with pitch and gaudily painted to represent the bad demon Nokolidyana; they also had one of the strangest which would be used in the dance of the Wood Ghosts, their faces and bodies being shrouded with the red bast material, while on their head would be worn light wooden cylinders with wings and each with a Sun or Moon face painted on the front; the woman dancer would have a long plait of palm pith.

MAYANS
The greatest and most exciting of the ancient civilizations of the New World, extending over Southern Mexico, Guatemala and Honduras, a people of great artistry. They had a form of hieroglyphic writing, and without instruments they had records of astronomy, as well as a calendar with solar and leap year correction that was more accurate than the Gregorian. The date for their emergence is about 100 BC and the 'Old Empire' lasted until the 10th century when they left the temple cities and built new fortified towns further to the east on the Yucatán Peninsula. The old cities such as Tikal, Copan and Bonampak were deserted and in the New Kingdom came Chicken-Itzá, Uxemal and Mayapan.

Their art forms embraced many crafts. With ceramics, there were fertility figures; and al-

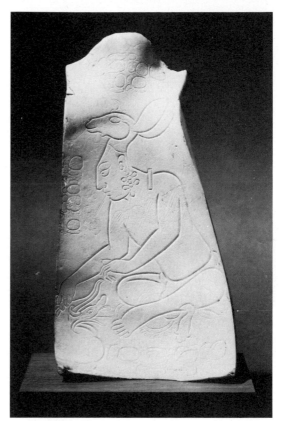

Seated Figure. Guatemala, Maya, 3rd–5th centuries AD. *(Courtesy The Cleveland Museum of Art)*

though some other pots might be plain there are instances of remarkable narrative-painting decoration. **Sculpture was sometimes in the round but gave the impression of being intended as a relief. Architectural frescoes were large genre scenes.**

MIXTECS
A people that flourished between 800 and 1521 in northern Oaxaca and southern Puebla in Mexico. **They were responsible for important examples of religious art, also extremely elaborate gold work.**

MOCHICA CULTURE (*c* 200 or 300 BC)
A people that lived on the north coast of Peru and **their art is characterized by attractive lively pottery with a strong sculptural feeling, showing advanced modelling** and on which sea motifs appeared quite often. The pots were formed either by hand-shaping or coiled. The stirrup handle was constantly used. **Their portrait jars with features fully modelled and painted have a remarkable realism.**

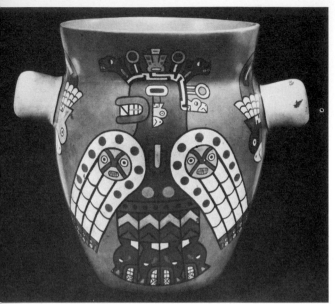

Tiahuanaco Culture Vase. The figure has a tail ending in four jaguar heads, representing the number of solar manifestations. The ten dots on each wing possibly stand for the 20-day cycle of the Mesoamerican Calendar. (Courtesy The Montreal Museum of Fine Arts)

TATTOO

(From the Samoan word *tatau*). A Body-decorating practice of considerable antiquity like body-painting and the rather vicious scarring and cicatrization, with the last the decorations are raised by cutting and inserting substances under the skin. With tattooing the colours are inserted into the skin layers with small sharp instruments. **One of the earliest known tattoos is that on a piece of skin from the right arm of an East Scythian man from the 4th century BC.**

Many peoples, in fact most, have had some form of this art, but with non-Europeans it is often associated with rituals, symbolism and deeper motives.

In New Guinea, girls were freely tattooed all over their bodies with geometrical symbols when they reached marriageable age. The Maoris made a speciality of facial tattoos, and the designs were characterized by swirling lines and spirals; their method of applying the tattoo produced grooving in the skin. The Indians of Bolivia also used symbolic facial tattoos. The young eligible Kayan ladies of Borneo had a threefold reason for the tattoos: firstly, as a mode of ornamentation; secondly, as a device for warding off and curing sickness; and thirdly, on account of the superstition that tattoo marks act as phosphorescent torches for the use of the deceased's spirit when making its long and tiring journey to the place reserved for departed spirits.

In Burma it appears to have been almost *de rigueur* for the self-respecting man to be tattooed from the knee to the waist in blue. The designs included ogres and tigers and were encircled with scroll lettering. On his arms would be red tattoos to ward off sword cuts, gunshot wounds or in contrast to bring success in love. It is probable that the greatest talent with body tattoos was shown by the Japanese, although in that country paradoxically it is a mark of low breeding and vulgarity. Generally it was only the coolies who, often stripped to the waist for work, had an over-all top decoration, which appeared as a highly decorative vest.

TEOTIHUACAN

The temple city of an ancient Mexican people active from about 200 to 800. The name means literally the 'Place or Seat of the Gods'. The principal buildings included two large pyramids, one of the Sun and the other of the Moon; that of the Sun being 200 ft (60·96 m) high, nearly up to the Egyptian Cheops. There is also the Temple of

American Indian Horn Ladle. (Courtesy The Montreal Museum of Fine Arts)

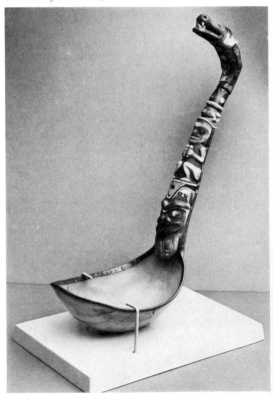

Quetzalcoatl, built like a stepped pyramid, liberally decorated with boldly carved images of the feathered serpent Quetzalcoatl and the rain god Tlaloc, both of which were probably originally polychromed. The Sun and Moon Pyramids are both faced with stone slabs and plaster.

TIAHUANACO CULTURE
Lasting from about 600 to 1000, the ruins of Tiahuanaco stand at over 12000 ft (3650 m) above sea-level beside the southern shore of Lake Titicaca in the Andes. The gateway has a relief frieze across the top, with the figure of Virachoca, one of the principal deities in the Peruvian mythology, in the centre of a repeating pattern of a short-winged figure. **The craftsmen of this culture produced tapestries of an extremely fine weave, generally with a figure worked into a repeat.** Pottery was decorated with bold motifs, it often simply portrayed fish.

TOTEM-POLES
An art form primarily of the North-west Coast Indians. The carved and coloured poles displayed the 'heraldic devices' also social standing and rank of the person outside whose house or hut they stood. The totem-poles also included representations of mythology, guardian spirits, animals, birds and grotesque forms. The Haidi Indians of the Queen Charlotte Islands (British Columbia) were celebrated for this art form; they also carved and decorated their canoes with the heraldry of their owners.

A totem-pole 173 ft (52·73 m) tall was raised on 6 June 1973 at Alert Bay, British Columbia, Canada. It tells the story of the Kwakiutl and took 36 man-weeks to carve.

WAMPUM
Beads made from white and dark mauve shells that would be used by the Iroquois, Huron and Algonquin tribes to decorate belts with symbolic and meaningful designs.

ZAPOTECS
Ancient Mexican culture lasting from about 300 to 1000. The people lived in the fertile valleys of eastern Oaxaca; **in one way they were similar to the Etruscans with a liking for large tombs with painted walls**; funerary urns were generally of red clay in the form of a ritually garbed figure sitting with crossed legs, and with a realistic and expressive face.

Chapter Five

Groups, Movements, Periods, Schools

ABC ART
(also minimal art, primary structure) Basic design structures associated with modern materials and construction. Encouraged by basic design courses in art schools, craftsmen had exhibited modern technology as applied to metalsmithing and joinery. **The name came into being in the 1960s** and the movement centred on Los Angeles and New York.

ABRAMTSEVO COLONY
It was founded in the late 18th century on a rail tycoon's estate near Moscow to encourage a revival of traditional crafts and challenged St Petersburg Academy, founded in 1754, for the control of artistic life.

ABSTRACT ART
An art form which eschews all natural shapes in its expression. In the early centuries the followers of Islam by their religious instruction were forbidden to represent any natural objects or shapes in artistic or decorative designs and from much of their early work many abstract ideas have grown. Abstract art as a definitive expression had its **beginning in 1910 with a water-colour by Kandinsky**.

ABSTRACT EXPRESSIONISM
A visual expression with considerable freedom, often using harsh, brash, bright colours. It was centred in New York about 1950. Much of the influence exerted on it came from the refugee artists such as: Chagall, Duchamp and Miró. Americans included Pollock, Tobey and Kline.

ABSTRACTION-CREATIONISTS
A group allied to Abstract Art who worked in Paris in the 1930s including Mondrian and Jean Helion.

ABSTRACTION LYRIQUE
In the period between 1947 and the late 1950s a new and completely unconscious art form was evolved. (Virtually identical developments are such as: Action Painting, Informal Art and Tachisme.) Two of the leading figures were Camille Bryen and Georges Mathieu.

ACADÉMIE CARRIÈRE
It was run by Eugène Carrière and included among the pupils: Derain and Matisse. It was started about 1898 and only lasted for a short time.

ACADÉMIE JULIAN
Founded by Rudolph Julian in 1860 in Paris, it became a preparatory establishment for the École Nationale. One of the principal teachers was Bouguereau and pupils included: Bonnard, Valloton and Duchamp.

ACADÉMIE RANSON
It acted as a centre for the young painters who called themselves the Nabis. **Founded in 1908 by Paul and France Ranson in Paris.** Teachers included: Maurice Denis and Sérusier; among

'Schlacht' by Wassily Kandinsky. (Courtesy Tate Gallery)

those that attended were: Bérard, Bonnard, Gromaire and Vuillard. Closed for the last war it was reopened in 1951.

ACADEMIES
The word academy is derived from a grove near Athens which was dedicated to the mythical Akademos; it was the meeting-place for Plato and his pupils. The first academy in line with the development today, was the *Accadèmia del Disegno* in Florence which was founded by Vasari in 1563; from this idea came the *Accadèmia di San Luca*, Rome, in 1593. In a like vein was the *Académie Royale de Peinture et de Sculpture*, brought into being in Paris in 1648 by Colbert and Lebrun. This was dissolved in 1793 and reinstated in 1816 as the *Académie des Beaux-Arts*.

The Royal Academy of Arts in London was founded in 1768 with 36 original members, the first President being Joshua Reynolds, who apart from his paintings left his celebrated *Discourses*, which remain as object lessons in artistic theory and codes for practice. The Academy was first housed in Somerset House, then for a short period in part of the National Gallery; in 1869 it started to operate from Burlington House, which it obtained on a lease of 999 years for a peppercorn rent. The Academy Schools started with a 'School for the Living Model' in 1769 with 77 students; painting and sculpture departments were added in 1815 and 1871 respectively. **The youngest RA** has been Mary Moser who was 24 when elected with the founding-members. **The oldest RA** has been (Thomas) Sidney Cooper who died in 1902 aged 98 and 136 days, and who between 1833 and 1902 exhibited 266 paintings. **The youngest exhibitor ever** has been Lewis Melville 'Gino' Lyons, who showed his 'Trees and Monkeys' in the 1967 exhibition when he was five.

In 1778 an academy of painting and sculpture was set up in Turin. Gian Galeazzo Visconti founded one for architecture in Milan in 1380. The Florence Academy which had fallen into decay was restored in the 18th century. In Russia the **Academy of St Petersburg was established in 1757** by the Empress Elizabeth at the suggestion of Count Shuvalov. John VI of Portugal founded in **1816 the academy in Rio de Janeiro**, which is now known as the *Escola Nacional de Bellas Artes*. In Madrid the *Accadèmia de Bellas Artes de San Fernando* was brought into being by Philip V. **The Swedish Academy was started in 1733 in Stockholm** by Count Tessin. The Austrians were earlier still with the institution being opened in **Vienna in 1705**. In America the National Academy of Design was set up in 1826,

and in 1906 absorbed the Society of American Artists.

ACTION PAINTING
(see Tachisme)
A method of painting which involves the splashing, dribbling and pouring on of the paints in a manner supposedly influenced by the unconscious. The practice is associated with New York and the **leading figure has been Jackson Pollock**.

ADA GROUP
A school of illuminators and miniaturists operating about 800, it took its name from the Abbess Ada of Mainz who ordered an Evangeliar from them. The manuscripts were sometimes written in gold on purple-grounded vellum.

AEROPITTURA
It was **an attempt to reproduce the sensations of environment and travel, especially flight**. In 1926 the first exploratory attempts were made. Three years later Marinetti produced the manifesto *L'Aeropittura* which was also signed by Balla, Benedetta, Deperio, Dottero, Filia, Prampolini, Somenzi and Tato. The adherents of Aeropittura after an exploratory exhibition in 1931, followed this up by a showing in Paris in 1932, and one in

Sketch by Mary Moser. (Photo by the Author)

Berlin in 1934, where they came under the patronage of Goebbels.

ALLIANCE OF YOUTH

A forward-looking group of artists started in St Petersburg, Russia, in 1909, it had a similarity to the Bubonovgi Valet association that came into being in Moscow the year after. Both looked towards the West and in particular to the artistic movements in Munich. In the first exhibition in 1910 all the pictures shown were by artists from St Petersburg, including: Pavel Filinov and Olga Rosanova. At a later exhibition the same year which was held at the Vladimir Izdebskiy Gallery, Odessa, Moscow artists joined in; among these were: Larionov, Gontcharova and the Brothers Burljuk. Kandinsky of the Munich school sent and showed 52 pictures, also from this school came works by Jawlensky and Gabriele Münter.

ALLIED ARTISTS ASSOCIATION

Led by Walter Sickert and formed in 1908, it set out as did the New English Art Club **to combat the strictures of the Royal Academy**. It organized exhibitions similar to the Salon des Indépendants. The ambitions of this association can be gauged from the fact that they held their first show in the Royal Albert Hall in 1908. The Allied Artists were absorbed into the influential Camden Town Group in 1911.

AMIENS SCHOOL

Active in the 15th century; it was closely influenced by the Flemish painters. Best-known master was Simon Marmion.

ANTIPODEANS

A group of Australian artists who rebelled against the obsessions of many with abstract art. They held an exhibition in 1959 and also brought out a manifesto which proclaimed for figurative art; among those associated with the idea were: Arthur Boyd, David Boyd and Robert Dickerson.

ANTROPOFÁGICO GROUP

An *avant-garde* association operating in São Paulo, Brazil in the 1920s.

ANTWERP SCHOOL

It flourished in the late 15th century and through the 16th; painters came to the town from Cologne, the Lower Rhine and Bruges. The leading figure was Quentin Matssys and others included: Patinir, Bles, Rubens, Seghers, Snyders and Van Dyck.

ARMORY SHOW

An international exhibition held in the 69th Regiment Armory, New York, between 17 February and 15 March 1913. It was to a high degree responsible for breaking down the barriers against modern trends. Primarily organized by Robert Henri and his associates. **The main idea was to publicize the new developments in American art**; although included were works from prevailing movements operating in Paris. One work that shocked was Duchamp's 'Nude descending a Staircase'.

'ART BRUT' (Raw Art)

A society formed in 1948 to study and appreciate the works of those who have had no artistic training or were mentally unstable. An exhibition of this form of primitive art was held in 1949 at the René Drouin Gallery, Paris. Jean Dubuffet was one who acclaimed this untrained, often explosive expression.

ART FOR ART'S SAKE

A term implying that the pure aesthetic qualities of a work of art should be considered without turning to such values as moral, political, story content or other such. It became current about 1870.

ART NOUVEAU

A type of artistic presentation and design that developed about 1895 in Belgium and France. In Germany it was called Jugendstil. It displayed an excessive use of flowing and curved lines based on plant forms; also motifs of La Tène. It spread rapidly all over Europe and the United States.

ARTS AND CRAFTS MOVEMENT

Largely peculiar to Great Britain, it was motivated by the theories of Ruskin, Pugin and William Morris; these **pointed to the necessity for combating the effects of over-industrialization** and the tasks often of great dullness thrust upon many workers.

ART STUDENTS' LEAGUE

Founded in New York in 1875 and incorporated three years later. It started off in an old piano factory in East 23rd Street with some 900 students. In 1893 it moved into the American Fine Arts Building on West 57th Street. A 50th anniversary exhibition was held in 1943 which showed works by past teachers and students including: Thomas Eakins, George McInnes, George Bellows and Robert Henri.

ASH-CAN SCHOOL
(see also Eight, The)

A term of derision aimed at a group of painters who chose for their subjects the often squalid realism of behind scenes in a great city, giving bold representation of the harsh, ordinary and environmental truths. Painters concerned included Robert Henri, John Sloan and Maurice

'*Moving Picture Theatre*' by *John Sloan (1871–1951)*. *Oil on canvas, 20 × 24 in (508 × 609 mm). (Courtesy The Toledo Museum of Art, Ohio)*

Prendergast. The name could be allied to the Kitchen Sink painters in Britain who had a vogue during the 1960s.

ATELIER GLEYRE

Marc Charles Gabriel Gleyre (1806–74) was a Swiss born at Chevilly in the Canton of Vaud. After studying and travelling in Italy and France he set up his studio in Paris, and is best remembered for the fact that Monet, Sisley, Bazille and Renoir spent some time with him in an atmosphere of *sympatico* rather than direct artistic influence. British painters who attended this atelier included: Poynter and Calderon.

AUTOMATISM

A method for **creating a work of art without conscious control**; the production of forms, lines and shapes, a free 'doodling'. The principal followers were the Automatistes, Canadian artists

who were rejecting the art round them and also other aspects of the contemporary pattern. They included: Fernand Leduc and Jean-Paul Riopelle. The peak of the feeling was between 1948 and 1951.

AVIGNON SCHOOL

A group of artists that gathered at the Papal Court when it moved in 1309 from Rome to Avignon in France. The Sienese painter Simone Martini was probably the leading figure, who with other compatriots helped to bring the Italian influence of this time to France.

BALLETS RUSSES

Serge de Diaghilev (1872–1929) one of the truly great impresarios who brought within the compass of ballet the genius of leading painters. During the period of his command such as the following were concerned with scenery and costumes: Léon Bakst, Gontcharova, Braque, Picasso, Matisse, Marie Laurencin, Utrillo, Ernst and Miró.

BALLETS SUÉDOIS

Founded by Rolf de Maré and Jean Borlin, it came to Paris, opening in 1920. Artists concerned with productions included: Chirico, Steinlen and Léger.

BARBIZON SCHOOL

A group of painters in the 19th century, seeking Realism and Naturalism, who worked together in the little village of Barbizon in the Forest of Fontainebleau. Their pictures are redolent with a feeling of freshness and close observation of nature. It is likely they could have received influence from the paintings by Constable, who himself gained first recognition in France when he was hung in the Salon in Paris. Members of the school included: Corot, Théodore Rousseau, Daubigny, Millet, Diaz and Troyon.

BATEAU-LAVOIR

A name coined by Max Jacob the poet for an

'Seated Woman' by Alexej Jawlensky. Oil on board, 28¼ × 20 in (717 × 508 mm). (Courtesy Marlborough Fine Art)

extraordinary collection of ramshackle studios in Montmartre. Tenants between 1890 and 1914 included: Picasso, Gris, Jaboc and Reverdy. As a meeting-place for discussing new ideas such as Cubism, many leading names attended: Matisse, Braque, Derain, Dufy, Laurencin, Lipchitz among them.

BAUHAUS

Among the most famous schools of architecture of this century, although in truth it was much more. It was under the direction of Walter Gropius and started at Weimar in 1919, moving to Dessau in 1925, and being closed in 1933 at the order of Hitler. The basic principle underlying the teaching was that **all aspects of the visual arts reflect man's creative ability**, and as such should be kept in separate compartments, but should nevertheless all work together; and with this there should be an inherent feeling for the artist and the craftsman that should be within the reach of everyone. The ideas of Ruskin and William Morris were taken further by the inspiration of Gropius and his teachers who included: Feininger, Kandinsky, Klee and Moholy-Nagy.

BENTVOGEL SOCIETY

A friendly club for the Flemish painters working in Rome at the beginning of the 16th century. As new countrymen arrived in the city they were admitted to the brotherhood by certain whimsical and bibulous ceremonies. These were liable to go on all night, and in the early morning the group walked to the outskirts of Rome to the Tomb of Bacchus. It is said that Raphael may have suggested the rites undertaken at the welcome to a new member.

BIEDERMEIER

A period in German art between 1830 and 1860 when there was a **rebellion against Romanticism**. Followers sought a more ascetic approach; in painting it evolved into a somewhat similar manner to the Pre-Raphaelites, a leading exponent being Adolf Menzel. The name came from a comic character, Papa Biedermeier, who featured in a poem in the magazine *Fliegende Blätter*. in design, animal, human and plant forms were used, also adaptions were made from Hepplewhite, Sheraton, and French Empire.

BIMORPHIC ART

A form of Abstract painting in which the artist uses organic forms of plant or other life rather than working from a geometric base only.

BLAST

A magazine founded by Wyndham Lewis in 1914, it set out to expound on Vorticism, attacking the many Establishment ideas left over from the Victorian period; it aimed at shocking the system.

BLAUEN VIER, DIE
Four artists who showed their work together in Germany in 1922 and in the United States and Mexico in 1924. They were Kandinsky, Klee, Feininger and Jawlensky.

BLAUE REITER, DER
A group of German artists who worked out a mild form of Expressionism. It was formed in 1911 in Munich by Marc and Kandinsky, members included: Klee and Macke. Many painters from other countries sent in pictures for the group shows, numbered among them were: Douanier Rousseau, Delaunay, Picasso, Vlaminck, Larionov and Gontcharova.

BLOK
The first Polish Abstract group, it ran close to the ideas of the de Stijl. Leading members included: Vladislav Strzeminski, Henryk Stázewski and the sculptor Katazyna Kobro. Later it was to include architects and then changed its name to the Praesens Association.

BOHEMIAN SCHOOL
A centre for painters, illuminators and sculptors that grew up in Prague at the Court of the Emperor Charles IV who became King of Bohemia in 1346 and reigned until 1378. Artists came in from neighbouring countries and produced a style international in its manners. **Sadly many of the works were destroyed during the Hussite troubles.**

BOLOGNESE SCHOOL
Founded by the Caracci Brothers, Lodovico, Agostino and Annibale. They opened the Academy of Painting in Bologna in 1598; here the trend was towards the manners of Correggio and Raphael and away from the dramatic realism, with its somewhat theatrical light and shade of the Caravaggismo fashion inspired by Caravaggio. An outstanding work by Agostino Caracci is the altarpiece 'The Communion of St Jerome' which was copied by Domenichino.

BROTHERHOOD OF ST LUKE
(see Nazarenes)

BRÜCKE, DIE
The moving spirits behind Expressionism were three architectural students from Dresden; Heckel, Kirchner and Schmidt-Rottluff. In 1905 all three gave up their technical studies and turned to painting. They formed Die Brücke (Bridge) and were joined by other painters including: Nolde, Mueller, Pechstein, Van Dongen and the Finn, Axel Gallén. Apart from painting the group gave an impetus to graphic work, notably bold woodcuts. In 1913 the group was disbanded after a dispute over the aims as set down by Kirchner in a Chronicle.

BRUGES SCHOOL
Flemish painters who were working in Bruges during the 15th century. Their work was characterized by an attention to detail, revelling in fine materials, furs, jewels, and interior furnishings. Some of the best-known artists were: Hugo and Jan van Eyck, Memlinc, Christus and Gheeraert David.

CABARET VOLTAIRE
The name of an establishment in Zürich at No. 1 Spiegelgasse which was started in 1916 by Hugo Ball and Emmy Hennings. There were exhibitions of work by such as Arp, Segal, Picasso, Modigliani, Marinetti, Kandinsky and Apollinaire. Recitals of poetry and music and song were also given. In March 1917 The Galerie Dada took over the running of the affairs.

CAFÉ GUERBOIS
A Montmartre café at No. 9 Avenue de Clichy which in 1868/69 became a haunt on Fridays of Manet, Monet, Sisley, Pissarro, Degas, Renoir. Here by repute the first rules and aesthetic principles underlying Impressionism were laid down.

CAMDEN TOWN GROUP
Formed by breakaway members of the New English Art Club. The leading personality was Sickert who had his studio in Fitzroy Square which became the centre of the group that included Spencer Gore, Harold Gilman, Augustus John, Robert Bevan, Charles Ginner and Henry Lamb. The work of the Camden Town painters

'Bavarian Farmers'. Woodcut by Karl Schmidt-Rottluff; 15·5 × 19·7 in (395 × 502 mm). (Courtesy Frederick Mulder Esq)

'View of the Basilica of St Mark's, Venice by Walter Richard Sickert. $14\frac{1}{2} \times 17\frac{1}{2}$ in (368×444 mm). Sold in 1963 for 2900 gns. (Courtesy Christie's)

was generally in low-tone, but with rich warm umbrous colours and often dramatic lighting, typified by Sickert's narrow upright canvas 'The New Bedford'. After several exhibitions the members joined with the circle of Wyndham Lewis and the London Group came into being in 1913.

CARAVAGGISTI
A collective term for painters who came to Rome in the early years of the 17th century and practised the manner of handling light and shade developed by Caravaggio. The dramatic effects appealed particularly to Dutch painters.

CAROLINGIAN ART
The period began about 800 and lasted through the reign of Charlemagne and his successors until the middle of the 10th century. During this time Late Antique forms were grafted on to the work being done in the North and were merged with the Northern traditions of art. It included the miniature-painters of Godescole and the Ada Group.

CHINOISERIE
A broad term used in the Arts to describe or indicate an influence related to Chinese ornamental design. Particularly in the 18th century it appeared in interior decoration, wallpapers, porcelain and furniture. At the peak of the vogue people were even landscaping in the manner with temples, pagodas and Oriental plants.

CLASSICISM
The classical period of the Arts of Greece and Rome lasted from the 5th century BC until the 4th century AD. Classicism implies styles used in later

'Ganymede and the Eagle' by Bertel Thorvaldsen. (Courtesy The Minneapolis Institute of Arts)

periods by painters or sculptors who adhered to the standards of these early artists.

COBRA
A movement that had roots in both Surrealism and Expressionism. Its members included: Appel, Constant, Corneille and Jorn. Influence lasted from between 1948 and 1951 and affected northern Europe from Belgium to Denmark.

COLOGNE SCHOOL
A group of painters who worked in Cologne from the last years of the 14th century until the mid 16th century. The best known of these was Stefan Lochner, most were nameless.

CONCRETE ART
A term coined by Van Doesburg in 1930 to be used instead of Abstract Art. Although Arp and Kandinsky used it for a time it has now become more or less obsolete. Abstract Art as a term to the purist is misleading, because the name can only be really applied to an intangible something.

CONSTRUCTIVISM
An abstract non-objective art movement largely linked to sculpture. It was first evinced in the work of Tatlin. It received an impetus when Antoine Pevsner and Naum Gabo returned to Russia in 1917. Ideas connected with the theory included a use of space and dynamism in the arrangement of the constituent members incorporated. The theory of the movement was put forward in 1920 *'Realist Manifesto'*.

COSMATI
Sculptors, mosaicists and marble-workers in Rome between the 12th and 14th centuries. Their *métier* was largely in decorative inlay stonework and coloured marble. The name was derived from craftsmen called Cosmas. The Cosmati signed their work which was a departure from the medieval practice of anonymity.

CUBISM
One of the most important movements in 20th-century painting in France. It was the years

between 1907 and 1914 that saw the greatest interest. Early leading experimenters included Braque and Picasso. The three steps of development were **Facet Cubism**, separate components of a composition being transformed towards geometrical shapes; **Analytical Cubism**, with objects being further broken down; and **Synthetic Cubism**, with the artist freed from disciplines of Naturalism. Cézanne said, 'You must see in nature the cylinder, the sphere, and the cone.' A dictum which can be recognized in his landscapes.

DADAISM

A term derived from the French word *dada*, meaning hobby-horse. **One of the strangest theory-based movements**, it originated in Zürich in 1916. Briefly it was all-out Nihilism. Protagonists set out to ridicule current civilization combined with a somewhat savage iconoclastic destruction and obliteration of all preceding artistic and aesthetic theories.

The credo for its followers was 'Everything the artist spits is art.' Leading figures associated, included: Duchamp, who shocked hard with his 'The Fountain' (a commonplace piece of glazed earthenware uprooted and put up as a work of art), and who also decorated a photograph of the 'Mona Lisa' with a moustache; and Ernst, Arp, Picabia and Man Ray.

DANUBE SCHOOL

Painters who worked during the first three decades of the 16th century. They expressed a strong feeling for nature; hence the romantic content of some of Altdorfer's work; others included Wolf Huber, Rueland Frueauf the Younger and Michael Ostendorfer.

DECADENT ART

(also *Fin de Siècle*)
A sneer term for some of the works by some artists operating during the last 20 years of the 19th century. One who came under some assault was Aubrey Beardsley with his exotic and near-erotic drawings, which were in fact exquisite examples of black and white with fastidious and economic use of the medium.

DE STIJL GROUP

Founded in Holland in 1917 by Van Doesburg and lasting until 1930. The ideas were connected with the Abstract movement in art and architecture and the use of primary colours and rectangles; a prime exponent was Mondrian.

DEVETSIL

A branch of the Surrealist movement that was founded in Czechoslovakia in 1920 by Jan Zrzavz.

DILETTANTI (DILLETANTE) SOCIETY

It came into being in 1734, and had its meetings at the Thatched-house Tavern in St James's Street, London. 'A body of noblemen and gentlemen who had studied Antique Art at home and abroad, and were anxious to spread its knowledge, and encourage a taste for objects which had contributed so largely to their intellectual gratification.' A term used now for those with education in the arts and a taste for beauty, quality and rarity without making a profession for themselves in any branch of the arts or their handling.

DIVISIONISM

A method of colour application worked up by Seurat and to a lesser extent by others of the Neo-Impressionists. **Pigments are not mixed on the palette, but colours are placed on the canvas as a series of small dots**; for example, yellow and blue interspersed, and the eye mixes optically and would see green.

DÜSSELDORF SCHOOL

Founded in 1767, the students' work reached prominence between 1830 and 1850. Landscapes had a romantic touch associated with the Nazarenes; also in the handling of genre subjects there was feeling of Biedermeier Art.

DUTCH CARAVAGGISTI

In the early part of the 17th century a number of Dutch painters were strongly influenced by the dramatic light and shade of Caravaggio. The leading painter in this vein was Hendrik Terbrugghen. Rembrandt did not belong to the group although he would have appreciated the manner. His own handling of chiaroscuro had the drama and power but possessed an innate quality of extreme subtlety and sensitivity not attained by others.

ECART ABSOLU

An exhibition of Surrealist art held in 1965 at L'Oeil Gallery in Paris. The artists set themselves towards a diametric revulsion from Establishment ideas.

ECLECTICISM

The picking out, borrowing, purloining of manner, style or composition by one artist from the works of others. The word is derived from the Greek *eklegein*. Signs of Eclecticism appear in the progress of art from as early as the Etruscans and they crop up in most periods and certainly in this century.

ÉCOLE DES BEAUX-ARTS

Originally known as the École Académique it was founded in Paris in 1648; its name was changed in 1793. Departments include painting, sculpture,

'Regatta at Cowes' by Raoul Dufy. (Courtesy Sotheby Parke Bernet)

graphic arts and architecture. All students can compete for the Prix de Rome which allows the winners to study at the Academy of France in Rome.

EIGHT, THE
The hard core of the Ash-Can School; they were: Robert Henri, George Luks, Arthur Davis, Ernest Lawson, William Glackens, Everett Shinn, John Sloan, Maurice Prendergast.

ELEMENTARISM
A theory of Van Doesburg based on Neo-Plasticism; to the order of the right-angled geometry he introduced inclined planes and a sense of fluctuation. Van Doesburg in the *De Stijl* review of 1926 wrote in part: 'Elementarism is partly a reaction against the too dogmatic application of Neo-Plasticism, and partly a consequence of Neo-Plasticism itself. What it seeks, above all, is a strict rectification of the Neo-Plastic ideas.'

ESPRIT NOUVEAU
A regular publication that started in 1920 and for some 20 years was the carrier of *avant-garde* theories. The editors were Le Corbusier and Ozenfant, and the periodical had considerable effect on thought in the schools of art and architecture.

EUSTON ROAD GROUP
In the immediate period before the Second World War a number of painters were in gentle rebellion against some aspects of modernism, they made a return to Post-Impressionism, to a sensitive, in some cases softly lit, Naturalism. They included: William Coldstream, Lawrence Gowing, Victor Pasmore.

EXPRESSIONISM
A collective term for a number of Modern Art movements that came after Impressionism and Post-Impressionism. The principal roots lie in Germany from where came Die Brücke and Der Blaue Reiter. A painter in the sense of the German, Emil Nolde, epitomizes the basic tenets of Expressionism, a combination of freedom, colour

power with sensitivity and strong emotional under-tides. Further back can be traced an influence that could have been born with Grünewald. The spirit of the movement spread to a number of countries: Edvard Munch, Norway; James Ensor, Belgium; Oskar Kokoschka, Austria; Georges Rouault, France, and Chaim Soutine born in Minsk, Russia. In the hands of the foregoing and others in a similar vein **Expressionism has proved to be probably the strongest and most lasting of manners in this century.**

FAUVES, LES
A group of painters who became revolutionaries in their handling of colour, whose work shattered most prevalent theories. Rich, powerful, fiery colour came off their palettes and on to canvas, producing effects and contrasts not attempted before. The name Les Fauves is derived from the French for the wild beasts, and was applied as a sneer by a critic at one of the first exhibitions of work by such as Henri Matisse, Albert Marquet, Maurice Vlaminck, André Derain, Raoul Dufy and Georges Rouault. It was one of the greatest explosions of colour in the history of painting. As with many movements, quickly it seeped across countries and was picked up by a number of artists, examined and used and then perhaps dropped as they passed on to the next theory. In Germany two painters who were strongly influenced were Emile Nolde and the Russian-born Alexej von Jawlensky.

FEDERATION OF BRITISH ARTISTS
An administrative and exhibiting centre with the Mall Galleries, London; including the following: Graphic Artists, Industrial Painters Group, Royal Society of Marine Artists, Royal Society of Miniature Painters, Sculptors and Gravers, Royal Institute of Oil Painters, Royal Society of British Artists, Royal Society of Portrait Painters, Royal Drawing Society, Society of Mural Painters, National Society of Painters, Sculptors and Printmakers, New English Art Club, Royal Society of Painter-Etchers and Engravers, Royal Institute of Painters in Water Colours, Pastel Society and the Society of Portrait Sculptors. The Secretary-General and moving spirit is Maurice Bradshaw.

FOLK ART
A widely embracing term for the arts and crafts of those who have had no specialist training; art forms that have come from native sources. Different areas or countries may have a particular use of pattern, colour relationships, or specific manner of carving, presentation of figures, animals, plants and flower forms.

FONTAINEBLEAU SCHOOL
This began as an answer to the need for decorating the Palace of Fontainebleau which was largely built under François I (1515–47) and Henri II (1547–59), which kings have the two great galleries named after them. A number of leading Italian painters were sent for, including: Francesco Primaticcio, Niccolò dell'Abbate and Il Rosso. Under the patronage of Henri IV the so-called Second School of Fontainebleau flourished.

FORMISM
A blend of Abstract, Futurism and Expressionism followed by Polish painters such as Chivistek and Wietkiewicz between 1917 and 1922; it was centred at Cracow, Poland.

FRONTE NUOVO DELLE ARTI
The first movement towards a 'New Art Front' in Italy after the Second World War was started in 1947. Members included: Renato Bitolli, Renato Guttuso and Emilio Vedova. The group lasted until 1948 and then was disbanded to be followed by the 'Otto Pittori Italiani'.

FUNCTIONALISM
A credo in the useful arts and particularly architecture that lays down that the purpose of an object or building should dictate its design before and to a degree regardless of decoration.

FUTURISM
A fierce, virile movement in Italy that flourished between 1909 and 1915; the followers sought to show the violence of the times, they were concerned with machinery and politics. These included: Carlo Carrà, Giacomo Balla, Gino Severini and Umberto Boccioni.

GIOTTESCHI
Painters who were followers or pupils of Giotto in the 14th century; among them, Bernado Daddi and Taddeo Gaddi.

GOTHIC
The period from the 12th to 15th centuries in European art. Within this time manuscript-illumination was to reach its highest execution.

GRAECO-ROMAN
Extending from the late 2nd century BC to about the 4th century, covering the marrying of the Greek to Roman art, in which the idealism of the Greek image turned into the realism of the Roman.

GUILDS
The peak of the guild system was reached in the Middle Ages, particularly on the Continent, in places such as Antwerp, Florence and Germany. The guilds acted as collective societies for

craftsmen, which would not only protect their interests but at the same time advance the craft and aim for the highest standards; to pass from apprenticeship to Master could take many years.

GUTAF
The Japanese group who introduced 'happenings' to their country, it was founded by Yoshihara in 1951. At one happening they hung kakemonos from captive balloons.

HAARLEM SCHOOL
Late 15th- and 16th-century painters centred on Haarlem. Geertgen tot Sint Jans was one of the early masters. A characteristic was the rendering of garden backgrounds, an idea reputedly first introduced by Dirk Bouts. It was the pattern for art schools to come in Holland in the 17th century.

HAGUE SCHOOL
Realist Dutch painters who worked during the latter half of the 19th century towards a sensitive involvement with landscape, a feeling for the delicate effect of light on a subject and the nuances created by the action of light on colour. Members included: Jozef Israels, Antoine Mauve and Jacob, Matthias and Willem Maris.

HALMSTADTGRUPPEN
A group of painters in Sweden working along Surrealist lines. Axel Olson, Stellan Mörner, Waldemar Lorenzton are among those who have found recognition in Stockholm's *Akademien För De Fria Kosterna* since as far back as 1932.

HEIDELBERG SCHOOL
A party of Australian painters who worked with and under the encouragement of Tom Roberts at Heidelberg, Victoria. The approach was *plein air*, a basic Impressionism suited to the Australian scene. The main period of activity was the last decade of the 19th century. Co-founders of the Heidelberg School were Arthur Streeton and Frederick McCubbin; a leading member was Charles Conder. Their method of work was rapid in an endeavour to catch quick impressions with the quick change light effects. A favourite support was a 9 × 5 in (228 × 127 mm) cedar panel, prepared from a cigar-box lid. Tom Roberts saw the possibility for a showing of these small paintings, and this was done in Buxton's Gallery, Melbourne, in August 1889, and was called the 9 × 5 Impression Exhibition. Roberts himself showed 62 pictures, Streeton 40, and Conder 46, other panels totalled 29. Critical reception varied from a thorough slating to warm support, resulting for the artists in a sell out.

HELLENISTIC ART
The period between 323 and 100 BC, the time when outside influence grew into the work of the Greeks, resulting in a fresh realism, particularly with portraits, a departure from the idealism.

HERAT ACADEMY
Found at the beginning of the 15th century by the Timurid ruler Baisunghar. The miniature-painters broke with earlier manners of having pictures and text together and produced the paintings as separate entities. The most important member was Bihzad.

HERMANNSBURG SCHOOL
John Gardner and Rex Battarbee, Melbourne painters, travelled in 1932 to Hermannsburg Mission Station, near Alice Springs in Central Australia. Here, while they were working they met Aborigines of the Arunta tribe, who became fascinated with the picture-making. Battarbee made two further trips in 1934 and 1936 when he came across Albert Namatjira (1902–59), a full-blooded Aborigine, who had been working as a general jobsman, carpenter, blacksmith and the rest. Battarbee gave him materials and lessons; and a number of the other tribe members practised and obtained a skill. Namatjira worked in oil with crayons and also in water-colour which was his most successful medium. His first show was in 1938 in Melbourne. Other Arunta painters include Enos Namatjira (son of Albert), Ewald, Gabriel and Keith Namatjira, Enoch Tranby, Ruben Pareroultja, Herbert Raberaba, and Joshua Ebararinja.

HLEBINE SCHOOL
This virile, pure, primitive Yugoslavian movement grew from the artistic group Earth, that was founded in 1929. The painters of Earth sought to express social ideas and problems, a course that brought them into collision with the police, and they were banned. So, in the village of Hlebine and its environs there came into being a group who looked for their subjects in the scenes around them: farms, gatherings of peasants, ceremonies, sports and portraits. Their work has an affinity with the earlier French painter Douanier Rousseau, although in many ways it is stronger and nearer to the rugged nature of their subjects. Ivan Generalić born at Hlebine in 1914 is one of the leading exponents of the manner. His work shows a studied observation, trees and components of the landscape being set down with care and a fastidious attention to detail. In contrast are the paintings of Franjo Mraz also of Hlebine and born in 1910; he works with a greater sense of freedom and more naturalistic colour. During the last war he was with the People's Liberation Struggle and showed his paintings at exhibitions by Partisan artists: in 1943 at Otočac, 1944 at Topuska and 1945 at Split. A third important

'Peasant with Flowerpot, 1962' by Ivan Generalic. Oil on glass. (Courtesy Mercury Gallery)

member is Mirko Virius born in 1889 at Delekovac near Hlebine. In the First World War he was taken prisoner by the Russians, and on his release in 1918 he was destitute and had a fierce struggle for life. After meeting Generalić and Mraz he started to paint; often using a low-tone warm palette, his figures massive and landscape details firm and strong. Sadly his output only spans five years, and after being captured his life ended tragically in 1943 in the Zemun Concentration Camp.

HUDSON RIVER SCHOOL
A group of American painters who worked in the area of the river and the Catskill Mountains from about 1826 to 1876. The leading members included Thomas Cole, Thomas Doughty, Asher Durand, John Kensett, Albert Bierstadt and Frederick Church. Outside influences included the work of the Barbizon School and John Constable. They specialized in imaginative and romantic renderings of the scenes before them.

ICONOCLASM
The breaking up of images and the destroying of works of art that are allied to veneration and worship. Sadly, in history there have been a number of periods when iconoclasm has burst forth, and the iconoclasts have mindlessly run wild destroying and damaging many masterpieces.

The term iconoclasm first came into prominence in the period 725 to 842, when the Byzantine World was split into the iconoclasts and iconophiles, the lovers of idols. This conflict lasted until a peace was found, and there was the Feast of Orthodoxy which commemorated the victory of the iconophiles. Emperors who initiated and encouraged the iconoclasts were Leo III, Constantine V, Leo V, Michael II and Theophilus. In the latter period Theodore of Studion and his monks, called the Studitae opposed the rulings of the emperors and saved a number of works.

In the Crusading times the Cathars and Paulicians carried the iconoclastic spirit all over Europe and probably assisted in giving it to Wycliffe. In 1389 a document was drawn up by the Wycliffites in defence of their past leader which reads in part: 'Hit semes that this offrynge ymages is a sotile cast of Antichriste and his clerkis for to drawe almes fro pore men . . . certis, these ymages of hemselfe may do nouther gode nor yvel to mennis soules, but thai myghtten warme a man's body in colde, if thai were sette upon a fire.'

IDEALISM

An art form which is based on a preconceived, perfect mental conception of the subject chosen by the artist, whether it be figures, landscapes or whatever. The painter might select parts from the models in front of him and so build the perfection he seeks by a type of pastiche.

The first illustration of true idealism comes when Socrates reacted against the pantheistic conclusions of early Greek philosophy, which was supported by Plato. Modern idealism has been fostered on its way by such as Galileo, Kepler and Hervey, and the thinking of Locke, Berkeley and Kant gave an impetus to idealism as it grew into the visual arts. Later idealism was to be theorized and explored by the various Academies and nearer to today by the different Neo-Classicism movements.

ILLUSIONISM

The technique of creating as closely as possible an illusion of visual reality. The artist uses – working on scientific rules – perspective, foreshortening, chiaroscuro, colour control and modelling; all subjects that can be taught by academies. With sculpture immense care is taken so that the material chosen will imitate the flesh or clothing, painting of the figure with natural colours may be done and real hair and clothing added.

Illusionist works were in fashion in the Hellenistic period and had much favour with the Romans. A prevalent use for this almost mechanically ruled art as concerned with painting has been to provide impressions of space. Signs of this occur with Pompeian murals, such as those in the House of the Vettii. In the Renaissance architects used the device to extend the appearance of their interiors. Murals often had illusionist frames with bold third-dimensioned modelling inducing the eye to sense depth. In the 'Mass of Bolsena' at the Vatican by Raphael this false framing is used.

There is the simulated apse in the Church of St Satiro, Milan, designed by Donato Bramante, 1514; actual depth worked on is 4 ft (1·22 m). while the illusion suggests 30–40 ft (9–12 m). The perspective murals of the Presenti Brothers carried out about 1580 in such places as the Palazzo del Giardino, Sabbinioneta, Mantua, challenge the eye to see where the illusion begins and ends. This aping of the third dimension was also carried into building with the same rules of perspective handling; in the Teatro Olimpico, Vicenzo, is a permanent setting by Vincenzo Scamozzi, 1582.

Illusionism crops up with the work of cabinet-makers, inlaying coloured woods and stone such as marble to simulate forms unrelated to the piece itself.

Perhaps it is the wax-modellers who take figure treatment to almost the limit; there is the jarring likeness of Charles II in the Westminster Abbey Collection. The discovery of fibre-glass has enabled the modeller to perfect even further the image. The American Duane Hanson produces quite extraordinary reality with such as his 'The Rocker', with real clothes, the figure stands apparently breathing and waiting. Allan Jones has used fibre-glass to make figures into furniture. In painting there has been in most periods a taste for the smaller trompe l'œil pictures, which have an obvious relationship to illusionism, but which are minor technical triumphs for the artist as opposed to the broader, larger spatial manipulations.

IMPRESSIONISM

A momentous movement in the latter part of the 19th century, painters sheared off many of the restrictions of academic ruling and there exploded on to the art world a new realism. Light was of primary concern, its effect on colours, on the whole scene in front of them. Then came the putting down of the paint, not so much as before with long, sweeping strokes, but staccato dabs so that the areas of colour were broken; generally they worked with a bright palette that contained the colours of the spectrum and no earth pigments. When paintings by the Impressionists are examined closely it can be seen how their method gives a slight subtle understatement, and this in turn presents to the eye of the viewer a chance for his imagination to work and enter the visual excitement and satisfaction of the painter.

The three leaders of the Impressionists were Claude Monet, Camille Pissarro and Alfred Sisley. The name was coined by a critic derisively reporting on an exhibition of their work in Paris in 1874, a work by Monet 'Impression: Soleil levant' providing the hack with his weapon. Contemporary with the three above and affected by the discoveries where Berthe Morisot, Pierre Auguste Renoir, Paul Cézanne, Edgar Degas, Mary Cassatt and Frédéric Bazille.

The realism that Impressionism produces on the eye and mind is different to Illusionism which is created primarily to a set of practical rules. Impressionism is far more subtle, it is to a large part the transfixing of a highly appreciated and worked-out mental vision into colours on a canvas. The feeling for the manner can be glimpsed in the magical brushwork of Frans Hals; from a distance a hand or face may appear to have every detail, yet as an approach is made the eye uncovers brush movements and marks which from a foot or so away give a feeling of abandon and do not produce the image that comes into being as a retreat is made. William Hogarth works in a similar manner in his 'Shrimp Girl'. Closer to root influences for the Impressionists is the work of Turner and Constable, with their ability to capture the elemental atmosphere by colour first and foremost.

It is a strange comment on the taste of a contemporary public that in their own day the like of Monet, Pissarro and particularly Sisley had a hard time winning a crust of bread. Their paintings fetched so little, often just double figures; yet today an Impressionist sale draws the highest of bidders, with prices reaching well up in six figures sterling.

INDÉPENDANTS, SALON DES
Held in 1874 for the first time in Paris, and intended for the work of those who had been thrown out by the official Salon. Among those concerned were Odilon Redon and the Pointillist Georges Seurat.

INTIMISM
An idea closely associated with Post-Impressionism that had its roots with the revolt of the Nabis at the end of the 19th century. Two of the leading figures were Pierre Bonnard and Edouard Vuillard. The painters set out to portray intimate, familiar and self-contained subjects, somewhat in the manner of the early Dutch genre low-life pictures.

ITALIAN SPATIALISTS
An ultra-freedom movement, founded in 1947 by Lucio Fontana, born in Argentina in 1899 and educated in Italy. Its principles were to emancipate painting and sculpture from aesthetic theories of the past, to: 'abandon the acceptance of known forms in art and begin the development of an art based on the unity of time and space' (from *The White Manifesto*, Buenos Aires, 1946). Other leading figures included Guiseppe Capogrossi and Alberto Burri.

JACK OF DIAMONDS GROUP
A Russian Futurist movement which was later incorporated into the Akkhr, that had been founded in 1910 with the Russian followers of Cézanne. One of the most important figures was Marc Chagall; in theory it was connected to the Fauves and Expressionists.

JUGENDSTIL
(see Art Nouveau)

KINETIC ART
Artistic constructions that rely on mechanical or other forms of movement to achieve their effect. For example coloured discs that change as they are rotated at different speeds.

KITCAT CLUB
A dining-club that was founded in London about 1700, its members included leading writers, artists and politicians of the Whig persuasian. They first met at a tavern and bakery near Temple Bar which was run by Christopher Cat – he baked a favourite mutton-pie nicknamed Kitcat, hence the name of the club. The Secretary of the club was Jacob Tonson. It was he who commissioned Sir Godfrey Kneller to paint the portraits of the members, which with one exception were all on canvases measuring 36×28 in (914×711 mm) and this size is still known today as a kitcat. Kneller's portraits executed between 1702 and 1717 now hang in the National Portrait Gallery, London.

KITCHEN SINK SCHOOL
A term levelled at English painters who have concerned themselves with what is called New Realism. Often their subjects have been drawn from sinks, wash-basins, dustbins. Protagonists include: John Bratby, Jack Smith and Edward Middleditch. The Americans with a fellow feeling were called the Ash-Can school.

LAETHEM-SAINT-MARTIN SCHOOL
Founded at the close of the 19th century on the River Lys not far from Ghent, Belgium. The moving spirit was Valerius de Saedeler who gathered round him such as: Alfred Servaes, George Minne and Gustave van de Woestyne. Their work was a form of Realism, simple, strong in colour and form that drew the criticism of academic circles. A second wave of artists followed a few years later, these included: Constant Permeke, Gustave and Leon de Smet and Hubert Malfaet.

LEAGUE OF BELGIAN ARTISTS
The members of this association were refugees in London during the First World War. The

'The Artist and His Model' by Marc Chagall. (Courtesy The National Gallery of Canada, Ottawa)

Chairman was Jean Delville, Professor at the Royal Academy, Brussels. The book *Belgian Art in Exile*, which claimed to show a representative selection of modern Belgian art owed its inception to the League. It was published in 1916 and was sold for the benefit of the Belgian Red Cross and the Convalescent Home for Belgian Soldiers. An important member of the committee was Frank Brangwyn.

LONDON GROUP
Founded in 1913 from the Camden Town Group, Vorticists and other small interested parties. The first President was Harold Gilman and early members included: John and Paul Nash, Eric Gill, C R W Nevinson and Jacob Epstein. The first London showing was in 1914 at the Goupil Gallery. Their manner followed Post-Impressionism and their advances brought them into conflict with the near-academism of the Slade and other bodies. Later members included: Marc Gertler, Duncan Grant and Vanessa Bell. The Group endured and was revitalized after the last war.

MACHINE ART
A general term applied to **art or craft methods that are produced from a pre-set pattern or control operated by a machine** in contrast to art and craft techniques that are entirely manipulated by the artist or craftsman.

MAGIC REALISM
A movement that commenced in Germany in the early 1920s, followers included: Otto Dix, George Grosz and Max Beckmann. Their work was typified by a bold, brash, at times brutal, stark realism; commenting on harsh conditions and cruelty of man to man. Versions of the theory appeared in Italy with Pittura Metafisica and in America when Beckmann arrived there in 1947.

MAIN À PLUME, LA
A strange and mixed-up, partly underground Surrealist splinter group which existed in Paris between 1941 and 1943; led by Noel Arnuad, members included Marc Patin and J.-F. Chabrun.

MANNERISM
A broad name for European Art which flourished between the end of the High Renaissance (*c* 1520) and the beginning of Baroque (*c* 1600). If there is any ruling to describe the general approach it must be that there was a tendency to disregard established rules for figure presentation and composition; exaggerated poses were adopted with distortion at times. There was considerable play with emotion, bright colours and forced perspective.

MASTER OF . . .
During the 14th and 15th centuries in particular, a great many paintings were anonymous and a custom has grown up to give them, if they display signs of an individual hand, a fabricated name for the executor . . . so have crept into art history such as these:
The Master of the Legend of St Barbara,
The Master of the Life of the Virgin,
The Master of Flémaille,
The Master of the Brunswick Monogram,
The Master of the Třeboň Altarpiece,
The Master of 1466.

METAPHYSICAL PAINTING
A theory and practice of ideas that came into being in 1917 in Ferrara, Italy, when Giorgio de Chirico met Carlo Carrà and that lasted until 1920. These two, joined later by Giorgio Morandi, left the main stream of Futurism and Surrealism to find a plane of expression that stemmed from the atmosphere of dreams and the unearthly presentation of at times quite ordinary objects.

MINIMAL ART
(see ABC Art)

NABIS, THE
From the Hebrew word meaning prophet, the members were a group of French artists. They came together in 1892 and were largely from the Académie Julian. They sought to bring Symbolism and decorative ideas to a manner of Post-Impressionism inspired by the work of Gauguin. The painters included Edouard Vuillard, Pierre Bonnard, Maurice Denis, and the sculptor Aristide Maillol joined them. The journal *Revue Blanche* gave them support, but when Bonnard and Vuillard left the Nabis broke up in 1897.

NAGASAKI SCHOOL
A realist Japanese school which grew up in the 18th century, influences included European painting and Chinese work as typified by Shên Nan-p'in.

NATURALISM
The painting of pictures in as lifelike a manner as possible with exact photographic detail. Primitive painters sometimes laboriously attempt this way. Pictures tend to appear stilted in contrast to those executed in the Realist manner which are alive and virile with atmosphere and emotion.

NAZARENES
A group of young German artists with an ascetic religious inclination who bonded themselves together in Vienna in 1809 (also known as Lukasbrüder – the Brotherhood of St Luke, who is the patron saint of painters). In 1811 they left

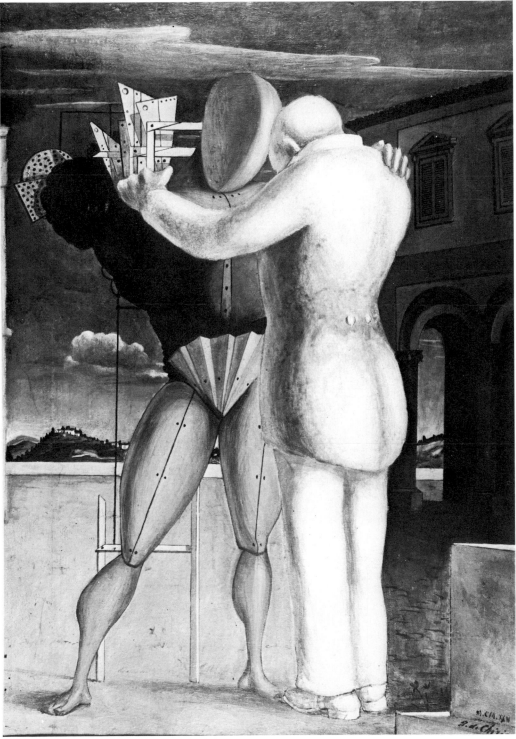

'*The Prodigal Son*' *by Giorgio de Chirico. Oil on canvas, 34 × 23 in (870 × 590 mm). (Courtesy Civico Padiglione D'Arte Contemporanea- Milano)*

Vienna to settle in Rome; here they wore monastic habits and grew their hair long. Some of the Romans mocked them with calls of 'Nazareni'. The young Germans desired to serve Christianity and the Church through the medium of their art. They wished to revive the 'Golden Era of the Middle Ages', choosing such subjects as the Childhood of Christ, legends of the Saints, and the Madonna. The showing of the Passion and martyrdoms were forbidden. The leading figures were Frederick Overbeck and Franz Pforr. Their drawings are considered more highly than the large communal frescoes such as those in Casa Barthody and the Villa Massimo.

NEO-CLASSICISM
A general fashion in painting, sculpture and architecture that grew up in the last part of the 18th century and reached into the first part of the 19th. **In part it was inspired by the French Revolution which created a reaction against the luxuries of Louis XV.** About the same time excavations were taking place at Herculaneum and Pompeii and this drew creative minds back to the Antique, and the Spartan design of the Greeks and Romans. In painting artists sought out mythological and heroic subjects. Winckelmann in his *History of Ancient Art* wrote in 1764 about 'noble simplicity and calm grandeur'; these were the qualities the painters and sculptors strove after. The peak of the movement varied in different countries. Principal painters included in France: David, Ingres, Prud'hon, Couture, Chassériau; in England: Alma-Tadema, Lord Leighton; in America: those working in the manner of Allston. Sculptors included the Italian Canova, the Dane Thorvaldsen and the Englishman Flaxman.

NEO-DADA
An offshoot movement started in 1955 in America as a protest against the excesses of Action painting. The two leading figures were Robert Rauschenberg and Jasper Johns.

NEO-IMPRESSIONISM
(see also Divisionism)
Developed by Georges Seurat and Paul Signac, a principle for colour application also termed Pointillism. One of the first paintings in this manner was Seurat's 'Bathers at Asnières'; it was shown in 1884 at the Salon des Artistes Indépendants and is now in the Tate Gallery, London.

NEO-PLASTICISM
Founded in 1920 it was a pure geometric-abstract method coming from Holland as a splinter idea from De Stijl. Its principle was the use of primary colours, non-colours with horizontal and vertical right angles.

NEO-ROMANTICISM
Romantic Realism in 20th-century Britain. Influences stem from Edward Calvert, William Blake and notably Samuel Palmer. Such as Graham Sutherland, John Piper, Paul Nash and John Minton demonstrate the manner.

NEO-TRADITIONALISM
(see Syntheticism)

NEUE SACHLICHKEIT (New Objectivity)
A type of Expressionism in Germany of the mid 1920s. The followers, who included Otto Dix and George Grosz, sought to depict, through representational acutely observed detail, social and political environment.

NEW ENGLISH ART CLUB
The origin of the club was as an opposite institution to the Royal Academy. It was founded in 1886 with Walter Sickert and Wilson Steer as two of the original members. Other members included: Augustus John, Lucien Pissarro, Ethel Walker, William Orpen and Spencer Gore. The club was associated with the Slade School of Painting attached to London University and with the work of Professor Henry Tonks who taught there. (See also London Group, Camden Town Group and Vorticists.)

NEW REALISM
(see Ash-Can School and Kitchen Sink School)

NKV
Neue-Kunster-Verlinigung (New Artists' Union), a society formed by Jawlensky and Kandinsky in 1908.

NON-FIGURATIVE ART
(see Abstract Art)
Also known as Non-Objective Painting, a term that was probably coined by the Russian Alexander Rodchenko and also used by another Russian Kasimir Malevich.

NORWICH SCHOOL
The Norwich Society was started in 1803 by John Crome 'for the purpose of an Enquiry into the Rise, Progress and present state of Painting, Architecture, and Sculpture, with a view to point out the Best Methods of Study, to attain the Greater Perfection in these Arts'. On Crome's passing in 1821 the presidency of the society passed to John Sell Cotman. The so-called Norwich School is a collective name for the East Anglian painters of the 18th century and early 19th. Apart from Crome and Cotman others connected with the scene include Gainsborough, Constable and Turner; and with the society, John Bernay Crome, Edmund and John Joseph Cotman and

'Great White Heron' by John James Audubon. (Courtesy Christie's)

David Hodgson. **What the society did most of all was to make great advances in the handling of water-colour** and also in the sympathetic treatment of English landscape.

NOVECENTO
Formed in Italy in 1926, its members set out to return to traditional methods and subjects. The followers included: Felice Casorati, Mario Sironi and Arturo Tosi.

NOVEMBERGRUPPE
Some of the leading German Expressionists formed the group in the period following the Social-Democrat revolution of 1918; one founder was the sculptor Rudolf Belling. Soon they were joined by some Dadaists. After a few years, the disenchantment began to overtake the freedom of some of the artists as the political scene shifted away, to the right. A splinter movement came into evidence, this was known as the New-Objectivity. The Bauhaus under the guidance of Gropius had been endeavouring to use the ideas of the Novembergruppe and the Workers' Council, also theories from further back akin to those of William Morris and John Ruskin; that is, to bring the artist and the public into a closer relationship.

NOVGOROD SCHOOL
One of the most important periods in Russian painting. It was active in the 14th and 15th centuries at Novgorod, and it begins with the work of Theophanes the Greek whose frescoes are in the Cathedral of the Transfiguration in the city. His handling of the figure has much in common with that of El Greco. The panel-paintings that followed from the school were the first to have a Russian style of their own. The most important of the pupils of Theophanes was Andrei Rublëv who was to become the leading genius in the art of the icon and a model for those painters working round him.

NUREMBERG SCHOOL
In 1662 Joachim Nützel, a Councillor of the City of Nuremberg, started a private academy of art with the intention that it should provide tuition in drawing and painting for under-privileged children. The academy was directed by Joachim von Sandart and Elias Gödler.

'Portrait of George Dyer Talking' by Francis Bacon. (Courtesy Marlborough Fine Art (London) Ltd)

Below: 'The Wedding Feast of Nastagio degli Onesti' by Sandro Botticelli. (Courtesy Christie's)

'Cupid crowned by the infant Apollo' by François Boucher (32 × 32 in (820 × 820 mm)). Sold in 1977 for £20 000. (Courtesy Sotheby Parke Bernet)

Below: 'Woody River Landscape' by Aelbert Cuyp. (Courtesy Christie's)

Icon of St Nicholas, Novgorod School, 16th century; 32¾ in (832 mm) high. (Courtesy Christie's)

OMEGA WORKSHOP

A project, not unrelated to ideas of William Morris and also certain departments of the Bauhaus, which was started by Roger Fry in 1913 and lasted until about 1919. As with the associates of Walter Gropius the followers held to the credo that the creative spirit of art should be instilled into everything made for use. Active in the workshop were Duncan Grant and Vanessa Bell.

OP ART

A term that was evolved in New York in 1963 to describe a form of geometric art, that was so created that the patterns or arrangements of lines and shapes reacted on the eye to produce a sense of movement, dazzle or even at times discomfort. One of the first deliberate exponents was Victor Vasarely; in England the leading figure has been Bridget Riley.

ORPHISM

An abstract movement founded in Paris in 1912, which in theory employed a 'musical orchestration' with colours; in a number of their pictures can be noted a rather off-beat separation of planes and colour areas. Guillaume de Kostrowitsky Apollinaire (1880–1918) gave the group its name and followers included Robert Delauney and Frantisek Kupka.

OTTONIAN ART

The first German movement, it developed from the Late Carolingian manner. It took its name from the German Emperor Otto the Great of Saxony (936–73), but the period lasted until the first quarter of the 11th century. Within this era was the Reichnau School which produced the 'Egbert Codex'. The cathedrals of Essen, Mainz and Worms were founded.

OTTO PITTORI ITALIANI

This group succeeded Fronte Nuovo delle Arti in 1952, and consisted of, among others, Afro Basaldella, Birolli and Morlotti. Their work can be compared to the more lyrical Abstractionists of the School of Paris. The critic Lionello Venturi termed it 'Abstract-Concrete'.

PARIS SCHOOL

A very general term to encompass art movements that followed after Impressionism. These included such as: Cubism, Fauvism, The Nabis and many aspects of Abstraction and Surrealism. Painters were by no means all French nationals. To the City of Paris were drawn artists from America and all over Europe. Uprooted ones and others found an atmosphere of sympathy for work, and there grew up the greatest concentration of artistic activity since the Renaissance in Italy; studios, galleries, dealers and critics proliferated.

An early meaning of Paris School implied the great school of Illuminators who worked under the inspiration of St Louis (1226–70). Later the Limbourg brothers were to work in Paris.

PERGAMENE SCHOOL

Greek sculpture that was wrought from the early years of the 3rd century BC and that carried on until the Roman expansion. It is characterized by a virile energy and the figures are generally shown in poses involving muscular tension. The peak of accomplishment was during the reigns of Atallus I, Eumenes II and Atallus II. Outstanding works include the 'Dying Gaul', in the Capitoline Museum, Rome; 'Achilles with the body of Penthesilea', in Geneva and the group with 'Menelaus carrying the dead body of Patroclus' at the Loggia dei Lanzi, Florence.

PITTURA METAFISICA
(see Metaphysical Painting)

POINTILLISM
(See Divisionism)

PONT-AVEN SCHOOL
A centre that grew up in the small town in Brittany around and inspired by Gauguin before he departed for the Marquesas Islands, Polynesia.

POP ART
A broad term to include the work of painters and sculptors, largely American, who have sought such as soft-drink bottles, beer cans, exploded and magnified strip cartoons, hamburgers and cast figures as vehicles for their expression. The artistic content leans towards commercial mannerism. Practitioners include in America: Larry Rivers, Roy Lichtenstein, Jasper Johns, Robert Rauschenberg, Andy Warhol and in England Peter Blake.

'Boy (Greystoke) and Palm Tree' by Peter Blake; 10 × 8 in (254 × 203 mm). Water-colour. (Courtesy Robert Fraser Gallery)

POPULAR ROMANTICISM
Painting and sculpture which unashamedly caters to a mass popular taste and in so doing tends to cheapen the subjects in handling and treatment; it is in direct contrast to the true virile Romanticism that came from such as Delacroix and Gericault.

Popular Romanticism had a considerable fashion in the latter half of the 19th century. English painters who followed the vogue included Orchardson and Fildes. The subjects generally leant heavily on literary content and over-dramatic emotional situations.

POST-IMPRESSIONISM
A term brought in by Roger Fry to describe the period in French art between 1885 and 1905. It encompassed the work of Cézanne, Van Gogh, Gauguin, Seurat, Degas and Renoir. Sculptors included: Bourdelle, Despiau and Maillol.

PRE-RAPHAELITE BROTHERHOOD
A mid-19th-century movement in English painting. The founders and most important demonstrators of the style were William Holman Hunt, John Everett Millais and Dante Gabriel Rossetti. At the outset they were inspired by high principles, rejecting to some extent the volume of sentimental narrative-painting of the Early Victorian times; they set themselves to find a return to the purity and compositional treatment of the early Florentine painters, hence the name Pre-Raphaelite. However, it was not long before their work, particularly that of Hunt and Millais, had fallen from their high ideals to a total Naturalism and photographic handling of their subjects.

They also tilted at academic rulings on colour, composition and drawing. Many of their paintings have a strange appearance, as detail in the foreground, mid-distance and distance is all in sharp focus as though viewed through a camera lens with a very great depth of focus. Although they may have escaped the worst of the sentimentalities of those painters round them, they were themselves Romantics and sentimental in their picture titles and handling. Holman Hunt in particular would go to almost endless lengths for what he saw as truth. His weird picture 'The Scapegoat,' which shows a lunatic goat stuck in the salt-pans which surround the Dead Sea, was painted at least in study form *in situ*; Holman Hunt travelling to the area, and it is reported that while he was painting he had a loaded gun beside him in case of a break-out by his unpredictable model.

PRIMITIVES
Today the term is applied to those such as Douanier Rousseau, Séraphine, Camille Bombois, Edward Hicks and Grandma Moses, Christopher Wood and Alfred Wallis; painters who have largely taught themselves and who work, often in a highly personalized manner, with naïve treatment of their subject. They are not bound by rules of composition, light, colour or perspective.

'Adoration' by Gerard David. (Courtesy Christie's)

'*Albert Einstein*' *by Hans Erni; born in 1909, he studied at the Lucerne School of Arts and Crafts, later in Paris (1928–29) where he worked at the Académie Julian; afterwards at the Arts Academy in Hardenberg-strasse in Berlin. Medium is tempera and measures 78·7 × 70·8 in (2000 × 1800 mm). (Courtesy of the artist)*

There is often a considerable relationship to some work of the young child.

An earlier meaning of primitives implied the work of those artists who were active prior to 1450 in Italy and the Low Countries.

PROLECULT

Founded in 1906 in Russia but not becoming active until after the Revolution in 1917; its aims were to bring in a marriage of proletarian art with industry, as a somewhat vague outline of the Bauhaus that was to come later in Germany.

PROTO-RENAISSANCE

The art that came into being during the Hohenstaufen dynasty (1138–1254) in Germany.

PSYCHEDELIC ART

An art form where the painter has been stimulated by colours, forms or lights. It is likely to be abstract only, often with swirling, swelling curves reminiscent of Art Nouveau. Colours applied may have a predominance of pale acid greens and purples, often very intense. The effects are said to be related to the mental receptions after taking drugs such as Mescaline and LSD.

PURISM

In 1916 Ozenfant and Charles Jeanneret (later to become Le Corbusier) founded a movement that led away from Cubism. It had ideas for a faultless precise semi-Abstract art. Their theories were published in a magazine edited by the two founders called *L'Esprit Nouveau*.

RAJPUT PAINTING

During the rule of the Islamic Mughal dynasty in India the Rajput princes were patrons of a school of miniature-painting, the artists were concerned to continue the old Hindu traditions (c 1500– c 1800). Influences on them can be traced to a style similar to the Ajanta frescoes which had been painted between the 1st and 7th centuries; also from Islamic miniaturists which in association with Persian painters flourished in India under the

'Te Pape Nave Nave' (Life-Giving Bath) by Paul Gauguin (28·7 × 36 in (730 × 920 mm)). (Courtesy Marlborough Fine Art (London) Ltd)

'Le Pont des Arts' by Auguste Renoire; £645833. (Courtesy Sotheby Parke Bernet)

Mughals. The Rajput style has continued through till the present time, and members of the Tagore family of artists have tried to revive it.

RAYONISM

A type of Abstract painting practised in Russia in 1911. It was evolved by Michael Larinov and was characterized by radiating lines of light. Locally it went under the name of Lutschism. The theory of Rayonism dealt with the penetration and dispersion of rays coming from objects. Natalie Gontcharova was concerned with it for her work for the Ballets Russes under Diaghilev from 1914 onwards in Paris.

REALISM

Literally, in painting, this implies the putting down of the actual as it appears to the artist when he looks at it. Applied specifically to a school of painting, Realism starts with the work of Gustave Courbet and those around him in the 19th century in France. Realism differs from Naturalism in that the Realist sets out to show an impression of what is before him.

REGIONALISM

An American art movement of this century that ran on the theory that the painter should show what he knows best, such as his immediate environment or one that he has lived in. The principal painters in this style included: Thomas Benton, John Curry and Grant Wood.

RENAISSANCE

A single word that brings into being the wonderful awakening, rebirth of the arts and living, the whole creative activity in general in Europe, but in particular in Italy during the 14th, 15th and 16th centuries. It is in truth the transition, the bridge, between the medieval and the modern world. During this period, notably in Italy, there lived and worked one of the greatest concentrations of genius the world has ever known; a tremendous concourse of painters, sculptors, craftsmen, architects, writers, musicians, scientists and engineers, as well as a number of outstanding patrons. This conception of a rebirth of the Arts was probably first discussed by Ghiberti in his *Commentarii*; he spoke of the leading painters and sculptors at the end of the 13th century and at the beginning of the 14th. Alberti (1404–72), the architect, wrote in *Three Books on Painting* of the rise of a new era in the Arts of Florence with the work of Brunelleschi, Donatello, Masaccio and

Luca della Robbia. Giorgio Vasari in his *Lives of the Painters*, published in 1550, was the first to actually use the word *rinascita*, which means renaissance.

This era can be subdivided into the Early Renaissance (c 1400–1500); the High Renaissance (1500–20) and the Late Renaissance (1520–1600).

Masters of the High Renaissance include: Leonardo da Vinci, Michelangelo, Raphael, Giorgione and Titian.

The influence from the various periods of the Renaissance spread outwards from Florence and the other great centres in Italy; crossing national barriers as though they did not exist, reaching other countries at slightly later dates: France during the reign of François I (1515–47); England with the arrival of Holbein; Germany under the guidance of Albrecht Dürer (1471–1528); the Dutch and Flemish painters with the work of such as Mabuse (real name Gossaert), also Lucas van Leyden and Scorel; the waves reaching Spain largely at the end of the 16th century.

REVUE BLANCHE

A literary review that was started in France in 1891 by two brothers, Alexandre and Thadée Natanson. It sought support from painters during its twelve-year life; some that participated were Guillaume Apollinaire, Octave Mirbeau and Tristan Bernard. Meetings took place at the publishing house where discussions were out-spoken and *avant-garde*. Such as: Bonnard, Denis, Ranson, Roussel, Sérusier, Toulouse-Lautrec, Steinlen and Vuillard attended and theories in

Study of a Lion. Drawing by Rubens. (Courtesy Sotheby Parke-Bernet)

Neo-Impressionism, Synthesism, Neo-Traditionalism, Naturalism, The Nabis, and Mystic art would be thrown around.

RHEIMS SCHOOL

A 9th-century group of illuminators in France that had a strong influence from the Byzantine evident in their work; for example, the 'Utrecht Psalter' and the 'Gospels of Epernay', the 'Evangelistary of Bishop Ebo of Rheims'. The cell of illuminators probably worked just outside the city of Rheims at Hautvillers.

ROMANISTS

Flemish painters who worked in Rome under the influence of Michelangelo and Raphael, they included: Gossaert, Van Orley and Franz Floris.

ROMANTICISM

A general term to describe works of art that are the opposite to Classicism. Under this manner the artist allows himself more freedom, and display of emotion and atmosphere. In France such as Delacroix and Gericault worked in this manner; in England, Constable and Turner often brought a feeling of Romanticism into their pictures.

ROSE-CROIX-KABALISTIQUE

A select and somewhat private group of artists who came into being in 1888; they were alongside the Symbolists. They felt themselves idealists; and included in their following such as Ferdinand Hodler, Émile Bernard, Odile Redon, Jan Toorop and later Vuillard, Sérusier. Denis and Rouault spared some thought for their credo.

RUBENS ATELIER

During the 17th century Peter Paul Rubens established a large and prosperous studio, that was in point a great picture factory. During Rubens's practising life, travelling round the courts of Europe as he did, he collected so many commissions that it would have been impossible for him to complete them all himself. In his studio at Antwerp he employed a great number of master artists, such as Seghers, Snyders, Van Dyck and others.

RUSSIAN FUTURISTS

A grouping of the followers in Russia, of Cubism, Expressionism and allied theories and movements.

SALIAN ART

German art during the reign of the Emperors of the Salian Frankish House (1024–1137). It continues the Ottonian.

SALON DES REFUSÉS

A special salon set up on the order of Napoleon III in 1863, to show the works of those who were refused by the principal Salon.

SALON, PARIS
The most important annual exhibition of painting and sculpture in France. The name comes from the fact that exhibitions were held in the Salon d'Apollon in the Louvre in the 17th century. The exhibitions were biennial after 1737 and annual since the Revolution. Since the last war the more usual place has been the Grand Palais.

SEVEN, GROUP OF
Canadian painters who were set on producing a nationalistic movement, they joined ideas and styles in 1913. In their painting came the force and strength of their own landscapes, from the environs of Toronto to the prairies, the Arctic regions to the towering Rockies. The members merged with the Canadian Group of Painters which came into being in 1933. The seven original faithfuls were: Franklin Carmichael, Lawren Harris, Arthur Lismer, A Y Jackson, J E H MacDonald, F H Varley and Franz Johnston.

SEZESSION
In the last years of the 19th century artists in Germany who were working in advanced styles found that traditional societies, galleries and academies would not accept their work, so a process of seceding began. Munich led off in 1892, Berlin with the encouragement of Max Liebermann did the same in 1899 and the Viennese in 1897.

SIENA SCHOOL
The town was the last medieval stronghold in Italy, and it was here that the final artistic expression of the spiritual emotion of the Middle Ages was produced during the 13th, 14th and 15th centuries. Leading figures included Duccio di Buoninsegna (1278–1319), Simone Martini (1283–1344), Domenico di Bartolo (c 1428–c 47) and Francesco di Giorgio (1439–1502).

SLADE SCHOOL
Next to the Royal Academy Schools in England, the Slade is the most influential. It was founded in London in the 19th century by Felix Slade, with Sir Edward Poynter as the first Slade Professor of Fine Arts. During the 1920s Henry Tonks was the leading teacher.

STURM, DER
A gallery and art magazine in Berlin that was for the progessive movements, notably the Expressionists, Blaue Reiter and Futurists. Both were the ideas of Herwarth Walden. The gallery lasted until 1924 and the magazine, which started in 1910 appeared until 1932.

SUPREMATISM
An Abstract art form founded in Moscow in 1913 by Kasimir Malevich. It was an extreme geometric form of Cubism, which relied on the circle, triangle, rectangle and cross for the main shapes. Another follower was Alexander Rodchenko, who in 1919 produced a painting 'Black on Black' as a companion to 'White on White' painted by Malevich in the same year. In his manifesto in 1915 Malevich described his theory: 'By Suprematism I mean the supremacy of pure feeling or perfection in the pictorial arts – the experience of non-objectivity. The movement was close to Constructivism and Neo-Plasticism.

SURREALISM
A 20th-century movement with roots deep in art history; elements being picked up in the 15th, 16th and 17th centuries with the work of Hieronymus Bosch, Pieter (Hell) Brueghel the Younger, Giuseppe Arcimboldo and Mathias Grünewald. Surrealists in the 20th century set out to liberate their thoughts and, if required, to dissociate objects from their natural placing and handle the whole in a diametrically opposite manner to the traditional or expected, and to explore their subconscious. Some practitioners claim their work is based on Nihilism, the Dialectical Materialism of Karl Marx, others include the thoughts of Sigmund Freud. The leading Surrealists include Max Ernst, Salvador Dali, René Magritte, Marcel Duchamp, Hans Arp, Man Ray, Yves Tanguy, Georgio de Chirico and Joan Miró.

The movement has had considerable support from writers and theorists and not least André Breton who brought out *Surrealist Revolution* in 1928; earlier he wrote a *Surrealist Manifesto* in 1924 and a second Manifesto in 1929; in 1942 *Prolegomena to a Third Manifesto of Surrealism or Not.*

SYMBOLISM
A literary movement that had a vogue in 19th-century France, the followers felt that poetry should relate to moral and physical acts by word symbols; poets primarily concerned were Baudelaire, Mallarmé and Verlaine. Painters who were a part of this thought process were: Puvis de Chavannes, Gustave Moreau and Odilon Redon.

SYN
A group of German painters who set out to find an integral art free from the disciplines and constricting theories of prevalent progressive movements. They became active in 1964 and members included Bernd Berner and Klaus Jürgen-Fischer.

SYNCHROMISM
(see also Orphism)
An approach to pure colour abstraction which started in 1913. It was the first American movement with thought-out rules and its own

'Barbares' by Max Ernst. (Courtesy The Redfern Gallery)

Manifesto. The two artists primarily concerned were Morgan Russell and Stanton Macdonald-Wright. In a visit to Paris before the First World War they had picked up influences from the Orphists.

SYNTHETISM
A type of Post-Impressionism worked on by Paul Gauguin in the last years of the 19th century. It employed a type of painting that used simple, flat, broad tones of colour; it was at variance with the basic 'broken colour' and other theories of Impressionism.

TACHISME
The name to express the freest method of Action painting; it is derived from the French word meaning stain or blot. When working in this manner the painter leans on accidental patterning and stippling of colours which can be applied as he desires: large brushes, pouring direct from a bottle, splashing, walking about or riding a bicycle over the wet paints or such uncontrollable methods as setting fire to oil- or spirit-based

colours. One of the first French exponents was Jean Dubuffet, others include Georges Mathieu, François Arnal and Jean Fautrier.

TEN, THE
A group of American artists who were influenced by the Impressionists. They showed together in 1895; the leading figure was probably Childe Hassam.

TENEBROSI
A school of painters led and influenced by Caravaggio (1573–1610), they were also known as Chiaroscurists. They made use of highly dramatic light and shade effects, often with a single source. Careful observation enabled them to produce effective studies in the effect of light, and reflected light on figures and objects at varying distances from the source. Rembrandt and Joseph Wright of Derby were influenced by their experiments and successes.

TEN O'CLOCK
A provocative and original lecture given at ten o'clock on 20 February 1885, by the American

painter James McNeill Whistler to, as he said, an audience of 'friends and enemies'. One of his main points was that the artist should be absolutely independent in his treatment and should not be under any restrictions from his patron.

TONALISM
A treatment evolved in the painting of Post-Revolution America. It was followed by such as George Innes, William Keith and Henry Ranger. Sunlight and its effects were one of their leading concerns, and they would often use numerous superimposed glazes to achieve the subtlety they sought.

TOURNAI SCHOOL
A 15th-century Flemish school that had centres at Ghent, Haarlem and Louvain; leaders included Robert Campin and Roger van der Weyden.

TOURS, SCHOOL OF
A 9th-century Carolingian group of illuminators and miniature-painters.

TUSCAN SCHOOL
A general name for the Florence school and including the schools of Arezzo, Lucca, Pisa and Siena.

291 GROUP
A group of American artists who took their name from the address of the gallery where they exhibited.

UMBRIAN SCHOOL
A 15th- and 16th-century Italian school; leading figures included: Raphael (1483–1520), Perugino (1446–1523), Piero della Francesca (c 1410–92) and Luca Signorelli (1441–1523).

UNIT ONE
A group of English painters, sculptors and architects, founded in 1933. Members included: Edward Burra, John Armstrong, Paul Nash, Ben Nicholson, Tristram Hillier, Colin Lucas, Barbara Hepworth and Henry Moore. They set out to express a 'truly contemporary spirit' in art.

UTRECHT SCHOOL
(see also Tenebrosi)
A group of painters in the early part of the 17th century who worked in Utrecht, Holland. They were native artists who had travelled to Rome and been to an extent influenced by Caravaggio. Largely through such as Gerrit van Honthorst and Hendrick Terbrugghen the influence spread through to Dutch painters such as Frans Hals, Rembrandt and Jan Vermeer.

VENETIAN SCHOOL
During the High Renaissance Venice was second only to Florence as a centre of the flowering arts, later she was to take over the leadership. The leading figure was Titian (c 1487–1576), others include Paolo Caliari Veronese (c 1528–88), Giorgio Barbarelli Giorgione (1475–1510) and Jacopo Robusti Tintoretto (1518–94). The pictures of the school were characterized by rich, full colouring and use of chiaroscuro.

VERISM
A theory of this century that a work of art need not have subject-matter based on beauty, nobility and other long-accepted standards. Followers claim that ugly objects, vulgar settings can possess inherent aesthetic qualities. (See Ash-Can and Kitchen Sink Schools.)

VORTICISM
This was started by Wyndham-Lewis in 1912. It was on a parallel course with the Italian Futurists, the name being derived from the founder's magazine *Blast*. The work of the group was predominantly abstract or non-figurative; the followers sought for freedom from Traditionalism and narrative-painting. The adherents included Gaudier-Brzeska and Jacob Epstein.

WANDERERS, THE
A group of Russian painters founded in 1872 by Nicholas Kramskoy; members included: Ilya Repin, Vladimir Makovsky, W Polenov and V V Vereshagin. They sought to enlighten the people through art and with their realistic pictures expressed criticism of the social scene and of the power of the Czarist régime.

WORLD OF ART GROUP
(also Mir Iskusstva)
A school of Russian artists led by Alexander Benois which was founded in 1890 in opposition to the ideas of the Wanderers. Exhibitions of their work continued until 1922. Members included Leon Bakst, Nicholas Roerich and Konstantin Somov (who was also associated with the pre-revolutionary Blue Rose Group). Main influence on the World of Art Group came from French Impressionism and they, as with others about this time, declaimed for 'Art for art's sake.'

X GROUP
An association of English Post-Impressionist painters which was founded in 1920.

Chapter Six

Gallery

Included here are painters and sculptors from many countries, representative of the story of art from the 13th century to the present time. From the great host, numbering tens of thousands, it is only possible to bring in the high peaks, and those of interest or incident, who can illuminate the breathless passage, often beautiful, often thoughtful, often savage, but seldom banal or dull of waxing creative sensation, the flowering that has come during the search for that illusive ideal expression for which every artist strives.

No one can truly point to a particular painter or sculptor and say that appreciation of him should be universal. Yet inevitably a select number raise themselves by some perhaps unsensed or unrealized genius to stand as indestructible marks to the results of an inner vision, a combination of a remarkable ability to manipulate paint or to wield a chisel with that something that could be within us all, and yet only they are granted that extra sight; but fortunately what they see and capture reflects back for all to study, to love, to reject, to learn from but seldom to ignore.

For each the choice will vary; but for most there will be some reception of the creative overcoming that will have gone into such as: 'The Agony in the Garden' by El Greco; 'David' by Michelangelo; one of the later self-portraits by Rembrandt; 'An Apostle' in limewood, Creglingen Herrgottskirche, by Tilman Riemenschneider, 'Snow Effect at Vétheuil' by Claude Monet. Yet it is intrusive to pick out; it is better to offer a selection. From these hundreds of men and women who have felt the humility that is the step to creativity, a pattern may emerge that will light up the history of art, a story at one and the same moment, complex and simple. This can provide a key to the often harsh and un-appreciated lives of those who have created memorable images from basic materials.

What pointers to public taste can a short survey present? Where possible evidence from the auction-rooms has been given, but this is not a real standard because values (monetary) can slide down and climb un-expectedly. Then it can happen, too, that the pundits are wrong and the artist was right all the time.

AACHEN, JOHANN VON (*c* 1562–*c* 1615) Born in Cologne and died in Prague. Studied with Barthel Spranger. Influenced by Tintoretto when on travels to Italy, which led him to be **one of the first to romanticize the stiff Gothic taste of his country.** Outstanding work 'Bathsheba Bathing' in Vienna. 'The Adoration of the shepherds' Asch 25000, 1969, Dorotheum.

ABBEY, EDWIN AUSTIN (1852–1911) American painter and illustrator, studied at Pennsylvania School of Fine Arts. Joined Harper Brothers, New York, as illustrator. Spent 12 years on murals based on 'Quest for the Holy Grail' in Boston Public Library. In 1902 painted official pictures of the Coronation of Edward VII. Last important work – murals for the State Capitol of Harrisburg, Pennsylvania.

ABILDGAARD, NIKOLAJ ABRAHAM (1743–1809) Born in Copenhagen, studied with his father Sören. Travelled to Italy and **copied pictures by Titian, Raphael and Michelangelo.** Painted themes from the Norse sagas. In 1789 he became Director of the Academy of Arts. Much of his best work destroyed in the 1794 fire at the Palace of

See end of chapter for exchange rates and abbreviations.

Christianburg. **He taught Runge and Thorvaldsen.**

ADAM, ALBRECHT (1786–1862)
Born at Nördlingen, Bavaria. A battle-painter, he went with the French and Bavarian army against Austria in 1809, and in 1812 to Russia with the Grand Army. In 1859 went with his son to the Italian campaign of France and Sardinia against Austria. Battle pictures include: 'On the Moscowa'; 'Abensberg'; 'Custozza'; 'Novara'. His last commission from King Maximilian of Bavaria was of the decisive charge of the Prussian cavalry at the Battle of Zorndorf.

ADAMS, ROBERT (1917–)
Born at Northampton and trained at the School of Art there. A non-figurative sculptor with architectural characteristics. Works in stone, concrete and metal, and has done an amount of wood-carving.

ADLER, JANKEL (1895–1949)
He came from Lodz, Poland. Early training as an engraver. He lived an oppressed and wandering life, going to Germany, Spain, France, Italy, the Balkans, Russia, Scotland, England. His work has the massive quality of Léger with a little of the poetry from Chagall.
'A still life' Sfrs 12000, 1969, Galerie Motte, Geneva.

AETSEN, PIETER (1507–73)
Born in Amsterdam. The son of a stocking-maker; he was known as Lange Peer because of his great height. Studied with Allart Claessen, joined Antwerp Guild in 1535. His subjects often included kitchens, cooking utensils, foods and the odd serving-girl. He also painted altarpieces, diptychs and religious panels such as the 'Adoration of the Magi' and 'Nativity'. **Many of his best works were destroyed by the iconoclasts in 1566.**
'Two Market Women' £2700, 1975, Southeby Parke Bernet, London.

AIVAZOVSKY, IVAN KONSTANTINOVICH (1817–1900)
Born in Feodosia of Armenian nationality, he studied at the Academy of Arts. A marine painter who worked nearly all his life round the Black Sea. His pictures have considerable sensitivity for atmosphere with light and elemental happenings. **His house in Feodosia is now a museum.**
'Full moon at sea' $8600, 1975, Sotheby Parke Bernet, New York.

ALBANI, FRANCESCO (1578–1660)
Son of a silk merchant in Bologna. At 13 he entered the academy of Denijs Calvaert. Here he met Guido Reni and they both continued studies with Lodovico Caracci, under the patronage of the Duke of Mantua. **He died in the arms of his pupils – with his brush still in his hand – in his 82nd year.** In the Church of San Sebastiano, Rome, is an altarpiece of the martyrdom of that saint and an 'Assumption' painted together with Reni.

ALBERS, JOSEF (1888–)
Born in the Ruhr, Germany, he became one of the teachers with the Bauhaus, concerned primarily with colour and form. He made a study of optical illusions and tried to work out a **physiology and psychology of painting.** When the Bauhaus was dissolved he went from there to Black Mountain College in North Carolina, and from 1950 until his retirement in 1960 he taught at Yale. 'Homage to the Square: two whites between two yellows 1958' L 7500000, 1975. Finarte, Milan.

ALEXEYEV, FEODOR YAKOVLEVICH (1755–1824)
The son of a watchman at the Academy of Science; he studied at the Academy of Arts from 1766 to 1773. Then he travelled to Venice for four years; on his return he worked as a theatrical decorator in St Petersburg. Principally a painter of townscapes, often with water, there is a slight influence from Canaletto. He travelled a number of times across Russia to paint views of: Kherson, Voronezh and Orel.

ALKEN FAMILY
One of the painter families that have caused and do cause considerable trouble to art historians and collectors as they try and **sort out which relation painted what.** The Alkens started with Sefferin (1717–82) and continued in production until Henry Gordon Alken died in 1894. Within this period male painting members included: Samuel, the son of the original Sefferin, and four of his sons, Sam, George, Sefferin and Henry. This Henry then had two sons, whom, to help matters, he christened Henry and Sefferin. All the Alkens painted in the same style, all were horse-painters and signed their work S. or H. Alken. Not quite all because the last Alken, who was christened Samuel Henry but was called Henry Gordon, sometimes signed his work H. Alken Junior, and at other times left out the Junior when he attempted to pass off his rather inferior work for that of his father.
Henry Alken Senior 'Hunting Scene' £5500, 1975, Christie's, London.
Henry Alken Junior 'Hunting Scenes' (four) $3750, 1975, Sotheby Parke Bernet, New York.

ALLSTON, WASHINGTON (1779–1843)
Studied first with Edward Malbone and Charles Fraser in America and later in London with Benjamin West. He travelled after this for further study to Paris, Florence and Rome. In the last city he met the poet Samuel Taylor Coleridge and formed a lifelong friendship. Nicknamed by some the American Titian, Allston painted in the grand Italian manner, grafting on his own individual personality. Subjects included: 'The Deluge', the Metropolitan Museum of Art, New York; 'Belshazzar's Feast', Detroit Institute of Arts. **His main pupil was S B Morse.**

ALMA-TADEMA, SIR LAWRENCE
(1836–1912)
Born at Dronrijp, Holland, he studied with Gustav Wappers. He became a British citizen in 1873, and was knighted by Queen Victoria in 1899. His art was somewhat similar to that of Lord Leighton, a sentimental classicism. His market so strong in his time, with figures like £6060 for 'A reading of Homer', shrank till in 1960 'Rose of Elgagabalus' only brought £105, when in 1888 the artist had been paid £4000 for it. But tastes can change quickly; for in 1975 'The Women of Amphissa' brought £11 500 at Christie's, London.

ALTDORFER, ALBRECHT (c 1480–1538)
It is likely he was born in Ratisbon (Regensburg), Germany, probably the son of the painter Ulrich Altdorfer. **He did much to develop landscape**, bringing a sense of romance and mystery particularly to woodland scenes. As an engraver on wood and copper he was close behind Dürer. He established himself not only as a painter in Ratisbon but also as an architect, and he was a City Councillor who, on at least one occasion, involved himself in a pogrom. He lived in some splendour as his Will shows. This mentions chests, pictures, weapons, gems, stuffs, coins, silver goblets and a horse with trappings. **His most amazing picture 'The Battle of Arbela'** displays a talented control with the handling of hundreds of armed figures, giving a sense of swirling movement to the columns by the directional placing of the great lances; in the background is a fantasy landscape, all prickly mountains, lakes and storm-torn sky. This picture is in the Pinakothek, Munich.

AMARAL, TARSILA DO (1900–)
Brazilian painter, sculptress and writer born in São Paulo. From 1917 onwards she came to Europe where she met Lhote, Léger and Gleizes and was influenced by them. Her painting reflected aspects of Impressionism. She returned to Brazil in 1925 and joined a progressive group the *Antropofágico*.

ANDREYEV, NIKOLAI (1873–1932)
One of the greatest Russian sculptors. He understood that simplicity of form gives strength, and could give a plastic quality to the hard stone he worked in. He spent the last years of his life in concentrated studies of Lenin known collectively as 'Leniniana'.

ANGELI, RADOVANI KOSTA (1916–)
Yugoslavian sculptor born in London. His tuition included a period at the Accadèmia di Brera in Milan and he had his first show in 1940. His *métier* is in the handling of portrait heads and busts, figure groups and nudes.

ANGELICO, FRA (c 1387–1455)
Born in Vicchio, died in Rome. Original name was Guido di Pietro, also called Fra Gio. da Fiesole and Il Beato. In 1407 he took the Domenican vows and was to become Prior of the Fiesole Monastery in 1449. He worked in the manner of Giotto, but introduced greater subtlety, refinement and depth of vision. **He was among the first to bring a feeling of joy into his work.** His arrangement of figures is an advance on Giotto, although the anatomy is still wooden and perspective is not yet understood. Although he may have technical shortcomings, Fra Angelico had a manner of creating a spiritual quality in the people he painted; they seemed to have an inner seraphic glow. The depth of his own consciousness comes through in the faces of his sitters. Among his finest works are the 'Annunciation', Prado, Madrid, and the 'Christ Glorified', National Gallery, London.

ANGUISCIOLA SISTERS
The most famous was Sofonisba, born in Cremona in 1535. She studied first with Bernardino Campi later with Bernado Gatti. She quickly built up a name as a portrait-painter and was invited to Madrid by Philip II. She arrived attended by three of her five sisters. She was married twice; first to Don Fabricio di Moncada, a Sicilian nobleman; then when he died she took a galley to Genoa and on landing married the Captain, Orazio Lomellini. There is a portrait of her drawn by Van Dyck when she was 96 included in his sketchbook at Chatsworth. Van Dyck records that 'she gave me many hints, such as not to take the light from too high, lest the shadows in the wrinkles of old age should become too strong . . .'.

Her sisters were: Elena, who after studying with the same masters as Sofonisba, entered the Convent of San Vincenzo, Mantua; Lucia who distinguished herself both with painting and music; Minerva who died young and Europa and Anna Maria who both painted religious historical subjects for churches.

ANNIGONI, PIETRO (1910–)

Born in Milan and trained at the Accadèmia Bella Arti, Florence. A founder of a group of modern realistic painting in Milan. He works principally in oil, tempera and fresco.
'Julie Andrews in My Fair Lady' £7000, 1975, Sotheby. Parke Bernet.

ANTONELLO DA MESSINA
(active 1456–c 79)

Born Messina, died in Venice. Reputed to have been a pupil of Colantonio del Fiore. He was strongly affected by a painting of Jan van Eyck he saw in the collection of the King of Aragon in Naples. He is supposed to have gone to Bruges to study the new manner with oils and to have returned to introduce it into Italy. There is no direct proof of this, but he would have had opportunities to study the style as Flemish paintings were coming to Italy; also Roger van der Weyden, Hugo van der Goes and Justius of Ghent worked in Italy in the middle of the 15th century. **His portrait command was particularly noted** as with the 'Condottiere', in the Louvre, Paris, and his self-portrait in the National Gallery, London. As one who has not appeared in a modern sale-room it is of interest to note: 'St Jerome' sold 1849 for £162 15s, and the same picture sold 1894 for £425 5s.

APOLLINAIRE, GUILLAUME DE KOSTROWITSKY (1880–1918)

Born in Rome, died in Paris. He had a Polish mother and a Swiss-Italian father. An inveterate practical joker, he liked to pass himself off as the son of an Italian prelate. His first interest was with Cubism, but later he moved to Futurism, publishing *The Futurist Antitradition* in 1913. He was close to Duchamp, Picabia and Chirico. He dabbled in the disturbing fields of dreams and the subconscious and their combination for Surrealism. **A forerunner of the Dadaists.**

APPEL, KAREL (1921–)

Born in Holland. As an Abstract painter his true affinity is with Abstract-Expressionism. His forms which can incorporate near-naturalistic shapes and ideas are freely laid in with a considerable impressionism of emotional strength; colours are strong, rich and bright, being laid beside each other to give the greatest effect from contrast.
'Figure, 1968' L 7000000, 1975, Finarte, Rome.

ARCHIPENKO, ALEXANDER (1887–1964)

Born in Kiev, Ukraine. Studied at Kiev Art School, thence to Moscow, and to Paris in 1908 for a short spell at the École des Beaux-Arts. A sculptor, he worked in clay, stone, bronze and iron; his expression was primarily through the Cubist idiom. 'Femme Assise' 1912 Bronze $12000, 1975, Sotheby Parke Bernet, New York.

ARCIMBOLDO, GIUSEPPE (1533–93)

A favourite painter of the Emperor Maximilian II and Rudolph II in whose service he was employed for most of his life. He is best recalled by **his almost Surrealistic manner in dealing with heads and figures; building them up from flowers, fruits and vegetables.**

ARDON, MORDEACAI
(formerly Bronstein) (1896–)

Born in Tuchow, Poland. He studied with Kandinsky, Klee and Itten and later himself taught at Itten's Art School in Berlin. In 1933 he emigrated to Palestine. As a painter he excels most as a remarkable colourist. There is a monumental triptych by him in the Amsterdam Stedelijk.

ARIKHA, AVIGDOR (1929–)

Born in Radautzi, Rumania. In 1944 he went to Palestine and studied with Ardon. He set out as a book-illustrator for writers such as Rainer Maria Rilke (Austria) and Samuel Beckett (Ireland), and then changed to Abstract art, evolving a mystical manner.

ARMITAGE, KENNETH (1916–)

Born in Leeds. He studied first at Leeds College of Art and then the Slade School, London. After the war he was Head of the Sculpture Department at the Bath Academy of Art. Bronze is his principal method of expression and his figures and groups have a strange emergent feeling, the hard alloy being given a malleable plastic look. 'Little Torso' (10 in 254 mm) £360, 1975, Sotheby Parke Bernet.

ARP, HANS (1887–1966)

Born in Strasbourg, died in Basle. In 1905 he went to study painting at the Weimar Academy of Art. In 1908 he went to Paris to join the Académie Julian. In the years immediately after this he was at Weggis in Switzerland painting abstracts and also beginning to model, and so he turned towards the third dimension working in marble and bronze. His forms not always entirely abstract at times lean on the torso and sinuous plant shapes. **One of the basic influences on Barbara Hepworth and her followers.**
'Silencieux, 1942' DM 30000, 1975, Karl Faber, Munich.

ARPINO, GIUSEPPE CESARI
(c 1560–1640)

Born in Naples, died in Rome, also known as Il Cavaliere d'Arpino. He studied with his father who earned a poor living painting ex-voti. When Arpino came to Rome he first earned his living preparing the colours for artists working in the Vatican under Gregory XIII. Once when the painters were away Arpino sketched some figures on a wall, which were seen by Fra Ignazio Danti, the Superintendent of the Vatican Works, who told the Pope. His Holiness put Arpino into the School of Pomarancio. Afterwards the young painter attained great prestige, which, to a high degree, he manufactured himself by using intrigue, deprecating the works of others and praising his own. His figures display a lack of anatomical understanding.

ASHFORD, WILLIAM (1746–1824)

He was born in Birmingham and moved to live in Dublin in 1764; he showed regularly for 60 years and was elected the **first President of the Royal Hibernian Academy** when it was founded in 1823. His landscapes have a subtle sense of atmosphere, with well-handled light and skies.
'An overshot Mill' £2000, 1975, Sotheby Parke Bernet.

'The Threatened Swan' by Jan Asselyn. (Courtesy Rijksmuseum, Amsterdam)

ASSELYN, JAN (1610–52)

Born in Diepen, near Amsterdam, died in Amsterdam. He studied with Esaias van de Velde; later travelling to Italy while still young and remained there for many years. **He had a strange manner of contracting his figures** which earnt him the nickname of Krabbetje. One of his best-known pictures is the 'Threatened Swan' in the Rijksmuseum.
'The Battle of Lützen' (1635) Sfrs 38000, 1975, Galerie Koller, Zürich.

AUDUBON, JOHN JAMES (1785–1851)

Born possibly in Haiti and died in New York. He was educated in France having some instruction from David. In 1804 he went to America and made a living as a hunter, naturalist, taxidermist and recorder of birds and wildlife. **His *The Birds of America* is a classic work of its type**, published in four volumes, 1827–38, it has 435 plates. This was followed by *The Quadrupeds of America* in 1845, with this he was assisted by his sons. There are numbers of his water-colours and sketches in the collection of the New York Historical Society at Harvard.

AUERBACH, FRANK (1931–)

Born in Berlin. Studied at the St Martin's School of Art, and with David Bomberg and after at the Royal College of Art. Generally paints in oil on board, employing a very heavy, rugged, sharp impasto.
'Jym in the studio II' £600, 1975, Sotheby Parke Bernet.

AVERCAMP, HENDRIK (1585–1634)

Dutch painter who produced a number of rectangular and round *Kermesses* on ice: small pictures alive with great numbers of skaters and onlookers. **Sadly through an unwise choice of materials, many of them have changed colour.**

AVRAMIDIS, JOANNIS (1922–)

Born in Batoum, Russia, of Greek parents. He studied at the Batoum Art School, later in Vienna. Between 1953 and 1956 he learnt sculpture in the studio of Wotruba. His forms have an affinity to those of Brancusi and the painted modelling of Schlemmer.

BACON, FRANCIS (1910–)

Born in Dublin he is entirely self-taught. A painter in the full-blooded stroke manner of Rubens, colours strong, rich and bright interplayed with sombre tones. In subject-matter often terrifying yet quite brilliant. Looking at many of his canvases is to experience a sensation of extreme shock; it is almost as though the sitters have been torn open and the darker emotions of their minds have been exposed. His work is entirely free of any pandering to fashion, nicety or attempt to hide an unpalatable feature. Although there have been a number who might have tried to emulate his style, he remains alone in his ability to stake savagery on to the canvas and evoke his particular glimpses of a netherworld. In the Tate Gallery, London are his 'Three Studies of figures for the base of a Crucifixion'.
'Study for a Pope 1954' £65000, 1975, Sotheby's, London.

'Reclining Man with Sculpture' £89890, 1976, Sotheby Parke Bernet, New York.

BAKHUYSEN, LUDOLF (1631–1708)

Born in Emden, died in Amsterdam. He studied with Everdingen and later with Dubbels. His father had intended him for a mercantile career, but this was argued against by Ludolf, although it did set him on his way as a marine painter of considerable talent. At first it was just pen and ink drawings of vessels lying in Amsterdam; these sold readily. Then he set himself to challenging in paint the anger of the sea and the trials of shipmasters riding out the steep-waved gales of the North Sea. **Often he would be in considerable danger as he hired fishermen to take him out on the stormy waters.** He is said to have given drawing lessons to Peter the Great while he was studying at Saardam. Among his better judged paintings are: 'Disembarkation of William III' in The Hague and 'The Tempest' in Brussels.
'A lugger in a fresh breeze' £4725, 1969, Christie's, London.

BAKST, LÉON (1886–1925)

Russian painter and stage-designer who had an almost unique quality particularly with costume. His choice of colours, materials and cut blended themselves perfectly with the characters and stars of the ballet, especially those for the great dancer Nijinsky. He was a member, with Diaghilev and Benois of the World of Art Group. Influences on him included late Persian art and the theories of colour being explored by the German painters in the period 1900 to 1920.
'Design for the décor, "Shéhérazade", water-colour, with pencil and gold paint' £13000, 1975, Sotheby Parke Bernet.

BALDOVINETTI, ALESSO (1426–99)

Born in Florence, little is known about his tuition; first firm mention being that he was admitted as a member of the Guild of St Luke in 1448. A great number of his paintings were either lost or destroyed. Of those existing these are of quality: 'Annunciation' in the Uffizi, Florence and a religious conversation piece in the Medici Villa at Cafaggiolo. Vasari records that 'he sketched his compositions in fresco, but finished them in *secco*, tempering his colours with a yolk of egg mingled with a liquid varnish, prepared over the fire'. This somewhat unwise procedure contributed to the deterioration of many pictures. Latterly he turned to mosaic and it proved to be his true *métier*. He restored a number of mosaics including those in the Baptistery, Florence.

BALDUNG, HANS (c 1475–1545)

(also called Grien or Grün, possibly from his habit

'*Mass of St Gregory*' by Hans Baldung. (*Courtesy The Cleveland Museum of Art*)

of wearing green). He was born at Gmünd, Swabia, died in Strasbourg. He was a painter, engraver and designer. Little is recorded of his youth. He knew Dürer and it is possible he may have had lessons from him. His most important work is the altarpiece in the Cathedral of Freiburg, which bears the inscription: 'JOANNES BALDUNG COG.GRIEN GAMUNDIANUS DEO ET VIRTUTE AUSPICIBUS FACIEBAT 1516'. It shows the Coronation of the Virgin, on the inside wings the Twelve Apostles and on the outside the Visitation, the Flight into Egypt, the Nativity and the Annunciation which Kugler attributes to another hand. As with other German painters of this period there often appears the sense of fantasy and threatening elemental happenings. He executed a large number of wood-blocks including the strange 'Bewitched Stable Boy'.

BALLA, GIACOMO (1871–1958)

Born in Turin, a leading figure with Futurism, as such **he was intent on the capturing of the sensation of movement** in a picture. He did this by repetitively brushing in a number of positions of limbs and objects as they would be in different stages of motion. Best-known work is probably 'Dog on Leash' which is in the Museum of Modern Art in New York.
'Balfiore' L 11000000, 1975, Finarte, Rome.

BANDEIRA, ANTONIO (1922–)

Born in Fortaleza, Brazil. He studied first in Rio de Janeiro and later in Paris at the Académie and Atelier de la Grande Chaumière. He went back to Brazil for a short time, then returned to Paris and began to paint abstract lyricisms.

BARBARI, JACOPO DE' (c 1450–c 1516)

(also called Jacob Walch and the Master of

Caduceus). Born probably in Venice, died in Brussels. In 1500 he went to Nuremberg and entered the service of the Emperor Maximilian and later the Archduke Frederick III of Saxony and Joachim I of Brandenburg. The year 1510 finds him as Court Painter to the Regent of the Netherlands, Margaret of Austria. **Among his graphic works was a large bird's-eye view of Venice which was engraved on wood and published by Anton Kolb,** the head of a German merchant family settled in Venice. His work is of interest as being a direct link between the Italian and German, generally it was a case of the Germans going to Italy, but Barbari reversed this.

BARKER, THOMAS (1769–1847)
(Known as Barker of Bath). Born near Pontypool, Monmouthshire, Wales, he died in Bath. His father was a lesser hand at animal painting. When the family moved to Bath, Thomas was helped in the practice of his art by Mr Spackman, a well-to-do coachbuilder. He spent four years copying Dutch and Flemish artists; at 21 his patron sent him to Rome, although he did little painting but absorbed the atmosphere and the works of the great. In his time he was among the better-off artists. One subject, for example, which he made two versions of, brought him 500 guineas for each from Mr Macklin and Lord Paulett. He had great popularity with all, his pictures and designs appeared on Staffordshire pottery, Worcester china, Manchester cottons and Glasgow linens. He painted a fresco 30 feet (9 m) long on the wall of his house, Sion Hill, Bath, entitled 'The inroad of the Turks upon Scio'. At one time he had large properties. Today his works are unlikely to command the prices he achieved in his lifetime.
'Wooded landscapes', France (a pair) £120, 1975, Sotheby Parke Bernet.

BARLACH, ERNST (1870–1938)
Born in Wedel, Holstein, the son of a country doctor, he studied at the Hamburg School of Arts and Crafts and then went to the Dresden Academy, and finally for a year to the Académie Julian, Paris. After two years in the army in the First World War, he survived to 1919 and became a Member of the Berlin Academy. In 1936, he was elected an Honorary Member of the Vienna Sezession, at the same time as the Nazis were ordering his works to be taken out of the Berlin Academy Jubilee Exhibition. He witnessed his monument at Kiel broken up and his memorial to the war at Güstrow cut up and melted down for scrap-metal. A final blow came when 381 of his works were banned from public showing. This was the year of his death.

Barlach was essentially a carver in the line of the German sculptors of three and four hundred years earlier. Whether working into huge blocks of wood or the clay preparatory to casting in bronze, his figures have an Expressionist power; they flow, they strain, they proclaim mutely. There is sense of dynamic power that he could tap with his chisel or mould with his fingers. At Luneberg there is a museum of the sculptor's works that have survived the latter-day iconoclasm of the Nazis.
'Das Wiedersehen' (bronze $18\frac{1}{2}$ in (470 mm) high) DM 35 500, 1975, Hanswedell & Nolte, Hamburg.

BAROCCIO, FEDERIGO (1528–1612)
Born at Urbino the son of a sculptor. After some study with his father he was placed in the studio of Battista Franco. He went to Rome to work in the palace of Cardinal Della Rovere for four years. Back again in his home town he painted 'St Margaret' for the Confraternity of the Holy Sacrament. This brought him to the notice of Pope Pius IV, who asked him to assist in the

Prince of Urbino, Federico (18 months) by Federigo Barocci. (Courtesy The Detroit Institute of Arts)

An Italianate River Landscape by George Barret. (Courtesy Christie's)

decoration of the Belvedere Palace in Rome. He painted 'The Virgin Mary and Infant Saviour, with several Saints', and a ceiling fresco showing the 'Annunciation'. While he was toiling in the Vatican, someone slipped him a powerful poison, which although it did not kill him, kept him from work for four years and thereafter weakened him. His figures are shown in a manner similar to Correggio. The German painter Mengs remarked that Baroccio's work lacked yellow tints, and Bellori felt he used too much ultramarine and vermilion.

BARRADAS, RAFAEL PÉREZ (1890–1929)

Uruguayan painter born in Montevideo. He began his career as a caricaturist. In 1912 he left South America for Europe and stayed there for 16 years. He had some affinity with the Cubists and Futurists both in Italy and France. After moving on to Spain he fell in with Torres Garcia. Here he painted harbour scenes, landscapes and towns and at the end of his stay turned to religious subjects. By 1928 he was back in Montevideo producing a series of what he termed 'Pictures of Montevideo', which were extra large water-colours.

BARRET, GEORGE (1732–84)

The son of a draper he studied under Robert West. He met Edmund Burke while still in his late teens and retained his friendship all his life. Barret spent many years working on the landscape round Dublin, in particular the Powerscourt Demesne. In 1762 he went to London and found popularity with his landscapes which were in the Italianate vein practised by Richard Wilson and others at this time. As with other artists he at times collaborated and in 'An extensive moorland landscape' he worked with the horse-painters Sawrey Gilpin and Philip Reinagle. It was sold for £3750 in 1975 by Christie's, London. In 1977 three of Barret's landscapes went for just over a £1000 each at a house sale in Ireland.

BARRY, JAMES (1741–1806)

Born in Cork, died in London. Also, as with Barret, he studied with Robert West and became a protégé of Edmund Burke, who financed a trip for him to Italy for four years; while there he was made a Member of the Clementine Academy at Bologna, painting as his diploma picture, 'Philoctes in the Isle of Lemnos'. Returning to England he became an Associate of the Royal Academy. He carried out a series of six pictures showing 'Culture and Progress of Human Knowledge', as a gratuitous offering to the Royal Society of Arts; the largest of these was a view of Elysium, 42 ft (12·80 m) long, and showed the portraits of the good and famous of all nations. **In 1783 he was elected Professor of Painting for the Academy Schools.** Sadly, Barry had a wayward and uncontrollable temper which often brought him into dispute with fellow Academicians, bringing his expulsion in 1799. Latterly he lived a squalid life in a ramshackle house in Castle Street, Oxford Market, London. Burke on one visit was reputed to have been left cooking the steaks while Barry went out to fetch supplies of porter.

'A warrior tortured by remorse' £1150, 1975, Christie's, London.

BARTOLOMMEO DI PAOLO, FRA (1475–1517)

He was also known as Della Porta because he lived in Florence near the Porta Romana (then known as the San Pier Gattolini). He worked in the studio of Cosimo Rosselli, and while there got himself involved with the fiery preachings of Fra Girolamo Savonarola, and became one of his most faithful disciples. In 1489 on Shrove Tuesday the people of Florence gathered in the square in front of the Convent of St Mark to hold an *auto-da-fé*, **Bartolommeo joined in fervently and consigned a number of his drawings and studies to the flames**. He took the Dominican vows in 1500 and for four years never painted or drew. Quality works include: 'The Virgin and Child, enthroned with Saints', Lucca Cathedral, and the 'Annunciation' in the Louvre. He also left us the profile portrait of Savonarola.

BARTOLOZZI, FRANCESCO (1725–1815)

The son of a Florentine goldsmith, the celebrated engraver studied with Ferretti in Florence and also with Joseph Wagner in Venice. In 1764 he came to England and was appointed Engraver to the King at a salary of £300 a year, and four years later on the founding of the **Royal Academy he was one of the original members. He was celebrated for his stipple engravings** after such artists as Cipriani and Angelica Kauffmann. In 1802 he accepted the invitation of the Prince Regent of Portugal to become Director of the Engraving School at Lisbon.

'The Four Seasons' (after Wheatley) (a set of four) F 1000, 1969, Hôtel Drouot, Paris.

BARYE, ANTOINE LOUIS (1795–1875)

Born in Paris, he studied sculpture with Gros. He was elected to the Academy in 1868 and received the Legion of Honour for the excellence of his work. He also painted in water-colour and etched a number of plates. Animals were among his favourite subjects.

BASSA, FERRER (c 1290–c 1348)

A Spanish painter of the Catalan school who worked for the court of Aragon. There are records of him working as a miniaturist, also executing altarpieces and wall-paintings. But the only surviving work that can be truly credited to him are frescoes in the chapel of the Franciscan Convent of Pedralbes, near Barcelona. Possible influence could have come from the Avignon school, with Simone Martini who was the leading figure.

BASSANO, FAMILY OF

(their name was adopted; proper name being da Ponte). A Venetian family of painters, a grandfather, father and four sons. They were:

Francesco, the Elder (1475–1530), after formal schooling in Venice, established himself at Bassano, a small town on the Venetian mainland. He may have been a pupil of Bellini, if not was certainly influenced by him. A leading work by him was 'St Bartholomew' in the cathedral of his home town. **Jacopo**, his son (1510–92) was the most important artist of the family. He trained first under his father and then in the studio of Veneziano; Titian had considerable influence on his later manner. Apart from religious and allegorical subjects he painted many portraits; including the Doge of Venice, Tasso and Ariosto. His studio later became a workshop with his sons helping to turn out a considerable output. One of his best pictures is 'Family Concert' in the Uffizi, Florence.

'The Israelites in the desert' £7800, 1975, Sotheby Parke Bernet.

Francesco, the Younger (1549–92), first son of Jacopo, studied with his father and then went to Venice where he worked on a series of historical paintings for the Doge. He was subject to hypochondria and in one attack he threw himself from a window to his death. **Giovanni** (1553–1613), the second son of Jacopo. He was the least talented of the family. He spent most of time copying his father's works, some of which copies at least are probably credited to his father.

Leandro (1558–1623), the third son of Jacopo is best remembered as a portrait-painter. He put the finishing touches to his brother Francesco's paintings in the Ducal Palace, Venice. **Girolamo** (1560–1622), Jacopo's youngest son, also copied his father's pictures and must have **added his share to the difficulties with authentification since**.

BATONI, POMPEO GIROLAMO (1708–87)

Born at Lucca, died at Rome. At first he followed his father's profession of goldsmith, then he went to Rome and studied with Conca and Massucci, and was also strongly influenced by the work of Raphael. He developed a popular style with tourists undertaking the Grand Journey; they left his studio either with a portrait or small figure studies of themselves inserted into ruined classical landscapes. The German Mengs was his principal rival, although much of the time he was at work in Spain. Such was the report of Batoni's work that **he was commissioned by no less than 22 sovereigns**.

'Portrait of the Earl of Hillsborough' £58 000, 1975, Sotheby Parke Bernet.

BAUCHANT, ANDRÉ (1873–1958)

Born at Châteaurenault, Indre-et-Loire. A market-gardener, he turned to painting when he was 46 and set out to portray mythological and biblical scenes. His pictures brim with a charming naïve beauty; flowers, trees, fruit, the open air are all set down with an attractive honesty.

'Repos des marins grecs' Sfrs 26 000, 1975, Galerie Koller, Zürich.

BAUMEISTER, WILLI (1889–1955)

Born in Stuttgart he became a leading demonstrator of Abstract art in Germany, and was to a degree influenced by Le Corbusier and Ozenfant. In his work come other roots including Aztec, Negro, Constructivism and Neo-Plasticism. During the war he was forbidden by the Nazis to teach. But like his countryman Emile Nolde he worked in secret. After the war he turned more and more to a pure ascetic abstract.

'Forms 1938' DM 5300, 1975, Hauswedell & Nolte, Hamburg.

BAYEU, FRANCISCO (1734–95)

Born in Saragossa, he studied with Luzan, in Tarragona. Best remembered perhaps as **the master of Goya**, who also married his sister, Josefa, in 1773. He spent some time working under Mengs, who commended Bayeu to Charles III who employed him. His brother **Ramón** (1746–93) worked with him and also with Goya.

BAZILLE, FRÉDÉRIC (1841–70)

Born at Montpellier and was killed in action at Beaune-la-Rolande during the Franco-Prussian War. He studied in the studio of Gleyre and met Monet, Sisley and Renoir. Bazille with his use of sunlight and colours was very much in the roots of experiment that was to blossom after his sad removal from the scene. 'Family Gathering' in the Louvre shows his quality and the influence he could have had.

BAZZI, GIOVANNI ANTONIO (1477–1549)

(generally known as Sodoma). He was born at Vercelli in Piedmont, and started his studies with a glass-painter, Spanzotti, from Casale. Later he went to Milan and came strongly under the influence of Leonardo, signs of whose style were to appear in the pictures of Bazzi for the rest of his life. At times he almost let his wit run away with him, including jollities not always associated with his subjects; such earned him the nickname of Il Mattaccio (the arch fool). Vasari spared no moments to show his dislike for Bazzi, raking up and inventing malicious stories about him. His finest work is probably 'The Life of St Catherine' in Son Domenico, Siena.

'The Holy Family' £2300, 1975, Sotheby Parke Bernet.

BEARDSLEY, AUBREY VINCENT (1872–98)

A master of black and white he was encouraged by Sir Edward Burne-Jones and Puvis de Chavannes. He brought a sensual quality to the medium that it had not had before. Works he illustrated include *Morte d'Arthur*, Wilde's *Salome*, Pope's *The Rape of the Lock* and the *Lysistrata* of Aristophanes. **His style influenced the evolution of Art Nouveau.**

BECCAFUMI, DOMENICO DI PACE (1486–1551)

The Italian painter and sculptor was born at Cortine, near Montaperto. He studied with Mecarino and was able to examine a large collection of master drawings owned by his teacher. His work shows the currently popular Mannerist style, using fresco and oil. Bronzes cast by him included the six angels holding lamps in the choir of the Cathedral at Siena. He also produced chiaroscuro woodcuts, etchings and engravings.

BECKMANN, MAX (1884–1950)

Born in Leipzig and died in New York. He was one of those lone figures who cared not to link himself too closely with any particular movement. He had a brief study time at the Weimer Academy of Art and a short visit to Paris. The carnage of the First World War marked him and

afterwards his work grew into a fierce, rugged attack on many aspects of social injustice, excesses and boorish cruelties. His style has a highly individual stark look, the drawing is angular and powerful, his colours strong and without subtle toning. He held within himself a tension that drove him which is epitomized by a retort made by him. 'I would worm my way through all the sewers in the world, suffer every humiliation and disgrace to paint.' Among his most striking works were nine large triptychs, which went under such titles as 'Departure', 'The Actors', and 'The Argonauts'.
'Die Anprobe' DM 162000, 1975, Hauswedell & Nolte, Hamburg.

BEERBOHM, SIR MAX (1872–1956)
English caricaturist and writer. His books of drawings include *Caricatures of Twenty-Five Gentlemen* and *Rossetti and His Circle*.

BEHAM BROTHERS
Hans Sebald (1500–50) born in Nuremberg, he was largely a graphic artist, working on wood and copper. Classed as a Little Master as his prints were generally small. **Bartel** (1502–40) also born in Nuremberg. He was Court Painter to Duke William IV of Bavaria, but he worked mostly in Italy where he had been sent by his patron.

BELLINI FAMILY
The celebrated Venetians, **Jacopo** (*c* 1400–*c* 70) was an assistant to Gentile da Fabriano. Later he himself taught in the Academy of Squarcione, and among the pupils was Mantegna who was to become Jacopo Bellini's son-in-law. He worked in several centres including Ferrara, Padua, Venice and Verona. **Gentile** (*c* 1429–1507) was trained by his father and in his time enjoyed considerable popularity. In 1469 he was made a Count Palatine by the Emperor and some years later dispatched to Constantinople to carry out portrait commissions for the Sultan Mehmet II. The greatest of the three is unquestionably **Giovanni** (*c* 1430–1516) who was one of the important figures in the rising Venetian school. Two of his great pictures are in the National Gallery, London. 'The Agony in the Garden' and 'Doge Leonardo Loredan'. Particularly with the former it can be noted how he has instilled a spiritual power into the composition and accentuated this with a subtly controlled handling of the light.

BELLOTTO, BERNARDO (1720–80)
Born in Venice, died in Warsaw. He was a nephew and pupil of Canaletto, whose name he adopted. His style reflects strongly that of his uncle, although latterly it became more individual and often had a greater feeling of atmosphere. He travelled to Munich, Dresden, Vienna, St Petersburg and finally to Warsaw where he was employed by King Stanislaus Poniatowski. He left besides paintings a great many well-observed drawings of cities, including Warsaw; and the latter **were of assistance in the reconstruction of the town after the hell-fire destruction of the last war.**

BELLOWS, GEORGE WESLEY (1882–1925)
American painter and lithographer who was a pupil of Robert Henri. **His art had a lusty reportage manner** that could catch the stark floodlight and sweaty atmosphere of the prize-fight or the crowds out on the loose in a city. He was elected a Member of the National Academy of Design at 27. Well-known pictures include 'Stag at Sharkeys', 'Up the River' and 'Dempsey and Firpo' (lithograph).
'Rhode Island' $4000, 1975, Sotheby Parke Bernet, New York.

BENOIS, ALEXANDER NIKOLAYEVICH (1870–1960)
Born in Russia, died in Paris. Of French descent he studied law in the University of St Petersburg and then attended classes at the Academy of Arts. His subjects included, historical, theatrical design and illustration. He was associated with Diaghilev and the World of Art (*Mir Iskusstva*).

BENTON, THOMAS HART (1889–   )
He studied at the Art Institute, Chicago and in Paris where he lived for five years. On his return to America he worked with the theatre and in the First World War on camouflage for the US Navy.
'The Music Lesson' $8000, 1975, Sotheby Parke Bernet, New York.

BERCHEM, NICOLAS (1620–83)
Son and pupil of Pieter Claesz Berchem. Afterwards he worked in the studio of van Goijen, de Grebber and Weenix. His manner had a strong feeling of Italy and his subjects usually included peasants, ruined buildings and an over-all warm tone associated with a number of the Dutch painters.
'A mountainous landscape with herdsmen' £17000, 1975, Sotheby Parke Bernet.

BERCKHEYDE, GERRIT ADRIAENZ (1638–98)
Born in Haarlem he also died there by drowning. He studied painting with the encouragement of his elder brother Job, also a painter. His speciality was recording the view of towns, the figures in the foreground sometimes being added by Job.

'View of Haarlem' £8200, 1975, Sotheby's, London.

BERMEJO, BARTOLOMÉ
(working about 1474–95)
Spanish painter and stained-glass designer. **His 'Pietà' signed and dated 1490 is among the earliest oil-paintings done in Spain.** 'St Michael overcoming Satan' in the Werner Collection displays a certain influence from the Flemish with the stark realism and intense concentration on detail.

BERNINI, GIOVANNI LORENZO
(1598–1680)
The son of Pietro, who was also a sculptor, was the **outstanding genius of the Italian Baroque**. From his father he received not only training but more important an introduction to patrons such as the Barberini and Borghese. He was at one and the **same time, painter, sculptor and architect**. The magnificent baldachino in St Peter's is his, and for a period he was the Architect to St Peter's in Rome. Marble sculptures include groups in the Borghese Gallery in Rome – 'The Rape of Proserpine', 'Apollo and Daphne' and 'David'. In 1665 he was invited to work out plans for the Louvre by Louis XIV. A portrait bust by him of Louis XIV is at Versailles.

BERRUGUETE, ALONZO (1480–1561)
Born at Parades de Nava, Castile. At a comparatively early age he travelled to Italy, and there he was considerably influenced by Michelangelo, and also made a friend of Andrea del Sarto. In Spain once more he came under the patronage of Charles V and later Philip II. Among his leading works is a series of panels commissioned by the Grand Inquisitor, Torquemada for the Convent of St Thomas, Avila. His royal patrons had him working in Madrid and also in the Alhambra, Granada. On his death he was buried with considerable magnificence at the expense of his sovereign.

BEWICK, THOMAS (1753–1828)
Born at Cherryborn in the parish of Ovingham some 12 miles (20 km) from Newcastle, England. At the age of 14 he was apprenticed to Ralph Beilby, a copper-plate engraver. The art of Bewick had taken root; **he was to revitalize the craft of wood-engraving that had to a large extent lost favour since the achievements of the 16th-century Germans.** Many of his blocks were of small size, but the intricacy of his tool cuts maintained its perfection; in blocks such as 'The Chillingham Bull' he exhibits the mastery of which he was capable, endless innovation with shading, pecking and hatching.

BINGHAM, GEORGE CALEB (1811–79)
American painter of the backwoods, frontier life and steamboats. He attended the Pennsylvania Academy of Fine Arts and later was Professor of Art at the University of Missouri. Well thought of paintings by him include 'The Emigration of Daniel Boone', 'The Jolly Flatboat Men', and 'Fur Traders descending the Missouri', which last is in the Metropolitan Museum of Art, New York.
'Washington crossing the Delaware' $260000, 1975, Butterfield & Butterfield, Washington, DC.

BISSIÈRE, ROGER (1888–1964)
Born at Villeréal, in the Lot-et-Garonne, France; he came to Paris in 1910 and at first earned his living as a journalist, while still forging ahead with his painting which followed in general the tenets of the Cubists. He taught at the Académie Ranson from 1925 to 1938. Latterly his work divorced itself from the severities of Cubism and became imbued with rich colour and movement.
'Nature Morte Cubiste' F 28000, 1975, Hôtel Rameau, Versailles.

BLACKBURN, JOSEPH (c 1700–c 65)
American portrait-painter who worked mainly from a studio in Boston from 1750. His clients included many of the New England families, the Lowells, Winthrops, Winslows and Bullfinches. Works that he did not sign may have been done by **John Singleton Copley, who was his pupil**.

BLACKLOCK, RALPH ALBERT
(1847–1919)
American painter of wide landscapes often including Red Indians. he attended the Cooper Union Institute in New York. Well-known pictures include: 'Indian Encampment', 'The Capture' and 'October Sunshine'.

BLAKE, PETER (1932–)
Born at Dartford, Kent, England, he studied at the Gravesend School of Art and the Royal College of Art. He works in acrylics, oils and mixed media collage **presenting an individual Pop Art**.
'Girl' £11000, 1975, Sotheby Parke Bernet.

BLAKE, WILLIAM (1757–1827)
Born in London, the second son of a hosier of Broad Street, Golden Square. At the age of ten he was sent to Pars' Drawing Academy in the Strand. At 14 he went to study engraving with Ryland; here the boy exhibited his extra-sensory perception, and left this master as he felt he would be hanged (which he was) and went to James Basire for seven years. With this master he was made to make drawings of London churches. He was often locked in Westminster Abbey alone with the

'Good and Evil Angels struggling for possession of a child' by William Blake. (Courtesy Tate Gallery)

tombs and effigies; it is possible he may have been present when Edward I was exhumed. Always, from his early years, when at the age of four he thought he had seen God 'put His forehead to the window', Blake had increasingly shown a sympathy with the supernatural; at ten he had seen a tree 'bright with angels' in Peckham Rye. All through his life he claimed to be talking with such as Moses, Homer, Dante and Milton. Blake's art owes much to an unbounded imagination and an individuality that was not breached by those around him. He was a painter in water-colour; he also used his own so-called fresco method, which he worked with glue and whiting on canvas or panel; he was a painter in tempera; and a graphic artist with many unusual techniques; and with all this a writer, and a poet; **all together one of the most original minds**.

'Job confessing his presumption' fetched £105 in 1903, and £7770 in 1949.

BLES, HERRI MET DE (1480–*c* 1550)

Flemish painter born at Bouvignes, and is identified as Herri Patenier, nephew of Joachim Patenier. He introduced scriptural subjects and themes into his landscapes, and instead of signing he would paint an owl in one corner, causing the Italians to nickname him Civetta.

'The temptation of St Anthony' £12600, 1975, Christie's, London.

BLOEMAERT AND SONS

The most important painter was **Abraham**, the father, (1564–1651) he was the son of an architect and sculptor and he studied with Joost de Beer, also he was influenced by work of Frans Floris. His work was largely concerned with landscape and animals, also mythology and flowers. Apart from his sons **his pupils included Cuyp, Honthorst, Both and Weenix**. As a graphic artist he had considerable talent with chiaroscuro printing. **Hendrik** was his eldest son and he essayed to be a portrait-painter, but never rose above the mediocre. **Frederick** was the second son and established himself as an engraver of quality. The third son was **Correlis** who became a celebrated engraver after studying with Crispyn van de Passe. The youngest Bloemaert was **Adriaan**, who, after tuition from his father, travelled to Italy, Vienna and Salzburg where he was killed in a duel.

'Italian Landscape' by Adriaan, Asch 1000000, 1975, Dorotheum, Vienna. 'The Holy Family' (woodcut by Abraham) £168, 1975, Christie's, London.

BOCCIONI, UMBERTO (1882–1916)

Born in Reggio di Calabria he was sculptor and painter. After meeting Marinetti he joined with Balla, Carrà and Severini in signing the Manifesto of Futurist Painters. In Paris he contacted the Cubists, notably Archipenko and Brancusi. His figures hovered between a moving Realism and near Abstract.

'Forme uniche di continuita nello spazio' £17850, 1975, Christie's, London.

BÖCKLIN, ARNOLD (1827–1901)

Born in Basle he studied at the Düsseldorf Academy. Afterwards he went to Rome and then settled near Florence. At first he concentrated on landscapes and then, as his imagination expanded, introduced all manner of fantasies into his compositions. His technique was immaculate, with great attention to brush-strokes and finish. In Basle his strange and powerful 'Island of the Dead' which is filled with a daunting macabre atmosphere.

'Triton et Naïade' Sfrs 20000, 1975, Galerie Fischer, Lucerne.

BOL, FERDINAND (1616–80)

Dutch painter, a leading pupil of Rembrandt and **one of the most prolific especially with regard to portraits**, that to a degree come quite close to the master, and thus have made research difficult for the art historian.

'Portrait d'homme en buste de trois quarts vers la gauche' F 38000, 1975, Palais Galliera, Paris.

BOLTRAFFIO, GIOVANNI ANTONIO (1467–1516)

(also spelt Beltraffio). A noble member of a distinguished Milanese family who was taught by Leonardo. His work may resemble the master's in outline, but he did not achieve the *sfumato*, the softness of modelling of Leonardo.

BOMBOIS, CAMILLE (1883–1970)

Born at Venavey-les-Laumes, Côte-d'Or; the son of a barge-owner, he started life as a farm labourer and by 16 was reputedly sketching. He loved the atmosphere of travelling circuses and fairs, so much so, that one day he went off with one and gave himself a plentiful supply of models. leaving his friends he earned his crust as a road-mender in Paris. A full stint in the trenches in the First World War and he still kept painting his strange little canvases, peopled by wide-hipped circus ladies and tough-shouldered wrestlers. His first exhibition was from a chair and the pavement. A passing critic wrote a piece and so his market grew. A natural primitive Bombois has entirely a style of his own.

'La Clownesse' F 30000, 1975, Galerie des Chevau-Légers, Versailles.

BONE, SIR MUIRHEAD (1876–1953)

Scottish painter and etcher who studied architecture at the Glasgow School of Art and then dropped it in favour of painting and etching; with the latter he worked with Meryon and Whistler. **He was an Official War Artist in the 1914 struggle**, and produced some notable plates, especially of air battles.

'Rainy night in Rome' (dry-point) £170, 1975, Christie's, London.

BONHEUR, ROSA (1822–99)

She was born at Bordeaux. Her father was Drawing Master and Director at the Free School of Design for Girls in Paris, but at first Rosa was intended to become a dressmaker. As she expressed great misery at the thought, however, her father relented and taught her himself. By 1845 she was in full production with no less than 14 paintings in the Salon, and the recipient of a medal. When her father died she took on his position; **the first time it was ever held by a woman**. 'The Horse Fair', 16 ft (4·87 m) long, was her greatest success, selling for £12000, 1886. But since her time her prices have dropped dramatically – 1861 'Spanish Muleteers crossing the Pyrenees' £1995, 1940 £147; 1887 'A highland Raid' £4059, 1949 £199·50.

Rosa had an arresting appearance, a very large head covered with shaggy white hair which she brushed up high giving her a lion-like look.

'Un Griffon' F 850, 1975, Hôtel Drouot, Paris.

BONINGTON, RICHARD PARKES (1802–28)

Born at Arnold near Nottingham, England, the son of the Governor of Nottingham Gaol. Father committed some indiscretions which took the family scuttling to France. Here Richard studied with Francia at Calais and in the studio of Gros, where his emergent talent was noticed by Delacroix, who was to become a close friend. Bonington's art flowered first in water-colour and then oils, landscapes and seascapes came from his brushes in quick succession, also lithographs, but time was running fast for the young genius. A malady struck at him and was aggravated by

sunstroke which brought on somnambulism and then brain fever. Delacroix said of him: 'I knew Bonington well and loved him much. . . . He developed an astonishing dexterity with water colours . . . no one in the modern school, perhaps no earlier artist, possessed the ease of execution which makes his works, in a certain sense, diamonds, by which the eye is pleased and fascinated, quite independently of the subject and the particular representation of nature.'
'A Sailing ship at Anchor' (15 × 12 in (381 × 305 mm)) £8500, 1969, Sotheby's, London.

BONNARD, PIERRE (1867–1947)
Born at Fontenay-aux-Roses he studied at the Académie Julian at the same time as Vuillard, Roussel and Sérusier. A withdrawn character he started simply with small, sparsely furnished interiors with figures that were suffused with light and warm colour. From this his art expanded to street scenes of Paris. **He was with the Nabis** and was closely associated with Vuillard. An unexpected honour came in 1940 when he was elected an **Honorary Member of the Royal Academy**.
'Nature morte à la fenêtre ouverte Trouville' £73000, 1975, Sotheby Parke Bernet.

BORDONE, PARIS (1500–71)
Born at Treviso he worked in the studio of Titian and later with Giorgione, whose work he imitated so well. But he returned to the influence of Titian and then imitated him so well, too, that **his paintings have been confused with Titian's**.

BORÈS, FRANCISCO (1898–1972)
Born in Madrid, he learnt much from copying pictures in the Prado. In 1922 he became associated with an *avant-garde* movement, the Ultraists. He left Spain for Paris in 1925.
'Femme au corset' F 45000, 1975, Palais Galliera, Paris.

BOROVIKOVSKY, VLADIMIR-LUKICH (1757–1825)
Born in the Ukraine, the son and pupil of an icon-painter. He was a painter of religious compositions, portraits and miniatures, and worked as a church-decorator in Mirgorod.

BOSCH, HIERONYMUS (c 1450–1516)
A unique painter, generally included in the Flemish school, born at 's Hertogenbosch (Bois-le-Duc) (also known as von Aeken). Of his early days or possible tuition little is reliably recorded. Apparently he lived and worked in his birthplace all his life, had a happy and harmonious marriage, and was in comfort with worldly goods. **Yet Bosch gave us some of the most extraordinary pictures ever painted.** His landscapes and scenes abound with spectres, devils, incantations, the strangest of goings-on and acts of vivid Surrealistic imagination. A satisfactory and complete reading of his compositions has yet to be made. Some point to the fact that he was telling morals aided by one of the most excited imaginations ever granted to an artist. Unfortunately only about 40 of his works have survived, many being destroyed by over-zealous clergy, fire or other dangers. Philip II of Spain greatly admired his painting and a number of his finest works were secured for Spain. Some have accused him of belonging to the heretical Adamite sect. For contrasting appreciation of him there is the 'Christ Mocked' in the National Gallery, London, and the 'Garden of Earthly Delights' in the Prado, Madrid.

BOTERO, FERNANDO (1932–)
Born in Medellin, Columbia. In 1952 he was in Madrid studying at the Academy San Ferdinando, also in the Prado, where he was affected by the works of Goya and Velázquez. Later he visited Italy to work on the fresco technique in Florence and to see the frescoes of Piero della Francesca in

'*Self-portrait the Day of my First Communion*' by Fernando Botero. Oil on canvas, 43 × 37 in (1092 × 940 mm). (Courtesy The Hanover Gallery)

Arezzo. Since 1960 he has lived in New York. His portrayal of his sitters possesses an extraordinary mixture of the naïve, caricaturist, whimsy and potent comment; colouring is bright and delightful; still lifes and flowers give evidence of acute observation.

'Siesta del Cardenal' £9975, 1975, Christie's, London.

BOTH BROTHERS
Andries (c 1608–52) and **Jan** (c 1610–52). They both studied with their father, Andries the Elder, who was a painter on glass, later they were in the studio of Abraham Bloemaert. They travelled to Italy and worked alongside each other first in Rome and then in Venice. Their landscapes had a pleasant glow and generally included ruins and small groups of figures, which were usually painted in for both of the Boths by Andries. Fate came in one night, when after a fair celebration they were on their way back to their lodgings in Venice, and Andries toppled overboard from a gondola and was drowned.

'La Tour au bord de l'eau' F 7500, 1975, Palais Galliera, Paris.

BOTTICELLI, SANDRO (c 1445–1510)
(properly Alessandro di Mariano dei Filipepi). Born in Florence, the son of a tanner, he was with his brother Antonio in a goldsmith's shop before going apparently to study first with Fra Filippo Lippi, then with Pollaiuolo and visiting Verrocchio's studio workshop. **Sandro was the great poetical lyrical master of the Renaissance in Italy.** There is a touch of lightness in his pictures that gives them a quality of joy; in some there is a sensitive spiritual atmosphere; and with all a splendid sense of colour.

His first major work 'The Adoration of the Magi' included a small self-portrait as well as several Medici family portraits: the first Mage was Cosimo, the second Piero, Cosimo's elder son, and the third Giovanni, his younger son. In 1478 Sandro was commissioned by Lorenzo the Magnificent to paint the effigies of the Pazzi conspirators on the walls of the Bargello. Later commissions from Lorenzo included: 'Mars and Venus' now in the National Gallery, London and 'Primavera' in the Uffizi. Between 1481 and 1482 he worked in Rome at the command of Pope Sixtus IV on decorations for the Sistine Chapel beside other artists including Ghirlandaio, Perugino and Pinturicchio; Sandro's contributions being the frescoes of the 'Life of Moses', 'The Destruction of Korah, Dathan and Abiram', and 'The temptation of Christ by Satan'.

Back in Florence he produced 'The Birth of Venus' in the Uffizi; **he was the first to paint tondi**, circular pictures of the Madonna and Angels; his 'Magnificat' also in the Uffizi, being thought by many to be his masterpiece. He may have done some engraving, and certainly supplied others with designs, including Baccio Baldini. For Lorenzo he illustrated a manuscript of the *Divina Commedia* in silver-point gone over with a pen and ink. Late in life he fell under the oratory of Savonarola, **and reputedly burnt a number of his own pictures with pagan themes**. His last painting of the Nativity is in the National Gallery, London; it is an exquisite composition full of fervent, joyous acclamation of the moment. It was bought by the Gallery in 1837 for £25 4s, **a rather shattering example of how cheap many of the great masters were even in the 19th century**.

BOUCHARDON, EDMÉ (1698–1762)
French sculptor born at Chaumont-en-Bassigni, son of the architect Jean Baptiste Bouchardon. He studied with Guillaume Coustou in Paris. In 1722 he won the Gold Medal of the l'Académie Royale with his 'Gideon Choosing His Soldiers by Observing the Way They Drank'. He copied many busts and statues from the Antique. The equestrian statue of Louis XIV, which is part of the Fountain of Neptune at Versailles is by him.

BOUCHER, FRANÇOIS (1703–70)
Born in Paris he studied under Lemoyne and also worked with Jean François Cars the engraver. In 1727 he went to Italy at his own expense and stayed until 1731. In 1734 he was admitted to the Académie and rose to be the Director in 1765. Boucher was associated with the tapestry manufactory at Beauvais, and on the death of Oudry became Inspector at the Gobelins. In 1765 he became First Painter to the King. He had much patronage from Madame de Pompadour, and decorated with erotic and idyllic subjects the boudoir at the Hôtel de l'Arsenal, where she met her royal lover. The art of Boucher was a part of the life that evolved in the warm scented air of the never-never scene of the French Court that floated over the fermenting masses that were to destroy it. As with Watteau and Fragonard, he gave what his clients wanted and wished to be. Boucher left behind by his own calculation **10000 sketches and drawings and 1000 paintings and studies**. Throughout most of his pictures voluptuousness is the main feature, never squalid or in bad taste. His values have shown strange irregularities. 'La fête des Bergers' £3200, 1844; £1600, 1884. 'Diana and Calisto' (a pair of ovals) £130, 1851, £12500, 1957.

BOUDIN, EUGÈNE (1824–98)

The son of a Honfleur pilot, he met the Barbizon painters and they encouraged him to work. He left behind a long series of memorable glances at the beaches, ports and fishing villages of France; fresh in colour and atmosphere. His 'Laveuses au Bord du Port de Trouville' sold at Christie's in **1942 for £84** and in the same rooms in **1976 for £17000**: also 'Rotterdam, La Meuse' for **£357 in 1927** and **£25200 in 1975**.

BOURDELLE, EMILE-ANTOINE (1861–1929)

French sculptor who studied at the École des Beaux-Arts, Toulouse and also in Paris. **He learned much from Rodin.** His work includes the large 'Hercules and the dying Centaur' and excellent portrait busts of Beethoven, Rodin and Anatole France. A bronze cast of the last selling for F 30000 in 1975, Palais Galliera, Paris.

BOUTS, DIRK (1420–75)

Born in Haarlem, his early life is only slightly recorded, but he appears to have settled in Louvain after a period of study with Van der Weyden. With the advent of oils his paintings used the strength and handling of the colours to give a somewhat puritanical view of his subjects. Faces tended to lack expression although he often achieved an over-all dramatic effect. In 1468 he is mentioned as Municipal Painter for Louvain for a **reward of enough money to buy a coat**. Fine pictures by him include the 'Adoration of the Kings' in the Pinakothek, Munich.

BOYD FAMILY

An Australian family that has been centred round the pottery workshop in the Murrumbeena suburb of Melbourne. The most internationally celebrated member being **Arthur Merric Boyd II** the painter. He was born in Murrumbeena in 1920. No formal training apart from six months' night school in the Drawing School of the National Gallery, Melbourne. The war effected his painting style: it moved from a gentle primitive manner to dark Expressionist compositions brooding with psychological disturbance. Later came a period of studies of mining towns and landscapes teeming with people. Working as a ceramic sculptor he made in 1955 a huge totem-pole 30 ft (9 m) high to stand outside the entrance to the Olympic Swimming Pool, Melbourne; this **weighed about ten tons, was made up of 260 glazed bricks**, and was fired in a specially built kiln. 'Crucifixion', A$ 7000, 1975, Christie's Sydney.

The rest of this artistic family includes: **Arthur Merric Boyd I** (1886–1940) Born at Dunedin, New Zealand. Water-colourist. Father of Penleigh and William Merric; grandfather of Arthur Merric II, David, Guy, Mary and Lucy. **William Merric Boyd** (1888–1959) Son of Arthur Merric Boyd I. He was the one to start the pottery workshop at Murumbeena. Leading Australian potter. **David Boyd** (1924–) Born in Melbourne, son of William Merric Boyd. Potter and painter. **Doris Boyd** (c 1883–1960) Wife of William Merric Boyd. Painter, also decorated her husband's pottery. **Emma Minnie Boyd** (1856–1936) Born in Melbourne, wife of Arthur Merric Boyd I. Painter. **Guy Martin Boyd** (1925–) Born at Murrumbeena. Sculptor and pottery-designer. **Hermia Boyd** (1931–) Born in Sydney. Ceramist, wife of David Boyd and decorator of his pottery, **Jamie Boyd** (1948–) Born in Murrumbeena. Painter. **Lucy Boyd** (1916–) Born in Murrumbeena. Ceramist. **Mary Boyd** (1926–) Born in Melbourne. Painter and pottery-decorator. **Theodore Penleigh Boyd** (1890–1923) Born in England, died in Victoria. Painter and etcher.

BRACKELEER, FERDINANDUS (1792–1883)

Born in Antwerp, he studied at the Academy there under Van Bree. A historical and religious painter, he spent three years in Rome painting subjects such as Esau and Isaac and the Healing of Tobias in the classical style of the Frenchman David.

BRANCUSI, CONSTANTIN (1876–1957)

Born in Rumania, he left home at eleven and spent seven years as a lone traveller. In 1898 he went to Bucharest Art School on a scholarship. By 1902 he was journeying again, Munich, Zürich, Basle and to Paris where he settled. He developed his sculpture and was showing publically by 1906. **He has had a formative influence on many of the young sculptors**, as an innovator of form, working in wood, plain and polychromed, marble, limestone and bronze.

'Négresse Blonde' $750000, 1974, Sotheby Parke Bernet, New York.

BRANGWYN, SIR FRANK (1867–1956)

Born in Bruges of Welsh parents, died at Ditchling, Sussex. Early on he spent a period in the William Morris Workshops and then had a period at sea. His murals earned high praise; they included work at Lloyd's Registry, Royal Exchange in London and the Rockefeller Center, New York. He was untiring with his art, working in oils, water-colours, etchings, woodcuts, lithographs and numerous designs for posters, furniture, ceramics and textiles. **There is a Brangwyn Museum in Bruges.**

BRAQUE, GEORGES (1882–1963)

Born in Argenteuil the son of a house-painter. When the family moved to Le Havre he joined the École des Beaux-Arts there in 1899. In 1900 he went to Paris where in reduced circumstances he began to paint and draw. Braque, very much one of the most influential artists of this century, is **credited with the evolution of Cubism with Picasso**. He went through many of the current phases including Fauvism. Primarily to many he is a colour experimenter; his palette at times was limited to the umbers and black with an acid yellow. Particularly with his still lifes there is an interesting relationship with Chardin, both men producing from simple objects subtle harmonies and exquisitely balanced compositions.
'Mandoline à la partition' £170000, 1975, Sotheby Parke Bernet.

In the 1930s and even up to 1950 Braque's works were unlikely to raise bids of more than hundreds of pounds.

BRETON, ANDRÉ (1896–1966)

A poet and critic who is included as he was the **principal moving force behind 20th-century Surrealism**. He was born in the Orne department of France and he died in Paris.

BRETT, JOHN (1831–1902)

English landscape and marine painter, strongly influenced by Ruskin in his early life, also the work of the Pre-Raphaelites. Notable paintings include 'The Stonebreaker' in the Walker Art Gallery, Liverpool.

BRIL BROTHERS

Matthys (1548–84) and **Paulus** (1556–1626). Both were born in Antwerp. Matthys went to Italy to be employed in the Vatican by Pope Gregory VIII; he was a landscape-painter with a good hand for animals. Paulus studied with Damiaen Ortelmans, and earned a living to start with by painting the lids of harpsichords. He too travelled to Italy and for a time worked with his brother.

BRODSKY, ISAAC (1884–1939)

A Russian Realist painter, he studied in western Europe and met Gorky while in Italy. One of his best-known paintings is 'Lenin in Smolny' also 'Voroshilov on Skis'.

BROEDERLAM, MELCHOIR (c 1380–1409)

Flemish school, born at Ypres. He was Court Painter and Valet to the Duke of Burgundy, Philip the Bold or Hardy. His work has a simplicity of colour and purity.

BRONZINO, IL (1503–72)

(more properly Agnolo Tori di Cosimo di Moriano). Born at Monticelli near Florence, he was adopted by and became the pupil of Pontormo. His style in handling figures and portraits is immediately recognizable with its firm modelling and great clarity of form combined with clean colour, qualities epitomized by his 'Venus, Cupid, Folly and Time' in the National Gallery, London.

BROUWER, ADRIAEN (1605–38)

Probably born at Oudenarde; his mother, a dressmaker, put him in Hals's studio, who according to some recorders used him badly – all work and no food for his labours. He left and went to Amsterdam where he began a brilliant career as an observer of low life in taverns and among the peasants. But to a degree his own liking for such goings-on must have taken an edge off his life as he rollicked with the best. In Antwerp he was flung into prison as a spy, and was only released by the good offices of Rubens who possessed a number of his works as also did Rembrandt.

BROWN, FORD MADOX (1821–93)

Born in Calais he studied in Antwerp with Baron Wappers. Later he worked for three years in Paris and then came to London. **One of his pupils was Rossetti.** Although Madox Brown subscribed to many of the ideas of the Pre-Raphaelites, he never joined the Brotherhood. His best-known work is 'The Last of England' in the Birmingham Art Gallery.

BRUEGHEL FAMILY

Pieter the Elder (c 1520–69) (also known as Peasant Brueghel and the Droll) was born near Bruges. He entered the studio of Pieter Coccke. Later took a short trip to Italy on which he was fascinated with the wild scenery in the Alps. His pictures reflect his love and interest of the simple life around him, village feasts, merry-making, snow and ice scenes; sometimes indulging himself with Bosch-inspired fantasies. Even when engaged on a serious religious subject such as the 'Adoration' in the National Gallery, London, he brings in some rather strange faces and expressions.

Pieter the Younger (c 1564–1638) (also known as Höllen Brueghel (Hell Brueghel) eldest son of Pieter the Elder, he worked with Coninxloo in Antwerp. His nickname came from his choice of subjects which included tortures, charnel scenes and mass suffering. Execution in general was inferior to his father.

Jan Brueghel (1568–1625) (also known as Velvet Brueghel for the apparent reason of his liking for the material). The younger son of Pieter the Elder, he was brought up and taught by Marie de Bessemers, later studied with Pieter Goetkint.

'Tower of Babel' by Peter Brueghel the Younger. (Courtesy Hallsborough Gallery)

'Mother and Child' by Jan Brueghel. (Courtesy Leonard Koester)

'Hôtel des Rochers' by Bernard Buffet, 38¼ × 51¼ in (972 × 1302 mm). (Courtesy Lefèvre Gallery)

He excelled in flower-painting, also decorative landscapes with exquisitely painted birds and animals. He was one who worked with Rubens in Antwerp.
'A river landscape' £73 000, 1975, Sotheby Parke Bernet.

BRUYN, BARTHOLOMÄUS (c 1500–56)
Born in Cologne his early manner resembled that of the Master of the Death of the Virgin, whose pupil he may have been. His greatest works were the wings for the Shrine in the Church at Xanten, the inside showed the Lives of the Saints Victor, Sylvester and Helena, and the outside the Virgin and Child, with Saints.

BRYULOV, KARL PAVLOVICH
(1799–1852)
Born· at St Petersburg, the son of an Italian sculptor Pavel Bryulov, he studied at the Imperial Academy under Andrey Ivanov. He spent some time in Italy, and while there produced a **huge canvas 21 ft (6·4 m) long entitled 'The Last Days of Pompeii' which, when on exhibition in Rome, was seen by Bulwer-Lytton and the idea emerged later as his novel of that name.**

BUFFET, BERNARD (1928–   )
Born in Paris, he has had considerable success with an individualistic style that gives his landscapes and still lifes a harsh linear look, most shapes being outlined in dark colours.
'Clown au train rose' F 59 000, 1975, Palais Galliera, Paris.

BURGKMAIR, HANS (1473–1531)
Born at Augsburg, the son of Thomas, also a painter, he was a pupil of Martin Schongauer and a friend of Dürer, being also employed by the Emperor Maximilian I. His paintings were influenced by the Venetian school; but he excelled more with woodcuts, of which **he made about 700.**

BURNE-JONES, SIR EDWARD COLEY
(1833–98)
Born in Birmingham, he went first to King Edward's School, thence to Exeter College,

'Starker Traum' by Paul Klee. (Courtesy Christie's)

Oxford, where he met William Morris in 1853. He met Rossetti in London in 1855, whose studio he worked in for a short time. He journeyed to Italy with Ruskin and was entranced with the work of Botticelli and Mantegna, traces of whose influence were to appear in his own work. Legends, historical valour fired his imagination but in the final judgement his canvases lacked that quality of true power and greatness. One of his best-known pictures is 'King Cophetua and the Beggar Maid' in the Tate Gallery, London. His value has been riding a switchback. 'Chant d'amour' in 1886 £3307 10s, in 1898 £3360, in 1930 £620. 'The Fountain of Youth' in 1919 £273, in 1975 £8400, Christie's, London.

BUTLER, REG (1913–)
Born in Buntingford, Hertfordshire. He trained first as an architect. He turned to sculpture in 1944,

first as a wood-carver and later he applied himself to bronze and wrought iron, the latter material giving him the chance to experiment with rough harsh textures. His maquette won the Grand Prize in the competition for a memorial to The Unknown Political Prisoner.

BYSTROM, JOHN NIKLAS (1783–1848)
Swedish sculptor who worked in the studio of John Tobias Sergel at Stockholm. He spent a period in Rome, modelling a half life-size work 'The Reclining Bacchante', which brought him immediate recognition. He also produced colossal statues of the kings of Sweden.

CAILLEBOTTE, GUSTAVE (1848–94)
A French naval architect who collected pictures from the Impressionists and also painted himself, joining in the 2nd Impressionist Exhibition in 1876 and the following four shows. His work started as

'Otter Hunting' by Sir Edwin Landseer (58 × 94 in (1473 × 2387 mm)). (Courtesy Sotheby Parke Bernet)

'Captain W H Rickets, RN, with the New Forest Hunt at Longwood, Hants' by Benjamin Marshall (1767–1835), English sporting painter, pupil of the portrait painter L F Abbott; but from about 1792 he turned almost exclusively to animal painting. Sold in 1977 for $72 500 (40·7 × 50 in (1035 × 1270 mm)). (Courtesy Sotheby Parke Bernet)

Realism and then turned towards Impressionism with a subtle individual touch of its own.
'River landscape' F 20000, 1975, Palais Galliera, Paris.

CALAMÉ, ALEXANDRE (1810–64)
A Swiss painter who had a masterful manner with the handling of his native scenery: the rugged snow-topped mountains and lakes. Such as 'Waterfall near Meyringen', 'Mont Blanc' and 'Jungfrau' demonstrate his quality. He was also Head of the Geneva Art School.
'Vue du Lac des Quatre-Cantons' Dkr 52000, 1975, Rasmussen, Copenhagen.

CALDER, ALEXANDER (1898–   )
Born in Philadelphia, he trained first as an engineer, then attended evening drawing classes in New York. In travels to and from Paris he became influenced by a number of the current vogues, notably Abstract-Creation, and met Arp, Mondrian and Léger. He turned at first to static construction and then, **in 1932, made his first Mobile**. A form of sculptural expression which allows for movement of the component parts; this may be just by the movement of the air, or by using water or some mechanical means.
'Red Hourglass' (painted iron mobile) $30000, 1975, Sotheby Parke Bernet, New York.

CALDERON, PHILIP HERMOGENES (1833–98)
English painter who worked with a strong moral motive in most of his pictures, many having underlying religious cautions. Titles such as 'By the waters of Babylon there we sat down', and 'St Elizabeth of Hungary', which was intended to show an episode in her career, aroused scathing comment and he changed the title to 'Renunciation' to make it anonymous.

CALLOT, JACQUES (1592–1635)
Born in Nancy, France, he was the son of the Herald of Lorraine. At the age of 12 Jacques ran away from home to Italy burning with a desire to study art. Unfortunately he met some Nancy merchants who took him home; again he tried but fell in with his brother in Turin and so once more was sent back to the parental roof. At last, against considerable opposition, he did manage to start tuition with the Royal Engraver, Dumange Crocq. Again he ran away, this time with a company of armed gipsies, and with them travelled to Florence, where he worked in the engraving studio of Canta-Gallina. Slowly his skill for etching and engraving became recognized; there was further study with Philippe Thomassin in Rome. Thence to Florence again to work in the Court of Cosimo de' Medici. Back to

Nancy, visits to Paris and much involvement with the battles and turbulence of the time. **Callot was one of the greatest hands with etching and particularly engraving.** Plates such as 'The Temptation of St Anthony' not only exhibit his skill but show an influence from Bosch with the weird beasts and happenings. **He worked about 1500 plates.**
'La Grande chasse' (etching first state) £660, 1975, Sotheby Parke Bernet.

CALVERT, EDWARD (1799–1883)
Born in Appledore, Devon he started out as a midshipman. Leaving the navy he studied art first in Plymouth and then at the Royal Academy Schools. He was a great admirer of William Blake, and he was associated with John Linnell and Samuel Palmer. Painter and graphic artist, his best blocks include 'Christian ploughing the Last Furrow of Life' and the 'Cider Press'.

CAMPIN, ROBERT (c 1378–1444)
Details of his early life are vague, but it is known that by 1406 he was already a Free Master, and in that year he was painting the portrait of the widow of the sculptor James Braibant. It seems he was a somewhat unruly character, as in 1428 he was fined and sentenced to make a pilgrimage to Saint Giles; also forbidden to hold any civic office. Four years later he was banished for leading a dissolute life, but at the intercession of the Countess of Hainault this was replaced by a fine. This early Netherlandish painter can be associated with the Master of Flémaille; his style has to a degree the stiffness of the Gothic, yet when an examination of the 'Portrait of a woman', in the National Gallery, London, is made it exhibits considerable sensitivity with the modelling of the face and in the head covering.

CANALETTO (1697–1768)
(Properly Giovanni Antonio Canal). Born in Venice he studied with his father who was a decorator and scene-painter. His early work was as a theatrical decorator. In 1719 he went to Rome and was possibly influenced by the work of Panini and also Carlevaris. Back in Venice once more he was to come under the patronage of Joseph Smith, who was to become British Consul; many of the Canaletto drawings acquired by this gentleman are now in the Royal Collection at Windsor. In 1746 he travelled to London and stayed for two years recording many views of the city and river and environs. **It is possible he was the first artist to use a camera lucida** for picture-making. Many of the figures in his paintings were put in by Giovanni Battista Tiepolo; also on close examination in some cases the projection lines used in preparing the perspective may be glimpsed. His

'A view of the Piazza del Campidoglio and the Cordonata, Rome' by Giovanni Antonio Canal, Il Canaletto; 20½ × 24 in (521 × 610 mm). (Courtesy Christie's)

nephew was Bernardo Bellotto whose style approached his uncle's causing confusion with attribution. In the National Gallery, London, are two excellent works by Canaletto. 'The Stonemason's Yard' and 'View of the Grand Canal, Venice'. His values have gone up and down, dropping at the beginning of this century to a few hundreds, but since the war he has risen sharply, evidence being:
'The Old Horse Guards and the Banqueting Hall from St James's Park' £105000, 1975, Christie's, London.

CANOVA, ANTONIO (1757–1822)
Italian sculptor born at Passagno, near Bassano; he settled in Rome from 1781. He was Neo-Classical in style and produced a number of impressive works under such titles as 'Cupid and Psyche' and 'Theseus and the Minotaur'. Canova was responsible for the Tomb of Pope Clement XIII in St Peter's. Later in life he did return to Venice. **Among his pupils were two English sculptors James Wyatt and Richard Westmacott.** He worked on commissions for Napoleon I and at one time was asked to come to London to expound on the artistic value of the Elgin Marbles.

CANTARINI, SIMONE (1612–48)
Born at Oropezza near Pesaro, he studied with Pandolfi and Ridolfi. He was impressed so deeply with the work of Guido Reni, that he almost forced his way into the master's studio and stayed there until he had to leave not only the studio, because of his insolence and malevolence, but also Bologna. Principally a portrait-painter, he is reputed to have met his end when he failed to catch a likeness of the Duke of Mantua, and gulped down a draught of poison.

CAPPELLE, JAN VAN DE (1626–79)
Dutch marine painter, the son of a dyer, who was influenced by the work of Simon de Vlieger. His canvases are well conceived and he achieves a soft atmospheric effect in the distance.
'A Coastal Scene' £15750, 1975, Christie's, London.

'La Femme Bleu' by Amedeo Modigliani. (Courtesy Christie's)

CARAVAGGIO, MICHELANGELO AMERIGHI (1573–1610)

Born at Caravaggio near Milan, the son of a mason, as a boy he prepared the plaster for the fresco-painters in Milan. Inspired by the work of the artists in front of him, he began to paint and was to a high degree completely self-taught. At the start he worked away at flowers and a few portraits. He moved to Venice and was influenced by the painting of Giorgione. From Venice to Rome, where need forced him to work for Cesare d'Arpino, who used his talent for flowers and ornaments. He began to develop his dramatic use of light and shade, that was to have far-reaching effects on not only Italian artists, but also the Dutch and Spanish and to a lesser extent the German. He liked to paint with priming, either veiled with a darkish raw umber, or with the umber actually in the ground.

Of a fiery temperament he had to do a quick flit from Rome when he killed a friend in a moment of anger; he went first to Naples and then on to Malta. Here he had a short spell under the patronage of the Grand Master of the Knights of Malta; but shortly after was flung into prison for quarrelling with another Knight. He escaped and fled to Syracuse, Sicily, from there to Messina, then Palermo and so back to Naples. Here through friends he received a pardon from the Pope for his killing. He took ship to Rome, but was taken prisoner by some Spaniards by mistake for another man. The Spaniards released him, but they left him stranded without property on a sun-drenched malarial coast between Naples and Rome. A combination of it all killed him when he had managed to stagger quite close to Porto Ercole. In the Cantarelli Chapel of St Luigi dei Francesi, Rome, is his 'The Vocation of St Matthew' which well illustrates his personal handling.

CARPACCIO, VITTORE (active 1490–c 1525)

Early records are very incomplete, but it is likely he was influenced by Gentile Bellini. He was apparently employed at the School of San Girolamo, Venice at the same time as Giovanni Bellini but all the works of this religious establishment have vanished. His manner has a greater stiffness than most of his contemporaries, as may be judged by 'The Dream of St Ursula' in the Accademia, Venice.

CARRÀ, CARLO (1881–1966)

Italian painter born at Quargnento, he was one of the original signatories of the Futurist Manifesto. He dabbled with Cubism and Metaphysical Painting; at one time his works had a very close relationship to those of Chirico. He won the Venice Biennale Grand Prix in 1950.

'Woman, Knitting' L 16000000, 1975, Finarte, Milan.

CARRACCI FAMILY

Lodovico (1555–1619) and his two cousins who were brothers, **Agostino** (1557–1602) and **Annibale** (1560–1609). Lodovico was born in Bologna and was the **founder of the Eclectic School.** He studied with Fontana who suggested he should give up the career of an artist; he was called the ox by his fellow pupils because of his slowness of mind and movement. He ignored them and painted on, until in 1589 with his cousins he opened an Academy which kept the three together until 1600, when the cousins went to Rome. Lodovico kept it going until 1619.

Agostino also studied with Fontana. When he went to Rome he worked with his brother on the decorations for the Farnese Palace there.

Annibale was the most original painter of the three. He had a short time of instruction from

'Atom Piece' 1964 by Henry Moore. (Courtesy of the Artist)

Lodovico and after that taught himself by the study of works of other masters, including Correggio and Parmigiano. His works have fluctuated considerably in value, in the 18th, 19th and early 20th centuries, with few exceptions they have stayed in the hundreds. But in 1969 a small self-portrait did fetch £5040 at Christie's, London.

CARRIÈRE, EUGÈNE (1849–1906)

Born at Gournay, France. An early influence on him was the pastel work of Georges La Tour. He

attended the École des Beaux-Arts; and while a prisoner in Dresden as a result of the Franco-Prussian War, he had the chance to study the Old Masters there. Later he was to open his Académie in Paris. His own work had a straightforward honesty, unpretentious and pleasing.

'Maternity' F 55100, 1975, Hôtel George V, Paris.

CASSATT, MARY (1845–1926)

Born in Pittsburgh, Pennsylvania, died in Château Beaufresne, Beauvais. The daughter of a

Above: 'Harvest Moon' by Samuel Palmer. (Courtesy Christie's)

'Femme à la Mandoline' by Picasso. Painted in 1908. Sold in 1974 for £273 000. (Courtesy Christie's)

Pennsylvanian banker, she spent her childhood in France, then returned to America for study at the Pennsylvanian Academy of Fine Arts, Philadelphia. In 1868 she returned to Europe and settled near Paris. She joined the Impressionist group and showed with them in 1879, 1880, 1881 and 1886. Sadly from 1912 her sight partially failed and grew worse until her passing away. Mary Cassatt had a wonderful touch with children, she could catch the softness of their features, and their gentle colouring, in a way few artists have managed.

'Two Women in a Canoe' £62500, 1970, Sotheby Parke Bernet, New York. **(Highest price for a woman artist).**

'Mère et enfant' £60000, 1975, Sotheby's, London.

CASTAGNO, ANDREA DEL (*c* 1423–*c* 57)

His early life was spent in grinding poverty, until he was helped by Bernedetto de' Medici and taken to Florence. Here he painted a number of frescoes, including an impressive 'Last Supper' in the Convent of Sant' Apollonia, which is **now a Castagno Museum.**

CASTIGLIONE, GIOVANNI BENEDETTO (1616–70)

(also known as Il Grechetto and in France as Le Bénédetto). Born at Genoa he studied with Battista Paggi, then entered the school of Ferrari and lastly worked with Van Dyck. He painted a number of rather dark stormy landscapes often with many animals. Owing to a dark ground, many of his colours have sunk and spoilt the over-all appearance of his pictures. The latter part of his life he spent in the service of the Duke of Mantua who treated him most generously.

'Abraham's departure for the land of Canaan' £26250, 1975, Christie's, London.

CAULFIELD, PATRICK (1936–)

Born in London, he studied first at the Chelsea School of Art and then at the Royal College of Art. He has adopted a linear method of treating buildings, using little in the way of light and shade, and using dark lines to hold the forms together.

CELLINI, BENVENUTO (1500–71)

Italian sculptor, goldsmith and writer, he studied with Michelangelo Bandinelli, the father of the sculptor Baccio Bandinelli and with the celebrated goldsmith Marcone. Florence cast him out for duelling. He came under the patronage of Pope Clement VII. Here he produced a wonderful array of medals, candlesticks, jewel settings and silver and gold plate. Later he worked at the Court of François I of France. Many of his pieces are lost, but there remains the superb gold and enamel salt for François I. Sculpture includes a bronze of Cosimo I and 'Perseus with the head of Medusa'.

CENNINI, CENNINO (Born *c* 1370)

Although he studied painting with Agnolo Gaddi, there are no pictures of his extant. He is better known as the writer of *Il Libro dell' Arte*, **a compendium of recipes and methods.**

CÉZANNE, PAUL (1839–1906)

Born in Aix-en-Provence, the son of a banker, he was intended for the Law and went to Collège Bourbon of Aix. Here he met Zola and made a friend of the writer. He managed to persuade his father to let him study art in Paris, but he was refused admission to the École des Beaux-Arts there. He was not discouraged and set himself to study the paintings in the Louvre. His early work showed an affection for Delacroix, Courbet and Daumier. But in the 1870s he met Pissarro who introduced him to the colour theories of Impressionism. Almost to the end of the century he worked in isolation, he was to a degree supported by his father's money after the banker died. **Cézanne, by some, has been seen as the founder of modern painting**; he strode over many of the sterile rulings from the academies and to all intents fathered his own manner, which was and has remained unique.

'Chaumière dans les Arbres, Auvers' $220000, 1969, Parke Bernet, New York. 'Environs du Jas de Bouffan' Yen 24000000, 1969, Sotheby & Co, Mitsukoshi, Tokyo. 'Sept Baigneurs' £160000, 1975, Sotheby Parke Bernet.

CHADWICK, LYNN (1914–)

Born in London, he trained as an architect and during the war served as a pilot with the Fleet Air Arm. In 1945 he started to feel his way towards third-dimensional expression with mobiles, but has gradually evolved a strange angular construction using iron and other materials. The figures have an emergent quality, at times threatening, and filled with a sensation of suspended motion.

'Paper Hat II' $2500, 1975, Sotheby Parke Bernet, New York.

CHAGALL, MARC (1887–)

Born in Vitebsk, Russia, his father watched over a herring depot, his mother had a small grocer's shop. Chagall is one of those brave ones who has within him that something to carry through regardless. In the early years about the only influence on him was Bakst with his lyrical colours. He failed the entrance to St Petersburg School of Arts and Crafts. A tiny scholarship enabled him to go to Paris and here he discovered the catalyst that could bring into being his creative

'The Presentation in the Temple' by Philippe de Champaigne. (Courtesy Christie's)

surge. He found himself alongside such as Apollinaire, Delaunay and Modigliani, and the work poured out of him. He returned to Russia at the beginning of the First World War, and in 1922 again sought out Paris. Ambroise Vollard, the dealer and collector, had him illustrate Gogol's *Dead Sands*, Chagall produced 85 etchings, but the work was not published until 1948. To Marc Chagall, canvases and paints are tools with which he can create strange and beautiful poetry, shaded at times, particularly since 1939, with sorrow and darkness. But all the time through his pictures runs a strong music, that reaches back into the Russia he knew as a child and up into the atmosphere around him in which surge those spirits he takes into his compositions.

'Le couple sur fond bleu' Sfrs 420000, 1975, Galerie Motte, Geneva.

CHAMPAIGNE, PHILIPPE DE (1602–74)
Born in Brussels, though a native of Brabant, he is considered as a French painter. He worked in the studios of Bouillon, Michel Bourdeaux and Fouquière, and then went to Paris to complete his studies with L'Allemand. After this Du Chesne,

'Grace before the Meal' by Jean-Baptiste Siméon Chardin. (Courtesy Boymans-van Beuningen Museum, Rotterdam)

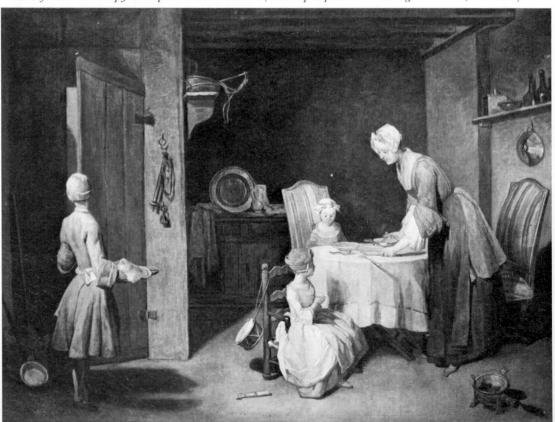

the Royal Painter, had him at work in the Luxembourg with Nicolas Poussin. The Queen Mother, Marie de' Medici, appointed him as Royal Painter, on the death of Du Chesne. An example of his skill is the triple portrait of Cardinal Richelieu in the National Gallery, London, and Louis XIII in the Louvre.

CHARDIN, JEAN BAPTISTE SIMÉON (1699–1779)
Born in Paris this most humble of painters studied with Nicolas Coypel, and was chosen to assist Van Loo in the restoring of one of the galleries at Fontainebleau. From here on he eschewed the fairyland atmosphere of the Court and set himself to paint simple genre pieces and most of all **some of the finest still life pictures ever executed**. A loaf of bread, a piece of fruit, a metal pot came alive under the brush of Chardin; his observant eye could see lights, nuances and colours where most would not think they existed. In 1786 one sees the 'Girl with Cherries' going for £5 5s and others about this time fetching not much more. It was not until 1912 that his merit started to be seen and appreciated. The Metropolitan Museum of Art in New York gave £13 225 for his 'Boulles de savon', and in 1951 the National Gallery, Washington DC, reached out with about £50 000 for a pair.

CHASE, WILLIAM MERRITT (1849–1916)
The American painter who studied under Karl von Piloty at the Munich Academy. When he came back to America he taught at the Art Students' League and the Brooklyn Art School. Later he opened the Chase School of Art and for a long time he headed the Society of American Artists. His work was principally genre with titles such as: 'Court Jester', 'Ready for the Ride' and 'Alice in Her Grandmother's Gown'.
'Hunter in Field, Shinnecock' $4500, 1975, Sotheby Parke Bernet, New York.

CHASSÉRIAU, THÉODORE (1819–56)
Born in San Domingo and died in Paris, he came to Paris when a boy and was set to study with Ingres. Primarily an historical painter he was also a good hand at portraits. His finest work was said to be the decorations in the Palais d'Orsay, Paris, but sadly they have been destroyed.

CHAVANNES, PIERRE CÉCILE, PUVIS DE (1824–98)
Born at Lyons he studied first with Henry Scheffer and then in the studios of Delacroix and Couture. He became the leading mural painter in France; working out his pictures on canvas, which was then marouflaged to the walls. His style had a pale, rather pastel-colour manner and compositions were monumental in size with subject-matter leaning back to the 15th-century Italian.

CHINNERY, GEORGE (1774–1852)
Born in London, he moved to Dublin in 1795 and worked there principally on portraits; he married and had two children. In 1802 under certain pressures he sailed for the East deserting his family. He worked first in Madras completing many portraits of the potentates for good money, from thence to Calcutta and on to Canton, finally to Macao in 1830 where he settled. His portrait output was prolific, and they were not always under his name, as well he produced many delightful and colourful small genre scenes and many thousands of pen and ink and pencil drawings. He also took a number of Chinese pupils into his studio.
'Goats in a landscape' (7 × 10¼ in (178 × 260 mm)) £1100, 1975, Sotheby & Lane Crawford, Hong Kong.

'A Little Boy standing in a landscape with a drum' by George Chinnery. Signed with monogram and dated 1797; 8¼ × 5⅝ in (216 × 143 mm). (Courtesy The Leger Galleries)

CHIRICO, GIORGIO DE (1888–)

Born in Volo, Greece, his father was an engineer from Sicily. He studied in Athens and later in Munich. His work hovers between the Metaphysical and Surrealism; in many compositions he shows strange empty streets, or geometrically constructed figures, and odd objects in unusual relationships. His approach is typified by 'Melancholy and Mystery of a Street' in the Museum of Modern Art, New York.
'Cavalli Antichi' L 28 000 000, 1975, Finarte, Rome.

CHODOWIECKI, DANIEL (1726–1801)

Born in Danzig, this Polish painter and engraver, after a hard start, became **one of the most celebrated graphic artists of his time.** In 1797 he was made Director of the Berlin Academy. His total output numbered 978 plates.

CHRISTUS, PETRUS (1410–c 72)

Early Netherlands painter at work around Bruges. He is known to have made copies of several pictures by Jan van Eyck, and was influenced by Van der Weyden.

CHURCH, FREDERICK EDWIN
(1826–1900)

American painter of landscapes that tended towards the spectacular in subject. He was a leading member of the Hudson River School.

CHURCHILL, WINSTON LEONARD
SPENCER (1874–1965)

With all his other activities he found a talent for painting. At the beginning he was encouraged by Sir John Lavery, but after that his own energy and impulse brought into being a surprising number of canvases of considerable quality; landscapes warm and glowing with Impressionistic colour harmonies, also still-life studies and flowers which betokened a sensitive and creative being.
'Blue Brass – La Capponcina' $10 000, 1975, Sotheby Parke Bernet, New York.

CHURRIGUERA, JOSÉ DE (1650–1725)

A family of Spanish architects and sculptors of whom the leading member was José. He made his mark by designing the catafalque for the funeral of Queen Maria Louisa. His main sculptures, with which he had the help of his sons, reflect the Baroque.

CIMABUE, GIOVANNI (1240–1302)

Of his life little has been recorded or is known. **Vasari claimed that it was he who first shed the light on painting, and he has been titled The Father of Modern Painting. His art does stand at a dividing line between the Byzantine and the Renaissance.** 'The Virgin-and Child with Angels' in the Louvre leans quite heavily towards the earlier icon-painters with the formal arrangement and lack of plasticity and emotion. His fine 'Crucifixion' in Florence was grievously savaged by the 1966 Flood.

CLAUDE (1600–82)

(properly Gellée, also called Claude Lorrain or Lorraine). French landscape-painter and etcher, he was born at Champagne, Lorraine. Not a hard-working pupil, his parents apprenticed him to a pastry-cook. He travelled to Rome where he worked in the house of Agostino Tassi, began to act as studio general help, and took some lessons from this painter. He travelled once back to his home, and then returned to Rome in 1627 where he stayed for the rest of his life. His painting life was marred by the behaviour of his pupil and assistant Giovanni Domenico who was with him for 25 years; those jealous of Claude's performance spread round the story that it was Domenico who was producing those marvellous sun-glowing paintings. Domenico very ungratefully joined the detractors and also sued for salary he said was owed to him. Claude paid, but for evermore shut his door to assistants.
It is with the handling of light that Claude is successful; it seems to permeate the whole picture, softening and enriching as it comes towards the viewer. The National Gallery, London, has a fine example with the 'Embarkation of the Queen of Sheba'.

CLOUET, FATHER AND SON

Jean (c 1485–1541) and **François** (1520–72). Both French portrait-painters of Flemish extraction. Jean had probably established himself at Tours prior to the accession of François I. He did become painter to this monarch who so understandingly patronized the arts, and he executed the incisive portrait of François which is in the Louvre. François was born at Tours and was also painter to François I and his *valet-de-chambre*.

COELLO, ALONSO SÁNCHEZ
(c 1532–c 88)

Born near Valencia he studied with Anthonis Mor; he became Court Painter to Philip II and was also Keeper of the Royal Collection. One of his finest portraits was that of Father Siguenza, a close friend; he also painted the likeness of Ignatius Loyola from casts taken 29 years earlier.

COLDSTREAM, SIR WILLIAM
MENZIES (1908–)

Born in Belford, Northumberland, he studied at the Slade School, and while there showed with the New English Art Club and the London Group, joining the latter in 1934. He was a **founder-**

member of Euston Road Group, and since 1949 Professor of Fine Art, University College, London. His observation is acute and his portraits are sensitive and incisive.

COLOMBA, LUCA, ANTONIO (1661–1737) Born at Arogno, Switzerland, he painted in oil and also fresco, helping to keep alive the medium. His manner was distinguished by careful design and delicate colours. He painted for some time in Germany where he was employed by the Duke Eberhard Ludwig.

CONSTABLE, JOHN (1776–1837) Born at East Bergholt, Suffolk, the son of a wealthy miller, he spent some years as a child working in the mill. It was not until he was 23 that with the help of that generous patron of the arts, Sir George Beaumont, he got to the Royal Academy Schools. Yet Constable was one of those who, having learnt some basic rules, was to be his own best teacher by his acute study of nature. He first showed with the Royal Academy in 1802, and was elected an Associate in 1819, and an Academician in 1829. Constable never had in his lifetime in England the acclaim he should have had for the breakthrough he made in the sheer appreciation of what landscape was about; it lay with the French to raise the plaudits when the 'Hay Wain' (now in the National Gallery, London; there is a study for it in the Victoria and Albert Museum) was shown in the Paris Salon in 1924. The style, handling of paint, composition made a mark on a number of his contemporaries over there.

COOPER, SAMUEL (1609–72) He was a miniature-painter, born in London, and he and his brother Alexander studied with their uncle John Hoskins, also a miniaturist. Samuel painted many of the famous of his time including Oliver Cromwell, John Milton and Charles II and his Queen, Catherine of Braganza.

COPLEY, JOHN SINGLETON (1737–1815) Born in Boston, Massachusetts, died in London. He may have had some training from his stepfather Peter Pelham, a portrait-painter and engraver. He left Boston in 1774 and set up his studio in London after a continental trip. He painted George Washington twice before he left America, other canvases of note are: 'Watson and the Shark', in the Museum of Fine Arts in Boston and 'The Death of Major Pierson' in the Tate Gallery, London.

CORINTH, LOVIS (1858–1925) Born in Tapiau, East Prussia, died in Zandvoort, Holland. He attended the Academy of Fine Arts, Munich, and later the Académie Julian. He developed a powerful Naturalism with broad attacking strokes, without preliminary drawing and working wet paint into wet paint. 'Weiblicher Halbakt am Fenster' £3700, 1975, Sotheby Parke Bernet.

COROT, JEAN BAPTISTE CAMILLE (1796–1875) Born in Paris he studied with Michallon and then Victor Bertin. He liked to work hard during the spring, summer and autumn out-of-doors and then in the winter develop the themes he had found. The individual look he gave to landscape had considerable popularity, also it **drew many forgeries.** Corot was a generous character, recalling his poverty in his youth, he donated some 25000 francs for needy painters during the Siege of Paris in 1870. **In value he shows all the signs of continuing to increase:** for example, 'Venus au Bain' in 1920 fetched £6750, in 1956 £27000. 'Le Pêcheur à la ligne' £33600, 1975, Christie's, London.

CORREGGIO, ANTONIO ALLEGRI DA (c 1489–1534) Born in the small village of Correggio, Italy, he worked in the studio of Ferrari and Mantegna, although an overriding influence, particularly in the handling of heads and anatomy, would appear to have been Leonardo. His pictures are rich in atmosphere with a subtle handling of light and shade, that gives considerable depth. One of his loveliest pictures is 'Mercury instructing Cupid before Venus' in the National Gallery, London.

COSWAY, RICHARD (1742–1821) Born in Tiverton, he came to London and worked under Hudson and in Shipley's School. His miniatures showed some influence from the style of Correggio.

COTES, FRANCIS (1726–70) Born in London he studied with George Knapton. His portrait work had elegance, and his crayon work was much admired. **A founder-member of the Royal Academy.**

COTMAN, JOHN SELL (1782–1842) Born at Norwich he was largely self-taught, but was assisted by Dr Monro, the collector and philanthropist. In 1807 he was made a member of the Norwich Society of Artists and the next year he sent in 67 works for the exhibition. **Cotman was a supreme master of water-colour,** knowing all the vagaries of that difficult and elusive medium. His 'Greta Bridge, Yorkshire', in the British Museum is a lesson in the purity of handling. He had two sons, Miles Edmund and John Joseph, both of whom were painters.

COURBET, GUSTAVE (1819–77)

Born at Ornams in France he started out studying law but gave it up for painting when he was 20. He did have some training with David d'Angers but his own impulse drove him on, regardless of early failures. He was the great demonstrator of Realism, a style that owed something to Rembrandt. He became involved politically and with the fall of the Commune in 1871 he was imprisoned for six months for encouraging the destruction of the Vendôme column. A painter who has probably yet to receive the appreciation due.
'Paysage, le torrent' F 110000, 1975, Palais Galliera, Paris.

COUSTOU, GUILLAUME (1677–1746)

One of the great French sculptors, brother of Nicolas also a sculptor. Guillaume studied with Pierre Legros. He worked alone and with his brother for Louis XV at Versailles and Marly. His work is at its best with the fine group **'The Horses of Marly' which once stood in the park at Marly, but can now be seen in the Champs-Elysées**, Paris.

COUTURE, THOMAS (1815–79)

French historical painter born ar Senlis, Oise, he worked under Gros and Delaroche. His subjects often harked back to the Greeks and Romans, and had a parallel in England with Alma-Tadema and Lord Leighton.

COX, DAVID (1783–1859)

Born near Birmingham, as a boy he was taught to wield a hammer as a smith, next he was apprenticed to a locket-maker, and thence to a night schoo! in drawing kept by Joseph Barker. He became a colour-grinder for the scene-painters at the local theatre in Birmingham, thence to painting scenery. He had some time as a drawing master in London and Hereford. Cox's manner had a lively broken look that at its best could capture the atmosphere of wind and movement in a landscape.

COYPEL FAMILY

The founder was **Noël** (1628–1707) who studied with Poncet and then Quillerier. He worked on the decorations of the Louvre, was elected to the Académie and then was appointed Director of the French Academy in Rome. He married in 1659 Madeleine Hérault, the sister of Charles Hérault, the landscape-painter, and she was herself an artist. From this marriage came a son **Antoine** (1661–1722) who among other chores decorated the ceiling of the Chapel of Versailles and those of the Palais-Royal. Antoine had a son **Charles Antoine**

(1694–1752) a weak portrait-painter. Noël on the death of Madeleine, married **Anne Françoise Perin**, also an artist. From this union a son was born, **Noël Nicholas** who was a minor hand at allegorical pictures. Father Nöel taught both his sons.

COYSEVOX, ANTOINE (1640–1720)

French sculptor who studied with Louis Lerambert. He worked with Le Brun and Mansart at Versailles. Distinctive works by him include: 'Louis XIV' at Versailles, 'Mercury' in the gardens of the Tuileries and the 'Tomb of Mazarin' in the Louvre.

COZENS, FATHER AND SON

Alexander, the father (1717–86) was born in Russia, reputedly the natural son of Peter the Great and an Englishwoman from Deptford, who was probably living with a shipbuilder employed by Peter. Alexander studied in Italy. **One method he developed was to use accidental blots as a basis for picture-making**, a method discussed by Leonardo. He worked largely in monochrome. His son John Robert was the pupil of his father. He travelled to Italy and made numerous sketches on

'Princess Sibylla of Cleves' by Lucas Cranach the Elder. Panel, 20 × 15 in (508 × 381 mm). (Courtesy Sotheby Parke Bernet)

the way, particularly in Switzerland, where the clear light and variety of scenery drew from him a pleasing Romantic style. In 1874 he became insane and was supported by the kindness of Dr Monro, the Royal Academy, Sir George Beaumont and others.

CRANACH, LUCAS, THE ELDER
(1472–1553)
Born in Kronach, it is likely that he was taught by his father, Hans. He became a painter and engraver with a distinctive style; the figures lack a live anatomical look but have a charm, set as they often are in woodland or rocky landscapes. There is but sparse record of where his early years were spent. In 1493 he may have visited the Holy Land with the Elector Frederick the Wise, after this he was Painter to three successive Electors of Saxony at Wittenberg. In the town as well as his studio workshop, he had a book-printing business and an apothecary's shop. **He was a close friend of Luther and Melanchthon.** His pictures are marked with a writhing serpent adopted from the winged serpent granted him as a crest by the Elector Frederick. He was often assisted by his two sons, Johann Lucas (1503–36) and Hans (c 1510–c 53). **Graphic work included about 800 prints, mostly woodcuts.**

CREDI, LORENZO DI (c 1458–1537)
Born in Florence he was working in Verrocchio's studio at the same time as Leonardo and Perugino. He stayed with his master until Verrocchio died, but was more influenced by Leonardo. As with others he got caught up with the frenzies of Savonarola, and **consigned to the flames a number of his pictures.**

CRIVELLI, CARLO (c 1430/40–93)
Probably born in Venice and early tuition could have been with Bartolommeo da Murano, after with Vivarini and possibly with the Academy of Squarcione. His manner was highly decorative, with great attention to detail as may be noted in 'The Annunciation' in the National Gallery, London, a picture full of fascinating detail, and using a somewhat forced perspective to give depth and carefully grouped figures to enhance the sense of spaciousness.

CROME, JOHN (1768–1821)
Born in a small public house in Norwich, he started life as an errand-boy for a local doctor, but soon apprenticed himself to Frank Whisler, a house- and sign-painter; he is **said to be the first artist to practise 'graining'** in imitation of the natural marking on wood. He studied the works of Dutch and Flemish masters in the collection of a Mr Harvey and he gradually made his way,

teaching and slowly increasing his sales. He was with the Norwich Society of Artists which has claimed John Sell Cotman as a member, and which is better known today as the Norwich School or East Anglian School. John Crome's work has a bluff strength combined with a sensitive feeling for landscape. The Tate Gallery, London, has one of his most striking pictures 'Slate Quarries'. He had the nickname Old Crome to distinguish him from his son John Bernay Crome (1794–1842), also a painter, whose works were very close in style to those of his father.

CRUIKSHANK, GEORGE (1792–1878)
Born in London, he was to become one of the leading caricaturists, aiming in general at social evils. Perhaps he is best known as a book illustrator with such as: *Grimm's Fairy Tales*, *Sketches by Boz*, *Don Quixote* and *The Ingoldsby Legends*.

CUYP, AELBERT (1620–91)
Born in Dordrecht and studied mostly with his father who was a portrait-painter. He developed a manner in handling light that earned him the name of the Dutch Claude. Cuyp's landscapes are generally fairly small and have an almost perfect sense of harmony held as they are in his fine golden light that brought even the sombre Dutch landscape to warm vital life. He excelled at sea pieces as well, and in the portrayal of animals. English collectors have in particular had a great liking for him and there are probably more of his works in Britain than any other country. **His value was not always high but is now climbing steeply.** Early in the 18th century his paintings could be had for as little as 30 florins and even at the end of that century were still well down in the £100s. Today five figures is more likely. 'Small view of Dordrecht' £25000, 1958.

DADD, RICHARD (1817–87)
The son of a chemist in Chatham, he went to the Royal Academy Schools. He started as a painter of genre subjects, and interspersed these with pictures based on ideas in Byron's *Manfred* and Tasso's *Jerusalem delivered*. In 1842 he left for a journey to Egypt, going through Italy, Greece and Asia Minor. On the way sunstroke appears to have brought on mental trouble and he returned to London. After a number of apparently normal water-colours he sent a strange cartoon to a competition at the Westminster Hall which pointed to his unsound mind. His father watched over him constantly. On 28 October 1843 he went to Cobham with his parent and there murdered him. Richard fled to France and near Fontainebleau suddenly attacked a fellow traveller, an incident that led to his arrest. He was committed

'The Sacrament of the Last Supper' by Salvador Dali. (Courtesy National Gallery of Art, Washington, DC, Chester Dale Collection)

to an asylum and continued to paint bizarre and eccentric pictures.
'The Haunt of the Fairies' £3380, 1975, Christie's, London.

DADDI, BERNARDO (*c* 1300–50)
Florentine painter and one of those who founded the guild of St Luke in 1339. It is possible he was a pupil of Giotto. His best work was probably the miracle-working picture of the Virgin in Or San Michele at Florence.

DAHL, JOHANN KRISTEN CLAUSEN (1788–1857)
Norwegian painter largely self-taught, he settled in Dresden in 1818 and in 1824 was made a Professor of the Academy there. His landscapes expressed the grandeur of the Norwegian scene, catching the intense atmospheric conditions often experienced.

DAHL, MICHAEL (1656–1743)
Swedish artist who had a period studying with Ehrenstral. He travelled through Europe and in Rome painted the portrait of Queen Christina of Sweden. Thence to London where he competed for popularity with Sir Godfrey Kneller; Queen Anne and Prince George of Denmark both sat for him.

DALI, SALVADOR (1904–)
Born at Figueras he worked for a time in the Madrid Academy, then rejecting the teaching went to Paris. **Here he developed one of the most outstanding techniques in the pure, non-impasto, handling of oils.** He seems to have been against Cubism and a number of the current vogues that were sweeping the scene. His own manner of Surrealism is quite original and has endless invention. He has made Surrealist films such as *L'Age d'Or* and *Le Chien Andalou* in collaboration with Luis Buñuel. Paintings like 'Persistence of Memory' in the Museum of Modern Art, New York, and the striking 'Crucifixion' in Glasgow Art Gallery, stay long afterwards in the mind.
'Pyramids at Gizeh' $45000, 1975, Sotheby Parke Bernet, New York.

DALMAU, LUIS (working 1430–60)
Spanish painter from Catalan. He travelled to Bruges and there became strongly influenced by the work of the Van Eyck brothers. The tempera painting 'Virgin and the Councillors' in Barcelona is one of the outstanding works of his still existing.

DANBY, FRANCIS (1793–1861)
Born near Wexford, Ireland, he was in the studio of a landscape-painter O'Connor in Dublin for a

period. In 1813 he, with his master and George Petrie, decided to look for fame and fortune in London; but their money ran out by the time they got to only Bristol. The others left him in Bristol where he made a living by selling drawings and giving lessons in water-colour painting. His 'Delivery of Israel out of Egypt' gained him an Associateship with the Royal Academy. But in 1830 he quarrelled with this body and went to Switzerland where he spent most of his time for eleven years boat-building and yachting. In 1831 he returned to England and set up a studio at Lewisham and concentrated on large compositions, often in a striking and imaginative vein. 'The Upas Tree' (the poison tree from Java), 'The Opening of the Seventh Seal', 'Rich and Rare were the Gems she wore'.

DANCE FAMILY

The father was George (1695–1768), an Architect to the City of London who designed in the manner of Wren; the Mansion House is credited to him. Nathaniel (1736–1811) was his elder son who studied with Hayman. He distinguished himself as a painter of portraits, conversation pieces and historical subjects. **He was a founder-member of the Royal Academy.** He spent a number of years in Italy not only studying art but also following the lovely Angelica Kauffmann around persistently smothering her in matrimonial propositions. When these failed he returned to England and settled for a widow, Mrs Dummer, who had a large fortune. He renounced his profession, took up politics and got a baronetcy and the name Dance-Holland. One of his better portraits is that of Captain Cook at Greenwich. His younger brother George (1741–1825) was an architect like his father. George's strongest work was Newgate Prison (pulled down in 1902).

DANNECKER, JOHANN HEINRICH
(1758–1841)
German sculptor who studied with Lejeune, Augustin Pajou and JV Sonnenschein and was certainly influenced by Canova, he may have had a period in his studio He became Director of the Academy of Stuttgart and was a lifelong friend of Goethe. Fine works by him include: 'Ariadne on a Panther', 'Girl Lamenting Her Dead Bird' and a bust of Schiller at Stuttgart.

DAUBIGNY, CHARLES-FRANÇOIS
(1817–78)
Born in Paris, he studied first with his father, who taught him to decorate clock-cases, boxes and other commercial items; and then with Paul Delaroche. He painted landscapes, often of a large size, in the open-air manner of the Barbizon School; he also was an accomplished etcher. He had a barge fitted out as a floating studio. 'Landscape at Sunset' Sfrs 14000, 1975, Auktionhaus am Neumarkt, Zürich.

DAUMIER, HONORÉ (1808–79)
Born at Marseilles, **he became one of the most brilliant French caricaturists**, also a supreme lithographer and painter. He was on the staff of *Le Caricature* in 1831 and made inflammatory attacks on the monarchy, which in the end got him six months in gaol. He levelled his brush and crayon at all classes of pomposity. Slowly as he grew older he turned more to painting which exhibited a lively brush manner working with a limited palette. His close friend Corot was to care for him when he went blind in 1877, and gave him a cottage at Valmondois, Seine-et-Oise.

'La Ratapoil' by Honoré Daumier. Bronze. (Courtesy Christie's)

DAVID, GÉRARD (c 1460–1523)
(also spelt Gheeraert) Born close to Gouda in Holland, he moved to Bruges in 1482 and was admitted to the Guild of St Luke there in 1484. His works have been confused with those of Memlinc; they have a potent sincerity although the anatomy is stiff and formalized; the landscapes in the background may have been executed by Joachim Patenir.

DAVID, JACQUES LOUIS (1748–1825)
Born in Paris, he was brought up by his mother, as his father lost his life in a duel, when the boy was quite young. He received some early instruction from Boucher and then entered the studio of Vien. This master, apparently in a fit of jealousy with his brilliant pupil, deprived him of the chance of getting the Prix de Rome in 1771. He travelled to Rome and there developed his ascetic classical style. In the Louvre are two of his paintings contrasting in subject; the lovely 'Madame Récamier' and the macabre 'Death of Marat'. Back in France he involved himself politically and ended up in prison after the fall of Robespierre. His art was a pictorial rejection of the splendour, riches and escapist work of the last days of the Louis' before the Revolution. The studio he ran became famous and the **pupils included: Girodet, Gros, Gérard, the elder Isabey, Léopold Robert and Ingres.**

DAVIE, ALAN (1920–)
The Scottish painter and goldsmith was born at Grangemouth, Scotland. Considerably influenced by Klee, Picasso and Jackson Pollock, he is an artist who to a degree relies on his subconscious mind to guide him. His pursuits include Zen-Buddhism, jazz musicianship and piloting his glider.
'Transformation of the Serpent No. 1' L 4400000, 1975, Finarte, Rome.

DAVIES, THOMAS (c 1737–1812)
A garrison officer in Canada, an amateur who left behind not only topographical work of interest but also a number of views of woods, waterfalls and Canadian scenery painted with a surprising understanding of colour and composition. **They had been 'lost' for a good many years, and were discovered in England** and are now in the National Gallery of Canada.

DEGAS, HILAIRE GERMAIN EDGAR (1834–1917)
Born in Paris he went to the École des Beaux-Arts and worked under Louis Lamothe, and met Ingres. He journeyed through Italy studying the early masters of the Renaissance. He showed with the Impressionists although he was not greatly attracted to landscape for itself. His art rested largely on the things and people close around him that could excite his imagination; thus his master works include ballet, theatre, race-meetings, café scenes. He was an immense enthusiast for working in many different media including: oils, pastel, water-colour, aquatint, etching and monotype. The Louvre has a number of his fine works including: 'Absinthe' and 'At the Races'; at the Glasgow Art Gallery is one of his best ballet compositions 'La Répétition'. As with a number of the 19th- and 20th-century French school his values have increased greatly.
'Deux Danseuses' £100000, 1975, Sotheby Parke Bernet.

DEINAKA, ALEXANDER ALEXANDROVICH (1899–)
Russian painter and engraver, who attended the Kharkov Art School and later was at the Moscow Art and Technical Workshops. His work concentrates on the modern Russian Factory worker. He has also illustrated a number of books both for adults and children. Honoured worker of the R.S.F.S.R.

DELACROIX, FERDINAND VICTOR EUGÈNE (1798–1863)
Born at Charenton-Saint-Maurice near Paris he was in the studio of Guérin and later with Gros. He shied away from the prevalent classical manner spear-headed by Jacques Louis David, and **joined with Gericault to bring into being the Romantic school.** His pictures had a rich passion in them full of emotion and involved feeling. Compositions such as 'Dante and Virgil in Hell' and 'The Massacre at Chios' both in the Louvre, underline the power of his feeling and the breakaway in style from the followers of David.

DELAROCHE, HIPPOLYTE (1797–1856)
Born in Paris, he set himself first at landscape with Watelet, then turned to historical painting under Gros. He produced a huge composition for the Ampitheatre of the École des Beaux-Arts for which he was to receive £3000, and it took him four years. He included almost a history of art, with the ancients, Apelles, Ictinus and Phidias, and a long concourse of painters and sculptors including, Claude, Titian, Velázquez, Rubens, Rembrandt, Dürer, Michelangelo, Raphael and the rest. The Louvre has his 'Death of Queen Elizabeth' which shows his somewhat theatrical and forced manner.

DELAUNAY, ROBERT (1885–1941)
Born in Paris he has been one of the main experimenters with quality of colour in the Abstract. He worked through the theories of Seurat, and applied himself to Cubism and

Orphism. He married Sonia Terk who was born in 1885 in Russia and went to Paris in 1905, and had been wed first to the critic Wilhelm Uhde. She showed 'Prismes électriques' at the Salon des Indépendants in 1914. Diaghilev had her do the costume designs for the ballet *Cleopatra*.

DELLA ROBBIA FAMILY

The Florentine workshop was started by **Luca** (1400–82) who was at first a goldsmith, his works later included a 'Madonna of the Roses' in the Bargello, and a balcony (*cantoria*) for singers in the Duomo. **He found a technique for enamelling terracotta with blue and white glazes.** Fine examples of this method are the 'Resurrection' and 'Ascension' also in the Duomo. **Andrea** (1435–1525), who was his nephew, worked in a similar manner, and among his successful works were the decorations for the Loggia dei Innocenti. Andrea had several sons who worked with terracotta not only in Tuscany but also in Naples, Umbria and Sicily. The finest craftsman was **Giovanni** (1469–1529) who is credited with a fountain at Santa Maria Novella, Florence. There was a grand-nephew of Luca, **Girolamo** (1488–1566) an architect and sculptor, who among other matters modelled a figure of Catherine de' Medici.

DELVAUX, PAUL (1897–)

Born in Belgium, he went through stages of Neo-Impressionism and Expressionism, and then with some influence from Chirico and Magritte he came to Surrealism. His compositions feature modest nude ladies set in strange settings, often with ruins and deep vistas, lit by an uncarthly light. 'Le Jardin Nocturne' £46000, 1975, Sotheby Parke Bernet.

DEMUTH, CHARLES (1883–1935)

American painter who was a pupil of William Merritt Chase at the Pennsylvania Academy of Fine Arts. His works include rather strange near-Cubist views of cities and flower studies. He illustrated Henry James's *The Beast in the Jungle* and *Nana* by Zola.

DENIS, MAURICE (1870–1943)

French painter, engraver and writer **who was a strong influence with the Nabis.** Among his expounded theories was the paragraph to explain modern art: 'Remember that a picture – before being a horse, a nude, or some sort of anecdote – is especially a flat surface covered with colours assembled in a certain order.'

DERAIN, ANDRÉ (1880–1954)

Born at Chatou, France, he studied at the Académie Carrière where he met Matisse and Vlaminck. He was a notable colourist, and working in London produced at least one memorable scene of the Thames loaded with shipping indicated with warm red strokes. He worked on stage designs and ballets, designing the sets for Diaghilev's production of La Boutique Fantasque'.

DESIDERIO DA SETTIGNANO (1428–64)

French sculptor who may have been a pupil of Donatello. His marble-carvings and those in wood have considerable quality. There is a good marble chimney-piece by him in the Victoria and Albert Museum in London.

DESNOYERS, AUGUST GASPARD, BARON (1779–1857)

French engraver, a pupil of Alexander Tardieu, whose speciality was producing plates of works by Leonardo and Raphael.

DESPIAU, CHARLES (1874–1946)

French sculptor born in Mont-de-Marsan who was an **assistant for a time to Auguste Rodin**, he himself studied with Hector Lemaire. His figure studies have a relationship to the work of Maillol.

DEVIS FAMILY

Senior member was **Arthur** (1711–87) who possibly studied with Peter Tillemans. He painted conversation pieces and portraits; the figures sometimes a little thin and wooden; the pictures are generally quite small. Arthur's brother **Antony** (1729–1816) was a landscape-painter mostly in water-colour. **Arthur William** (1762–1822) was the son of Arthur, Senior and had some instruction from his father. In 1782 he was appointed as a draughtsman by the East India Company and sent with Captain Wilson in the *Antelope* for a voyage round the world. Then he went to Bengal but came home in 1795. 'A family group in a garden' by Arthur Devis, Senior £58000, 1975, Sotheby Parke Bernet.

DIAZ DE LA PEÑA, NARCISSE VIRGILE (1808–76)

Born in Bordeaux, his parents were of Spanish extraction. He was placed at the age of 15 with a porcelain-manufacturer, but left him to study with Sigalon. He became a member of the Barbizon school with an individual rubbed style, in which fairly thick impasto was taken down to allow underpainting to ghost through. From his earliest years he liked to study nature in the woods and when young lost his left leg by sleeping on the damp grass: he died at Mentone from the bite of a viper.

DIX, OTTO (1891–1969)
Born in Unterhausen, Thuringia, he studied at the School of Arts and Crafts and the Academy at Dresden, later he was a Professor at Düsseldorf Academy. An artist using his talents in social and political criticism, particularly against the cruel obscenities of war. His work is parallel with that of Grosz. He was greatly influenced by the earlier Germans such as Dürer, Baldung and the great Spaniard Goya.

DOBELL, SIR WILLIAM (1899–   )
Born in Newcastle, New South Wales, Australia, he was apprenticed to Wallace J. Porter, thence to Julian Ashton's School, Sydney, and then to England for a time at the Slade School, London, and later to The Hague, Holland. He was away for ten years and brought back to Australia a considerable influence from outside. In the war he worked on camouflage. Since then his talent has flowered, a personal style of perceptive portrait work in particular has borne fruit. But as with so many artists he has had to watch others reap the main harvest. Not so many years ago pictures he had been paid £50 for were having the hammer dropped at about £7000.
'King's Cross' (on board $9\frac{3}{8} \times 6\frac{1}{4}$ in (238 × 159 mm)) A$ 12 000, 1975, Christie's, Sydney.

DOBSON, FRANK (1886–1963)
British sculptor who was in the line of Maillol and Gaudier-Brezeska, his brave use of form and the plastic quality he could give stone brought him considerable plaudits in the 1920s and 1930s.

DOBSON WILLIAM (1610–46)
Born in London he was apprenticed to Robert Peake. Early he had some material help from Van Dyck and after this master he was appointed as Sergeant-Painter to Charles I. One of the best existing works is the well-found portrait of Endymion Porter in the Tate Gallery, London.

DOESBURG, THEO VAN (1883–1931)
Mainly important for his influence with the de Stijl and his insistence that all the visual arts should be related.

DOLCI, CARLO (1616–86)
Born in Florence and worked in the studio of Jacopo Vignali. His output was mostly small well-finished pictures of Our Saviour, the Virgin and other religious subjects. **His pupils Loma, Mancini and his daughter made numerous copies of his pictures.**

DOMENICHINO (1581–1641)
Born at Bologna, he worked with Dionysius Calvaert, then to the academy run by the Caraccis. Later he assisted Annibale Caracci with the decorations at the Farnese Palace. He had a well-developed sense of landscape which gave considerable strength to his compositions.

DONATELLO, DONATO DI NICOLO (1386–1466)
Born in Florence, he was in the workshop of Ghiberti. **He emerged as a new force in sculpture**, whether working in marble or the clay to be cast into bronze. **Perhaps his peak is the beautiful figure of David in the Bargello**, Florence, with his shepherd's hat, as he stands holding the great sword he has taken from Goliath and with his left foot on the giant's severed head. The quality of the figure that emerges from the hard bronze is quite perfect. Florence is the place, as it is with so much, to see Donatello; his 'St Louis of Toulouse' in the Museo dell' Opera di Santa Croce, or 'Jeremiah' in the Museo dell' Opera del Duomo and the fine 'Judith and Holofernes' in the Piazza della Signoria.

DORÉ, GUSTAVE (1832–83)
French painter and illustrator born at Strasbourg, he modified his name from the German Dorer. He was showing pen and ink drawings with the Salon in Paris and a year later he was drawing for the *Journal pour Rire*. He illustrated many books including: *Contes Drolatiques* by Blazac, *Inferno* by Dante, *Idylls of the King* by Tennyson, *Fables* by Fontaine and Poe's *Raven*, his last work.

DORMIDONTOV, NIKOLAI
A Russian painter who brilliantly captured the savaging of cities like Leningrad in the last war; working with just India ink and pencil he captures the futile destruction.

DOSSI, DOSSO AND BATTISTA
Painter brothers, **Dosso** (1479–1542) **Battista** (birth not placed – 1548) who studied with Lorenzo Costa. To a degree they worked together afterwards. Dosso was good at figures, Battista was good at landscape.

DOU, GERARD (1613–75)
Born in Leyden, he worked with Bartholomeus Dolendo. Later **he was one of the first of Rembrandt's pupils.** He imitated his master's chiaroscuro at first and then achieved a successful manner of his own for highly finished small pictures of intimate and genre nature.

DRYSDALE, SIR GEORGE RUSSELL (1912–   )
Born in England, he emigrated to Australia when a child. Studies started with the George Bell School, Melbourne, thence to Grosvenor School, London and Grand Chaumière, Paris. The wide outback has been one of the inspirations for studies

in paint, crayon, line and wash; the deserted areas, bleaching bones, twisted trees, ghost towns, the latter inhabited by spindly legged figures, the colours raw, warm and harshly dark.
'Bob and Maudie' A$ 20000, 1975, Christie's, Sydney.

DUBUFFET, JEAN (1901–   )
Born in Le Havre, he made some rather desultory efforts at painting, but then gave up and settled down as a wine merchant. In 1944 he turned back once more to painting, he also tried collage and relief. **His is an art that sets out to catch the emergent primitive, a return to Child Art.** In 1948 he founded the Société de L'Art Brut.
'Composition, 1958' L 36000000, 1975, Finarte, Milan.

DUCCIO DI BUONINSEGNA (c 1278–1319)
Early records are sparse. An early work for the Fraternity of Santa Maria, Florence, 'Virgin and Child, with Saints' for their Chapel in Santa Maria Novella and an 'Annunciation' for Santa Trinità have disappeared, also a 'Madonna' for the Chapel of the Palazzo Pubblico of Siena is missing. The year 1308 saw him working on his altarpiece for the Cathedral of Siena; this took him three years and it was carried to its place with great ceremony. It was 14 ft (4·26 m) long, 7 ft (2·13 m) high and worked on both sides, the front showed the Virgin and Child with various Saints and Angels, and the reverse 27 subjects from the life of Christ. So much gold and ultramarine were used that the original price came out at about 3000 gold florins. Small paintings from this work have been removed at unknown dates and are now in national galleries in three countries. The remainder of this work is in the Cathedral Museum.

DUCHAMP, MARCEL (1887–1968)
Born in Blainville, near Rouen, the brother of the sculptor Raymond Duchamp-Villon and Suzanne Duchamp, and half-brother of the painter Jacques Villon. The painter became involved with Cubism and then was **one of the most unpredictable and explosive figures with Dadaism.** In America he made his name at the Armory Show with 'Nude descending a Staircase', now in the Philadelphia Museum of Art. From here Duchamp attacked the art establishment, exhibiting found objects, and sparing no pains in his efforts to shock.

DUFY, RAOUL (1877–1953)
Born at Le Havre, he studied at the École des Beaux-Arts in Paris. He was attracted to Les Fauves, and was influenced by Cézanne. But before the First World War he had begun to develop his own style which is one that is bright and colourful, sketchy in details, and often witty. He did much design work for tapestries and textiles. His values have increased considerably since 1950.
'Deauville, les régates' £21 500, 1975, Sotheby Parke Bernet.

DUGHET, GASPARD (1613–75)
(commonly called Poussin) Born in Rome, his parents were French who had settled in that city. They took under their roof a young student Nicolas Poussin, who in time married their daughter and taught their son. When Gaspard left Poussin's studio he came under the influence of Claude. His greatest strength lay in landscapes under storm conditions.

DUNOYER DE SEGONZAC, ANDRÉ (1884–1974)
Born at Boussy-Saint-Antoine, France, he attended the École des Beaux-Arts and the Académie Julian. He worked at an individual style, particularly with water-colour, which had a limited palette and employed a vigorous brushwork.

DUQUESNOY, FRANÇOIS (1594–1643)
Flemish sculptor, also known as François Flamand who lived and worked in Rome from 1618 until 1643. **He was one of the first to be able to simulate the anatomy of children with marble and terracotta.** It is interesting to compare Duquesnoy's figure of St Andrew with 'St Longinus' by Bernini that stands opposite to it in St Peter's, Rome.

DÜRER, ALBRECHT (1471–1528)
Born in Nuremberg, he was the 3rd of 15 children of a Hungarian goldsmith. In 1486 he was bound apprentice to Michael Wolgemut and of this time Dürer was to remark 'God gave me diligence so that I learnt well.' He did indeed for few artists have so successfully employed so many techniques; **Dürer mastered oils, water-colour, tempera, gouache, silver-point, pen and ink, charcoal, crayon, woodcut, wood-engraving and metal-engraving.** In 1494 he married Agnes Frey, the daughter of a rich musician, and he went to Italy for the first time and later again in 1505; here he found the influence of the Renaissance, particularly Mantegna and the Bellinis; he was to bring their manners back to Germany and they are reflected in his own work. Of the second trip to Italy, which took in Venice, Vasari remarked that Dürer went to that city to defend his copyright against one Marcantonio Raimondi who was plagiarizing some of his blocks and plates and using Dürer's monogram; if this is so it did not do much good as later Raimondi and others were to

commit wholesale piracy on the luckless artist. His output was prodigious, water-colour studies of such humble subjects as pieces of turf, a bullock's nose, a bird's wing; altarpieces included: 'Adoration of the Trinity' in Vienna, 'Four Apostles' in Munich and the 'Adoration of the Magi' in Florence; and graphic work, **one of the finest pieces of engraving 'Knight, Death and the Devil'**; the 15 woodcuts that go to make up the 'Apocalypse' which abound in invention, and which were followed by the masterful 'Little' and 'Great Passion' series.

Dürer was employed by the Emperor Maximilian from 1512 until the Emperor's death in 1519, and among the works he did at this time was what must surely be **the largest woodcut ever**; it showed a giant Triumphal Arch, and was **made from 92 blocks and the print measured 10 ft 6 in (3·20 m) high and 9 ft (2·74 m) wide**. It showed with allegorical presentation the triumphs, glories and might of the Emperor, as well as his family tree. Maximilian was extremely interested in the progress of the print and it is recorded one day he visited Dürer's studio-workshop and a number of cats came running into his presence, whereupon the Monarch gave rise to the proverb 'A cat may look at a king.'

Perhaps it is in his self-portraits Albrecht Dürer tells most about himself. When he was 22, he depicted himself with a young sensitive face, the right hand clutching a sprig of sea-holly or eringo, this is now in the Louvre; five years later the slightly more confident young man of fashion, in the Prado, Madrid, and then that done, in 1500, a mature man of nearly 30, friend of the Humanists, an intellectual who has reached some junction of the spirit and accomplishment, and perhaps sees it as his mission to reform German art, away from the barbarisms and savagery, to something more sublime. As to his values, as he is one of those artists whose important works are unlikely to leave their galleries, so any suggestion would only be guesswork, but it would certainly be very high indeed.

DYCE, WILLIAM (1806–64)

Born in Aberdeen, he studied at the Royal Academies of Edinburgh and London. Influences on him included the Nazarenes and Pre-Raphaelites. **In 1837 he published a pamphlet on art education, which led to him being placed as Head of the Government schools of design throughout the country.** He produced for the Houses of Parliament a series of frescoes based on the story of King Arthur.

EAKINS, THOMAS (1844–1916)

Born at Philadelphia he studied at the Penn-sylvania Academy of Fine Arts and later in Paris at the École des Beaux-Arts under Jean Gerome and Léon Bonnat. He ranks high in the American school of Realism as a painter and sculptor. Paintings include 'Turning Stake Boat' in the Cleveland Museum of Art, 'The Biglen Brothers Racing' in the National Gallery, Washington, DC; sculpture is well seen with his horses on the 'Soldiers and Sailors Monument', Brooklyn.

EARLE, RALPH (1751–1801)

American historical and portrait-painter. In 1775 he painted four scenes of the Battle of Lexington, which are understood to be **the first historical pictures done by an American artist**.

EASTLAKE, SIR CHARLES LOCK (1793–1865)

Born in Plymouth and died in Pisa, Italy. He studied first at Plympton Grammar School, where an earlier pupil had been Sir Joshua Reynolds; afterwards he was one of the first pupils of Prout and then with Haydon. He was elected an Associate of the Royal Academy in 1827 and a full Member in 1830. In 1841 he was appointed Secretary to the Royal Commission for decorating the Houses of Parliament and for the promotion of the Fine Arts. **In 1850 he was elected President of the Royal Academy and Director of the National Gallery in 1855.** During his tenure he was instrumental in obtaining many of its finest paintings. He wrote an **excellent textbook** *Materials for a History of Oil-Painting.* In 1865 he left England for his annual acquisition tour for the National Gallery and after being taken ill in Milan died at Pisa and was buried in the English Cemetery in Florence, later to be reinterred at Kensal Green.

ECKERSBERG, CHRISTOFFER-WILHELM (1783–1853)

Danish painter who studied first with Nikolaj Abilgaard and later with Jacques Louis David. He was on close terms with the sculptor Bertel Thorvaldsen and painted the portrait of him which is in the Academy of Fine Arts, Copenhagen. **He is considered the moving spirit behind the Danish modern school.**

ELSHEIMER, ADAM (1578–1610)

Born in Frankfurt and died in Rome, he was in the studio of Philipp Uffenbach, but soon outshone his master. He travelled to Italy and developed a manner with landscapes which are based on biblical or mythological subjects, and were carried out with a notably successful handling of light.

ENGLEHEART, GEORGE (1750–1829)

Born in Kew, he studied first with George Barret

'The Visitation' by Sir Jacob Epstein, bronze. (Courtesy Tate Gallery)

and then Sir Joshua Reynolds, many of whose works he copied in miniature. He was an outstanding performer in this way, and **left behind a Fee Book which gives the names of all those he painted between 1775 and 1813, some 4853 sitters**. His brother Francis (1775–1849) was a line-engraver having learnt the skill from Joseph Collyer and as an assistant to James Heath. A second brother was Thomas (1745–86) who was renowned for his sculptures in wax which were often shown with the Royal Academy. He also worked with Flaxman for Josiah Wedgwood. He had two sons, Timothy and Jonathan, who were both engravers. Henry (1801–85) was the son of George and was good with water-colour and architectural drawings. John Cox Dillman (1783–1862) was a nephew of George and also his pupil. Timothy Stansfield (1803–79) was the son of Francis and was a line-engraver.

ENSOR, JAMES (1860–1949)
Born in Ostend the son of an English father and a Flemish mother, he retained his British nationality until 1930 and then he became a Belgian. He attended the Brussels Academy, and at first confined himself to stormy dark seascapes and landscapes. Then his imagination expanded and drew into his compositions weird, mystical and often macabre subjects. One of his most remarkable pictures is 'The Tribulation of St Anthony' which was in the Cologne Museum but was banned by the Nazis; today it is in the Museum of Modern Art, New York.
'Fruits, fleurs, lumières effeuillés' FB 1 300000, 1975, Galerie Campo, Antwerp.

EPSTEIN, SIR JACOB (1880–1959)
American sculptor of Russian-Polish background; he studied at the Art Students' League, New York, and then at the École des Beaux-Arts in Paris. In 1905 he settled in London. When working in bronze he had the capability to evoke not just the likeness of his sitter but seemingly caught the personality too. In another vein some of his stone-carvings of mythological inspiration drew down on his head biting criticism. Best known of his works are 'Rima' commemorating W H Hudson which is in Hyde Park; figures of Night and Day on St James's Park underground Station and the 'Oscar Wilde Memorial' in Paris.

ERHART, GREGOR (*c* 1469–1540)
The Swabian sculptor who worked in wood and stone. He progressed from the Late Gothic manner of Multscher to the Renaissance style. There are a number of fine works by him extant including: the well-preserved altarpiece in wood with original polychrome at Blaubeuren Convent

Church, near Ulm; this includes five impressive figures, St John, St John the Baptist, St Benedict and St Scholastica, who surround the Madonna.

Another celebrated work he did was the 'Kaisheim Madonna', and according to records he was assisted in some of his works by Hans Holbein the Elder; in the Louvre is his 'Mary Magdalen', known as La Belle Allemande.

ERNST, MAX (1891–1976)

Born at Bruhl, near Cologne. A philosophy student at first, it was not until 1913 he became interested in art. A Dadaist show he put on in Cologne was short-lived as it was closed by the police. He went to Paris just after the First World War and was there linked with the moving spirit behind Surrealism, André Breton. Ernst experimented with collage, and then went on to use frottage and this technique had a formative effect on his work. The Nazis black-listed him in Germany and this with other influences darkened the mood of the artist. **His works have supernatural imagery of considerable force and mind-penetrating energy.** At the beginning of the last war in 1940 he was flung into a concentration camp in France, but escaped and fled to America, where he settled.

'La Forêt' (oil and frottage) $120000, 1975, Sotheby Parke Bernet, New York.

ETTY, WILLIAM (1787–1849)

Born in York, the son of a miller 'like Rembrandt and Constable' he liked to remark; he was apprenticed for seven years to Robert Peck a letter-press printer at Hull. But he had 'such a busy desire to be a painter – I counted the years, days, weeks, hours till liberty should break my chains and set my struggling spirit free'. Assisted by his uncle he broke free and after a series of introductions met John Opie and Henry Fuseli and thus to the Royal Academy Schools, followed by a stint in the studio of Sir Thomas Lawrence. **William Etty was one of the most outstanding anatomy-painters England has produced**, his subjects being treated with an honesty and purity. The small picture 'The Bather' in the Tate Gallery displays his attractive free treatment. He seems to be a painter who has yet to reach his full appreciation, certainly in the sale-rooms his values seem to have remained constant at about £500 since his own time.

EVERDINGEN, ALLART VAN (1621–75)

Born at Alkmaar, Holland, he studied first with Roeland Savery and Pieter Molyn. He travelled through Scandinavia and brought to Dutch landscape-painting the contrasting ruggedness of the mountainous and wild scenery.

EWORTH, HANS (c 1520–c 73)

The Flemish painter from Antwerp who came to England about 1550 and is known for his portraits such as *Queen Mary*, Society of Antiquaries.

FABRITIUS, CAREL (1622–54)

Born at Delft, he was a **pupil of Rembrandt**. He had considerable talent and those works existing show great quality. Sadly he was killed when the powder-magazine in Delft blew up in 1654, and a number of his works went with him. **He would have been one of the stylistic influences on Vermeer.**

FADRUSZ, JOHN (1858–1903)

Hungarian sculptor who worked in the studio of Vilstor Tilgner in Vienna. He had a fine expressive manner and notable works include: the 'Mozart Monument' in Vienna and equestrian statues of King Mathias I and Queen Maria Theresa.

FAED BROTHERS

John (1820–1902) and **Thomas** (1826–1900) both were born in Kirkcudbright, Scotland. John was self-taught and was practising as a miniaturist in Edinburgh in 1841. He went to London in 1862 and had some success, remaining there for 18 years, he then returned to his native place. Thomas joined John in Edinburgh when their father died, and at 15 he entered the Art School there. He became an Associate of the Scottish Academy at 23, and then followed John to London. Here he had considerable fortune with pictures that caught

'Phlox Blancs dans un Vase' by Fantin-Latour. (Courtesy Christie's)

the fancy of the Victorian collectors, such as 'Faults on Both Sides', Tate Gallery, London, 'A Wee Bit Fractious', 'Worn Out' and 'In Time of War'. Sadly he so strained his eyes with his minute painting manner that for the last seven years of his life he was blind. In the family there was also a younger brother James who was a line-engraver.

FALCONET, ÉTIENNE MAURICE
(1716–91)
French sculptor who worked in the studio of Lemoyne. To a degree he echoed the exquisite grace of the figures by the painter Fragonard, although his versatility encompassed a huge equestrian bronze of Peter the Great as well as modelling for Sèvres china. Unhappily many of his works were destroyed in the savagery of the Revolution.

FANTIN-LATOUR, IGNACE HENRI JEAN THÉODORE (1836–1904)
French portrait- and flower-painter, who studied with his father and was influenced by Courbet. Above all **he is one of the greatest flower-painters**, capturing the quality of their softness, colour and construction.
'Roses Aimé-Vibert' £36750, 1975, Christie's, London.

FEDOTOV, PAVEL ANDREYEVICH
(1815–52)
Graduated with the 1st Moscow Cadet Corps, and while a Guards Officer studied in evening classes at the Academy of Arts. He retired in 1843 from the army and devoted himself entirely to painting. He worked under Sauerweit and counselled with Bryullov. His work was largely in the vein of the British painters, Frith and Leslie with a feeling of Hogarth in his satirical manner. Pictures such as 'The Major's Courtship' and 'The Young Widow' both in the Tretyakov State Gallery, Moscow, illustrate his style.

FEICHTMAYER FAMILY
Sculptors and stuccoists from Wessobrunn, Bavaria. The two leading members were **Johan Michael** (c 1709–72) and **Joseph Anton** (1696–1770). Their skill epitomizes the Baroque extravagances to be seen in a number of the churches of Bavaria.

FEININGER, LYONEL (1871–1956)
Born in New York of German immigrant musicians, he went to Germany when only 16 to study music, but gave that up and took to painting, thereafter studying in Hamburg and Berlin and later in Paris where he developed a highly personal type of Naturalistic Cubism. He was part of Die Blau Reiter movement. He was invited by Gropius to be **one of the first artist-teachers at the Bauhaus**, where he stayed until it was disbanded by the Nazis. He left Germany for New York in 1937.
'Mellingen II' Sfrs 215000, 1975, Galerie Koller, Zürich.

FELU, CHARLES
He was born in 1820 without arms, but overcoming the disadvantage he taught himself to draw with his right foot and managed so well that he was admitted to the Academy in Antwerp, where he progressed to painting. He made several quite good portraits and also copied a number of great pictures from the past.

FERNÁNDEZ, GREGORIO (c 1576–1636)
Spanish sculptor who worked in Valladolid. He was concerned with extreme Realism on religious subjects in carved and painted wood, abandoning the early manner of much gilding and brash colours; he insisted on representational treatment which can give quite a shock with its almost total reality of suffering.

FERNÁNDEZ NAVARRETE, JUAN
(c 1526–79)
Commonly called El Mudo, from being deaf and dumb, he was born at Logroño, Spain. He worked first under a monk of the Order of St Jerome, and then went to Venice where he was in Titian's studio. He stayed in Italy for 20 years. Philip II heard of the prowess of El Mudo and had him return to Spain to work on the decoration of the Escorial.

FEUERBACH, ANSELM (1829–80)
German historical painter, the son of a well-known archaeologist was born at Spires. He studied at Düsseldorf under Schadow, then Munich with Genelli who introduced him to Classicism; after this to the Antwerp Academy and thus to Paris where he was under Couture. His paintings have a relationship to those of Lord Leighton.

FLAXMAN, JOHN (1755–1826)
English sculptor, who was **modelling terracotta figures at 12** and joined the Royal Academy Schools at 14, later working for Wedgwood for 12 years. He was more successful with relief work than standing figures. He was appointed the **first Professor of Sculpture at the Royal Academy**.

FLORIS, FRANS (c 1516–70)
Flemish painter who was in the studio of Lambert Lombard. he went to Rome with his brother Cornelis, an architect and sculptor. In 1541 he was apparently at the unveiling of Michelangelo's 'Last Judgement' in the Sistine Chapel; the

masterpiece inspired him greatly and on his return to Antwerp he did much to introduce the influence to·Flemish artists; he maintained a large studio with many pupils.

FOLEY, JOHN HENRY (1818–74)
Irish sculptor, he studied at the School of the Royal Dublin Society and the Royal Academy. Among his works are: 'Oliver Goldsmith' outside the front of Trinity College, Dublin; the large 'O'Connell Memorial' in O'Connell Street and the bronze statue of Prince Albert on the 'Albert Memorial', he also did the Asia Group on the same place. Other Irish sculptors working on the memorial were Patrick MacDowell (1799–1870) who did the Europe Group, John Lawlor (1820–1901) and Samuel Ferris Lynn (1834–76) who worked with Foley on the project.

FOUQUET, JEAN (c 1420–c 80)
French miniature-painter and illuminator born at Tours. He was employed by Charles VII and Louis XI. His output was high and varied; portrait of Pope Eugene IV, a 'Book of Hours' for Étienne Chevalier, the 'Virgin and Child' in the Musée des Beaux-Arts, Antwerp – stories have it that the Virgin is a portrait of Agnes Scorel Charles VII's mistress.

FRAGONARD, JEAN HONORÉ (1732–1806)
Born at Grasse he worked first with Chardin and then in the atelier of Boucher, whose manner he was to carry on and add to. After winning the Prix de Rome in 1752 he continued his studies with Carle van Loo. His brush was used in a broad manner and he developed an exquisite light touch, whether used as with 'The Schoolmistress' in the Wallace Collection, London, or 'The Swing' in the same collection. He was it seemed equally adept in the handling of not only oils but also water-colour, pastel, gouache, etching and crayon work.

FRANCESCA, PIERO DELLA (c 1410–92)
Born at Borgo San Sepolcro in Umbria, it is not known whom he would have studied with. It is possible he worked with Domenico Veneziano on the frescoes of San Egidio, Florence. He worked in the service of Sigismondo Pandolfo Malatesta, Duke of Rimini and decorated for him the Chapel of the Relic in San Francesco with a picture showing 'Malatesta kneeling before the enthroned St Sigismund of Burgundy'. Soon after he started the cycle of frescoes on the Legend of the Cross in San Francesco, Arezzo. The frescoes he did for the Duke of Borso in his palace at Ferrara have been destroyed. **He wrote a Treatise on Perspective**, one copy being in the Vatican.

FRANCIS, SAM (1923–)
American painter born in San Mateo, California. **An Action painter who works in a free uninhibited manner, applying colours with sponges, by pouring and dribbling.** His murals can be seen in the Chase Manhattan Bank, New York and in the Sogetsu Floral Art Centre, Tokyo.

FRANCKEN FAMILY
A long line of Flemish artists who worked in the 16th and 17th centuries in Antwerp and France. They included **Nicholas** (c 1520–c 1600) a painter of whom little is known; **Hieronymous** (c 1540–1610) a pupil of Floris and who worked at Fontainebleau; **Frans** (c 1542–1616) also a pupil of Floris; **Ambrose** (1544–1618) also worked at Fontainebleau; **Hieronymous II** (1578–1623) and **Frans II** (1581–1642).

FRIEDRICH, CASPAR DAVID (1774–1840)
German painter who studied in Copenhagen and Dresden. He was a leading figure of the German Romantic school in landscape. A friend of Goethe he had a melancholic outlook. His pictures bring into being rugged coasts, fogs, moonlight, sunrise and sunsets that are shown with crisp details; in the foreground there are generally a few figures scattered about who appear totally static and immersed in contemplation.

FRIEDRICH, JOACHIM-CARL (1904–)
German water-colourist born in Berlin. He studied at the Hochschule für Bildende Künste in Berlin and Munich. He lived for many years in the Tyrol but returned to Berlin in 1954. He works with a limited palette principally composed of yellow, red and blue with at times the addition of violet. His water-colours are broadly painted and capture the atmosphere of his subject, especially those worked high in the ragged fir-topped heights of the Austrian Tyrol.

FRINK, ELIZABETH (1930–)
English sculptress born in Thurlow, Suffolk, she studied at the Guildford School of Art and the Chelsea School of Art. She reacts to the hard surface of bronze, whether it is in the manner of the slim 'Crucifixion' in the Liverpool Metropolitan Cathedral or the animal strength of her 'Wild Boar' at Harlow New Town. Her work pulses with a harsh vigour and a feeling for the media she is using.
'Horse and Rider' (bronze 20 in (508 mm)) high) £1900, 1975, Sotheby Parke Bernet.

FRITH, WILLIAM POWELL (1819–1909)
Born at Aldfield, near Ripon, an English genre-painter who studied at the Royal Academy Schools. A great recorder of the Victorian scene

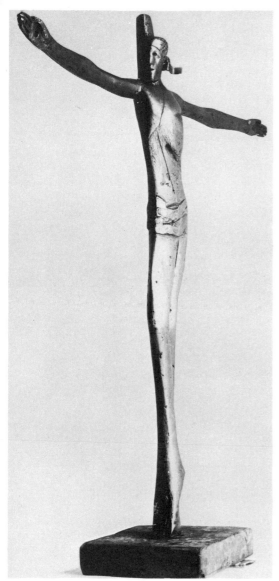

Crucifix on the High Altar, Metropolitan Cathedral, Liverpool. (Courtesy Metropolitan Cathedral Liverpool, Committee)

with such as 'Derby Day' in the National Gallery, London and 'Railway Station' in the Royal Holloway College. His prices have moved around, starting in the mid 19th century in the mid-hundreds, going to thousands in the last quarter of the 19th century and now back in the hundreds.

FUSELI, HENRY (1741–1825)
(Johann Heinrich Füssli) Born in Zürich, died in Putney, London. He had to leave his home town because he exposed a dishonest magistrate. He went to Berlin and there met the British Ambassador who persuaded him to go to England and gave him an introduction to Sir Joshua Reynolds. He travelled to Italy where he stayed for nine years and became a great admirer of Michelangelo. He never really mastered the basics of painting but his pictures succeeded for the sheer force of his imagination. A man of extremely eccentric habits, he was a friend of William Blake and another eccentric Benjamin Haydon. A fine demonstration of his style is 'The Nightmare' in the Kunsthaus, Zürich. Another version of this picture was sold in 1975 for Sfrs 65 000 at the Galerie Koller, Zürich.

GABO, NAUM (1890–)
Born at Briansk, Russia, he took the name Gabo to distinguish him from his brother Antoine Pevsner. Friendship with Die Blaue Reiter aroused his interest and he took to sculpture in 1914. He taught at the Bauhaus. Back in Russia in 1917 he met Tatlin and Malevich and joined with them in experiments. In the 1930s he was in England and **helped Ben Nicholson edit the *Circle*, a book on Constructivist art. He had considerable influence on a number of sculptors, not least Barbara Hepworth.**

GADDI FAMILY
The senior member of this Italian family was **Gaddo** (c 1260–1332) born at Florence and a close friend of Cimabue. Principally a craftsman in mosaic, among the works he did was the 'Coronation of the Virgin with Saints and Angels' in the portal of Santa Maria del Fiore, Florence. His son was **Taddeo** (c 1300–66) who was a pupil of Giotto, whom he worked with for 24 years. Important frescoes by him include scenes from the Life of the Virgin in the Baroncelli Chapel in Santa Croce, Florence. One son of Taddeo was **Agnolo** (c 1335–96) who, when his father died, was put in the studio of Jacopo del Casentino and Giovanni da Milano. He too painted frescoes in Santa Croce, eight on the Legend of the Cross. He had a number of assistants including Cennino Cennini. The other son of Taddeo was Giovanni, a somewhat lesser hand than his brother; frescoes which he did included 'Dispute of Christ with the Scribes in the Temple' in San Spirito, but they were destroyed when the church was rebuilt.

GAINSBOROUGH, THOMAS (1727–88)
Born at Sudbury, at an early age he was in the studio of the French engraver Gravelot, and he had contact with the St Martin's Lane Academy and Francis Hayman. Later he returned to Sudbury where he set up as a portrait-painter. In 1760 he

moved to Bath and had considerable popularity, charging 40 guineas for a half-length and 100 guineas for a whole-length. He was **one of the original members of the Royal Academy**. In 1774 he left Bath for London and set up his studio in Schomberg House in Pall Mall. From 1783 he stopped showing with the Royal Academy because of a dispute over a hanging of the picture of the three Princesses. **Gainsborough was a supreme colourist and was equally at home with portraits or landscapes.** 'Cornard Wood' and 'View of Dedham' are strong in atmosphere and show his influence as having come from some of the Dutch school, pictures by whom he would have seen in homes in East Anglia. With portraits he has an easy grace in handling far removed from the pompous formality of many of his contemporaries. Evidence of his quality is the steady climb of his values up to £130000 in 1960 for 'Mr and Mrs Andrews in a Park'; also in 1972 £280000 was given for his *'Mr & Mrs Gravenor and their daughters Elizabeth and Dorothea'* at Sotheby Parke Bernet.

'The First Marquess of Donegal' by Thomas Gainsborough. Full-length portrait c 1770. (Courtesy Sotheby Parke Bernet)

GALLEGOS FERNANDO (1475–1550)
Born at Salamanca, it has been said he may have studied with Albrecht Dürer; there is certainly a Germanic look with some of his work.

GAUDIER-BRZESKA, HENRI (1891–1915)
He was born at Saint-Jean-de-Draye, France. He secured a scholarship to London and Bristol, the latter sending him to study art in Germany. Then Paris where he started his career as a sculptor. In London he was associated with the Vorticists and showed with the London Group. He was killed while serving in the French Army on the Western Front.

GAUGUIN, PAUL (1848–1903)
Born in Paris and died in the Marquesas Islands. After the Franco-Prussian War he started out with some success in a stockbroker's office in Paris. Painting began as a part-time leisure pursuit, that quickly grew to a means of expression. He was almost entirely self-taught. The year 1886 found him in Pont-Aven, Brittany, and later at Pouldu; for a time he had some peace and with a party of followers he found what was necessary, to paint some of his finer works. In 1887 he took a trip to Martinique and back to an upsetting time with Van Gogh in 1888 in Arles. In 1891 he left France for Tahiti, where, except for one short trip home, he remained until his tortured and tempestuous life burnt itself out. In the last years he painted frantically pictures full of rich colour and barbaric atmosphere. His 'Hina Maruru' brought $275000 in 1965, $950000 in 1975. Sotheby Parke Bernet, New York.

GEERTGEN TOT SINT JANS
(c 1465–95)
A Dutch painter who may have studied with Albert van Outwater. His 'Nativity' in the National Gallery, London has great charm with its simple but utterly sincere rendering of the subject.

GENTILE DA FABRIANO (c 1360–c 1428)
He was probably born in Fabriano and could have been in the studio of Allegretto Nuzzi. During a stay in Venice between 1408 and 1414 he worked on frescoes in the Doge's Palace, which were completed by Pisanello and are now among the host of the missing. Jacopo Bellini studied with him while he was in Venice. Very little of Gentile's work still exists, having suffered considerably from losses. There is the Madonna from the 'Quaratesi Polyptych' in the British Royal Collection and the delightful 'Adoration of the Magi' in the Uffizi, Florence.

GENTILESCHI, ORAZIO (c 1563–c 1647)
(properly Orazio Lomi). Born in Pisa with his half-brother Aurelio Lomi and his uncle Bacci

'The Glorification of the Virgin' by Geertgen tot Sint Jans. (Courtesy A Frequin, Den Haag)

Lomi. In Rome he made friends with Caravaggio and developed a similar but less bold and dramatic handling of light and shade. Later he travelled to Paris and then at the invitation of the Duke of Buckingham came to England where he stayed for the rest of his life receiving handsome commissions from Charles I; there were nine paintings by him in the Royal Collection. 'Joseph and Potiphar's Wife' at Hampton Court gives a good idea of the influence on him of Caravaggio.

GÉRARD, BARON FRANÇOIS-PASCAL (1770–1837)

French historical and portrait-painter born in Rome, and studying with the sculptor Pajou and then with Brenet and finally with Jacques Louis David. He managed to steer clear of involvement in the Revolution and later political intrigues. During Napoleonic times and the Restoration he had much portrait work with varied sitters: Madame Récamier, Napoleon, Talleyrand, the sculptor Canova, Louis XVIII, the Emperor of Russia, the King of Prussia and Charles X.

GERICAULT, JEAN LOUIS ANDRÉ THÉODORE (1791–1824)

French painter of animals, landscape and history of considerable invention and character, **regarded as the leader of the French Romantic movement**. He was born at Rouen, the son of a well-to-do advocate. At the Lycée Impériale he showed clearly his coming talent, during holidays he spent much time studying animals in the circus, and horses pulling the varied transport in the streets. In studios he studied with Vernet and Guérin, and in the Louvre he absorbed the methods of the Masters, notably Rubens. A lively young blood he joined the Jockey Club and when the day's work was over, lived a somewhat dissipated existence. One of his outstanding pictures is the forceful 'Raft of Medusa' in the Louvre, which shows the survivors of the frigate *Medusa* which came to a disastrous end in 1815. His health became weakened from his way of life, and after a nasty fall from a horse he died, sadly so young.

GERTLER, MARK (1892–1939)

Painter of still life and figures, he was born in humble circumstances, nevertheless he managed to attend the Slade School with the help of a scholarship from the Jewish Educational Aid Society. He was a member of the New English Art Club. He committed suicide in 1939.
'Portrait of Natalie Denny' £670, 1975, Sotheby Parke Bernet.

GHIBERTI, LORENZO (1378–1455)

He started off as a goldsmith, then became a painter. In 1401 the Guild of Cloth-Finishers in Florence organized a competition for a pair of bronze doors for the Baptistery; against stiff opposition, including Brunelleschi, Ghiberti won; the completing of this work took him 23 years. In 1425 he was asked to make a second pair of doors, this took him until 1452. Time and to a degree neglect caused considerable darkening, and many had forgotten that the decorations on the doors had been originally gilt. During the Second World War the doors were stored for safety and then cleaned, and again treated after the disastrous flood. Today they stand with the gilding giving them the appearance they would have had in the 15th century.

GHIRLANDAIO, DOMENICO BIGORDI (c 1449–94)

His full name seems to have been Domenico di Tommaso Curradi di Doffo Bigordi. In the Florentine dialect Ghirlandaio comes out as Garland-Maker. He worked in the studio of Alessio Baldovinetti, and was influenced by Verrocchio and Masaccio. In 1485 he painted the frescoes in the Sassetti Chapel in Santa Trinità, Florence, and after this Giovanni Tornabuoni, a banking associate of the Medici, commissioned him to decorate the choir of Santa Maria Novella.

'Diego, ou l'homme au blouson' by Alberto Giacometti. Bronze, signed. Height: 22 in (559 mm) Executed in 1954. (Courtesy Sotheby Parke Bernet)

There is the painting 'Old Man and Child' in the Louvre, which has a touching sincere charm and timeless quality. **His most famous pupil was Michelangelo.**

GIACOMETTI, ALBERTO (1901–66)

This Swiss painter and sculptor was born in Stampa in the Bregaglia Valley; his father, Giovanni, was a painter following the lines of the Impressionists. Alberto studied at the École des Arts et Métiers, Geneva; after a short trip to Italy he was in Paris in the studio of Bourdelle. His figures have a strange elongated appearance, slender and rough textured.

'Femme, épaule cassée' (dark brown patinated bronze) $60000, 1975, Sotheby Parke Bernet, New York.

GILL, ARTHUR ERIC ROWTON (1882–1940)

He studied at the Central School of Arts and Crafts and from 1903 began to earn his living as a designer and cutter of letters. Sculptor, engraver, writer and typographer, despite his talents with the first two, he is perhaps best known for the last, as printing types based on his designs are in common use.

GILLRAY, JAMES (1757–1815)

One of the most famous English caricaturists and satirists. Apart from the fact that he was of Irish descent, little is known, but it is possible he may have been the son of a sexton of the same name at the Moravian Cemetery in Chelsea; he set out as a letter-engraver. Bored with the task, he ran away and joined a party of strolling players, and then he somehow got into the Royal Academy Schools. His output 'savaged' not only the English Court but also France. He travelled with De Loutherbourg in 1792 to France to gather material. It is recorded that he had such a sure touch and sense of what he wanted that he could needle or engrave his ideas straight on to the copper plate, without making previous drawings. Celebrated subjects include: 'A New Way to pay the National Debt', 'The Middlesex Election', 'The Loyal Toast' and 'Fatigues of the Campaign in Flanders'. Late in life he went insane.

'He steers his flight' (two coloured aquatints) £190, 1975, Sotheby Parke Bernet.

GILPIN, SAWREY (1733–1807)

Born in Carlisle the son of an Army Captain who gave him a little instruction. In London he worked for some time with a ship-painter. Under the patronage of the Duke of Cumberland he started to paint animals and soon became proficient, doing the likenesses of His Grace's racehorses. At times he worked with Barret, he would do the horses in Barret's pictures, and Barret would do the landscapes for Gilpin. His son William Sawrey (1762–1843) was a water-colourist of quality and became the first President of the Water Colour Society. William, born in 1724, was Sawrey's brother, and he became a vicar with a fair hand as a landscape-draughtsman.

GIORDANO, LUCA (1632–1705)

The Italian painter was born in Naples the son of an obscure artist, whom apparently he surpassed by the time he was eight years old, this accomplishment came to the notice of the Viceroy of Naples who placed him in the studio of Ribera. Then he travelled to Rome and studied with Pietro da Cortona. Giordano began to be in demand for paintings and sketches whereupon his father came to Rome and lived well from his son's labours. Luca was called on for murals and ceilings; those in the Escorial, Madrid and the Palazzo Ricordi in Florence evidence his skill.

GIORGIONE DEL CASTELFRANCO
(1475–1510)
Also known as Big George and as Zorzo da Castelfranco, he was born in the old city of the Trevisan March from which he takes his name. He went to Venice and there entered the studio of Giovanni Bellini. His art advanced swiftly and it is recorded by 1500, when he was only 23, he was commissioned to portray the Doge and one of the better known *condottiere* Consolvo Ferrante. With subjects such as the 'Concert Champêtre', in the Louvre, there is an affinity with Manet's 'Le Déjeuner sur l'Herbe' in the same museum. In his lifetime he enjoyed enormous popularity and he was imitated by many of the Venetian painters.

GIOTTO (c 1267–1337)
(properly Giotto di Bondone). **The painter who was to do more than any other to found the Florentine school** was born at Colle, in the Commune of Vespignano, a few miles north of Florence. Records of his life are sparse. Legend has it that Cimabue discovered him as a shepherd boy making a drawing of one of his father's sheep on a rock. Giotto's power lies with his deep sincerity and innocence; he appears uninfluenced by fashion or others around him. His largest work was the series of frescoes in the Arena Chapel in Padua, executed in 1303; it shows the Life of the Virgin and the Life of Christ in 38 scenes. In the Uffizi, Florence there is his 'Virgin and Child with Angels', a masterpiece of simplicity and a model for countless similar Madonnas to come. One of his last works was as an architect; he designed the graceful Campanile which stands beside the Duomo in Florence, the lower reliefs of the Creation of Man and his occupations were probably worked from his designs by Andrea Pisano. **The painter's numerous pupils were known as Giotteschi.**

GIRARDON, FRANÇOIS (1628–1715)
French sculptor, a pupil of François Anguier, who worked under Charles Le Brun in the Tuileries and at Versailles. His largest work was probably a huge bronze of Louis XIV in costume as a Roman, but this met with destruction at the time of the Revolution; In the Church of the Sorbonne is the Tomb of Richelieu also by Girardon.

GIRODET-DE-ROUSSY-TRIOSON, ANNE-LOUIS (1767–1824)
French historical painter and writer, he was born at Montargis. An early orphan he was adopted by an army surgeon. First tuition was from Luquet, he then joined the studio of Jacques Louis David.

He lacked power of colour, but his modelling and drawing were excellent.

GIRTIN, THOMAS (1775–1802)
English water-colourist who was born in Southwark, he was first apprenticed to Edward Dayes, and then he was set to colouring prints for John Raphael Smith. He had a period sketching together with the young Turner, also copying works from Dr Monro's collection. As with Turner, he travelled much in Britain, picking his subjects around the kingdom. **His techniques advanced the manner of water-colour considerably in strength and style.** Turner so admired his work that he once said, 'If Tom Girtin had lived, I should have starved.'
'Caernarvon Castle' (water-colour and pencil, 12 × 20 in (304 × 508 mm)) £6090, 1975, Christie's, London.

GLACKENS, WILLIAM JAMES (1870–1938)
American painter, a member of the Group of Eight, a student of the Pennsylvania Academy of Fine Arts. He did much illustration work for periodicals and after a short period in Paris brought back Impressionist influence.

GLEIZES, ALBERT (1881–1953)
Born in Paris, he commenced painting in the Impressionist manner but then passed to Cubism and was associated with the first Cubist show at the Salon des Indépendants. His manner is one using flat geometric planes.

GOES, HUGO VAN DER (c 1440–82)
He was probably born at Ghent, but there are few details about his early life and his master has been lost sight of; he was known to have been admitted to the Guild of St Luke in 1467. **He stands as one of the great masters of his period**; his finest work is the so-called Portinari alterpiece 'The Adoration of the Shepherds', now in the Uffizi; this he painted for Tommaso Portinari, a Medici agent at Bruges.

GONÇALVES, NUÑO (working 1450–70)
Court Painter to King Alfonso of Portugal. He may have studied in Italy and there is also evidence of Flemish influence in his work. Fine works include 'The Veneration of St Vincent' in the Museu de Arte Antiga, Lisbon and 'Christ at the Pillar' in the Convent of the Trinity in the same city.

GONTCHAROVA, NATALIA (1881–1963)
Born in the Russian province of Tula, she studied in Moscow and was exhibiting by 1900. She was associated with Rayonism through Larionov. Latterly she was heavily committed to the Ballets Russes, designing décors and costumes for *Le Tsar Saltan*, *Le Coq d'Or* and *Noces*.

'St Vincent' by Nuño Gonçalves. The saint is usually shown as a beautiful young man robed as a deacon. The figure to the saint's right with a round black hat is Prince Henry the Navigator. (Courtesy Portuguese State Tourist Office)

'Portrait of the Infante Don Luis de Bourbon' by Goya. Painted c 1783–84. (Courtesy The Cleveland Museum of Art)

GOSSAERT, JAN (c 1478–1533)

(known as Mabuse). Born near Utrecht, his early life is not recorded other than he was admitted as a Free Master to the Guild of St Luke at Antwerp in 1503. His paintings were mostly of religious and mythological subjects, although he did a number of portraits after travelling to Italy. Influences on him include Albrecht Dürer, evidence of this can be noted in his 'Adoration of the Kings' in the National Gallery, London; the little dog in the right-hand bottom corner is a lift from work by Dürer.

GOTTLIEB, ADOLPH (1903–   )

Born in New York, he studied at the Parsons School of Design there and then the Art Students' League. His manner is to use large masses of pungent colours beside each other with a stimulating effect.

'Trajectory' $9500, 1975, Sotheby Parke Bernet, New York.

GOUJON, JEAN (c 1510–c 66)

French sculptor who worked under Pierre Lescot, the architect of the Louvre. His work includes the Fountain of Diana at the Château of Anet, and decorations in stucco for Fontainebleau.

GOYA, Y LUCIENTES, FRANCISCO JOSÉ DE (1746–1828)

The great Spanish painter and graphic artist was born at Fuentetodos in Aragon. He worked first in the studio of José Luzán y Martinez in Saragossa. Unfortunately he spent more time fighting in the streets and on amorous pursuits than painting. After one incident he fled to Madrid, where he worked for Francisco Bayeu, whose sister he married, and settled in the city. As an artist he is a man of many parts. There is the splendid portrait of Doña Isabel Cobos de Porcel, and then there are the somewhat satirical Royal portraits, the delightful 'Nude Maja', the savagery of 'The Firing Party, 3rd May 1808', with his supernatural side epitomized by 'Witches Sabbat', there are paintings from bull-fights, and the large number of aquatints, etchings and lithographs which include his brutal assessments in 'The Disasters of War' with 65 etchings. Latterly he dropped from his palette the bright rich colours and his work became at times almost monochrome.

GOYEN, JAN VAN (1596–1656)
Born at Leyden, he studied with Jan Nicolai and later worked with Esaias van de Velde. An early Dutch landscape-painter, the father-in-law of Jan Steen.
'Winter landscape with Skaters' £63 000, 1975, Christie's, London.

GOZZOLI, BENOZZO (1421–97)
(also known as Benozzo di Lese). The son of a doublet-maker, he was born in Florence and for a time worked with Ghiberti. Later he was the leading assistant to Fra Angelico in Rome and Orvieto. After he had left this master his skills and talents flowered. Certainly for most **his greatest achievement was the decoration of the walls of the Chapel in the Palazzo Riccardi**, the theme being the Journey of the Magi. The chapel is quite small, 20 × 25 ft (6 × 7·62 m), and here on the walls is set out one of the most interesting pictorial statements, even if it has been slightly chopped around by vandalistic builders, an extra door cut through and a corner remodelled, but nevertheless it is largely as Gozzoli left it. Incorporated in the composition is a youthful portrait of Lorenzo the Magnificent.

GRAF, URS (c 1485–c 1528)
The Swiss painter and craftsman, born at Solothurn, he started out working for a goldsmith in Zürich. He was a wild and reckless cháracter and served as a mercenary in the war in Lombardy. He was a skilled wood-engraver and woodcutter, leaving behind some 327 blocks. For some he is best recalled for his drawings and prints of fantastically garbed and armoured soldiers with splendid moustaches and beards, well-muscled calves and massive weapons.
'The Sorrowing Virgin and St John standing before Christ on the Cross' (woodcut $9\frac{5}{8} \times 8\frac{1}{2}$ in (244 × 215 mm)) £9450, 1975, Christie's, London.

GRAFF, ANTON (1736–1813)
Born at Winterthur in Switzerland, the portrait-painter to be who studied with Johann Ulrich Schellenberg in his home town. Later he left for Augsburg and then Dresden, where he was appointed Court Painter in 1766. His portraits include: 'Frederic Augustus of Saxony', 'William II', 'Schiller', 'Gluck' and 'Herder'.

GRANT, DUNCAN (1885–)
Born at Rothiemurchus, Inverness, Scotland, he studied at the Slade School and later in Italy and Paris. A member of the close-knit Bloomsbury circle, he showed with the New English Art Club and the London Group. Influences on him include les Fauves and Cézanne.

GRECO, EL (1541–1614)
(more properly Domenikos Theotokopoulos). Born in Crete, he died in Toledo; he had primary tuition from monks on the island, but there are no pictures existing that can be placed at this time. The miniaturist Giulio Clovio writes that when he went to Venice El Greco was a pupil of Titian, yet in examining his work there are more signs of influences from Tintoretto and possibly Bassano and Michelangelo. In 1570 Clovio apparently gained him an introduction to Cardinal Farnese and he went to Rome and spent six years there. By 1577 he had left Italy and was established in Toledo; it is likely he hoped for the patronage of

'The Annunciation' by El Greco. (Courtesy The Toledo Museum of Art, Ohio)

Above: 'The Religious Procession in Kursk district' by Repin. (Courtesy Novosti Press Agency)

'Lenin receives peasant delegates' by Serov. (Courtesy Novosti Press Agency)

Philip II, but it was not forthcoming in the manner he might have expected, as the Monarch was not too impressed with what he saw, so the painter did not set up his studio in Madrid.

El Greco as a painter is almost unique in that his pictures have that rare quality of being visually timeless. His style could have been worked today or a hundred years ago. To examine 'View of Toledo' in the Metropolitan Museum of Art, New York, is to almost enter the landscape as El Greco must have viewed it nearly 400 years ago. The whole treatment is so fresh, the choice of cool and clean colours, the broad brooding sky, the handling of the distant town on the hill. But generally the painter reserved his creative power for religious subjects. In the London National Gallery is 'The Agony in the Garden', an amazing composition of masses and swirling forms and rhythms. Here can be seen the spiritual power which must have been one with the artist in his creation. As to the value a collector would place on his work, this is impossible to judge because pictures by him are unlikely to come on the market. The last evidence was in 1960 with his 'Christ healing the blind' which is now also in the Metropolitan Museum of Art, when the figure was spoken of as being considerably in excess of £100000.

GREENOUGH, HORATIO (1805–52)
An American sculptor who studied with Bertel Thorvaldsen in Rome and later in Florence with Lorenzo Bartolini. **Best-known work is the huge statue of Washington, worked out as Zeus, which is now in the Smithsonian Institution,** Washington, DC. He had a brother Richard Saltonstall Greenough (1819–1904) who was also a sculptor.

GREKOV, MITROFAN BORISOVICH (1882–1934)
From Moscow, he attended the Odessa Art School under Kasatkin and later the Academy of Art where he had the benefit of tuition from Ilya Repin. During the First World War he fought as a volunteer with the Red Army cavalry. He is best known as a painter of battle pictures.

GREUZE, JEAN-BAPTISTE (1725–1805)
Born at Tournus, he studied art first at Lyons and then in Paris. A genre-painter, he had considerable success in his early period; this started with 'A Father explaining the Bible to his Children' which was in the 1775 Salon. But as time went on he quarrelled with the Academy, and with the Revolution he found his market was shrinking as the Classicism of David triumphed. Greuze's handling of children was his forte; with his compositions of young ladies they appear too

strongly overlaid with sentimentality and a trace of insincerity.
'Sainte-Marie l'Egyptienne' F 5800, 1975, Hôtel George V, Paris.

GRIS, JUAN (1887–1927)
(real name José Gonzales). He was born in Madrid and died in Boulogne-sur-Seine. His studies started at the School of Arts and Crafts in Madrid, and later were continued in the studio of an elderly academic painter Moreno Carbonero. Tiring of what he felt was an outdated approach, he sold everything he had and went to Paris; Here he met Picasso and became strongly influenced by his fellow countryman. His manner that emerged was a Cubist treatment largely of still lifes.
'Bol et paquet de cigarettes' $53000, 1975, Sotheby Parke Bernet, New York.

GROGAN, NATHANIEL (c 1740–1807)
Irish landscape-painter from Cork, he was first a wood-turner and then served in the army during the American War. On his return to Cork he set out to make a living teaching drawing and painting. His best-known works are 'An Irish Fair' and 'An Irish Wake'; his style was broad and free and had a touch of caricature, perhaps best seen in his small figure studies of old fisherwomen or soldiers.

GROMAIRE, MARCEL (1892–1971)
He was born at Noyelle-sur-Sambre, and at first studied law in Paris and then went to art classes in Montparnasse. He was wounded in 1916 on the Somme. After the war he kept himself aloof from groups and theories. His colouring tends to be somewhat heavy and leans on the earth pigments, whether in landscapes or figure studies.
'Paysage aux trois clochers' F 70000, 1975, Hôtel du Barry, Versailles.

GROS, BARON ANTOINE JEAN (1771–1835)
French painter born in Paris, who studied first with his father who was a miniaturist, and then with Jacques Louis David. He found a friend in Josephine who introduced him to Napoleon, whose Official Artist he became. A series of battle pictures started with 'Bonaparte at the Bridge of Arcola', then 'Napoleon on the Battlefield of Eylau' and the shock realism of 'The Plague-stricken at Jaffa', all three of which are in the Louvre.

GROSZ, GEORGE (1893–1959)
German satirist, who with brittle pen and ink line drawings tore into the warmongers, war-profiteers, the night-club scene of the 1920s, and vice in general and pointed to the plight of the little man. His work was held up to scorn by the

Nazis in their Degenerate Art Exhibition. He emigrated to America in 1933, where some of the bite came out of his style.

'Der apokalyptische Reiter', Sfrs 28 000, 1975, Kornfeld und Klipstein, Berne.

GRÜNEWALD, MATTHIAS (c 1460–1528)

Born in Würzburg, it is likely that he was taught in Alsace. His major accomplishment is undoubtedly the Isenheim altarpiece, which is now in the Unterlinden Monastery, Colmar; this multi-panelled work, comprising the Annunciation, Nativity, Angels, St Anthony and St Paul in the Desert culminates in the Crucifixion, the figure of Our Lord displayed on the Cross with total and unremitting realism. This ability for painting the bleak and harsh facts can be seen again in 'Christ Mocked' in the Pinakothek, Munich. But in another mood Grünewald shows a gentle and sensitive nature with the delightful 'Madonna' in Stuppach Church.

GUARDI, FRANCESCO (1712–93)

Born in Venice, his father was a minor painter Domenico Guardi. He was strongly influenced by Canaletto, although his pictures were more free in style, perhaps the architectural perspective may not have been quite so accurate, but the over-all effect had a greater sense of life and atmosphere.

'The Rialto Bridge, Venice' £27 300, 1975, Christie's, London.

GUERCINO, IL (1591–1666)

(more properly Giovanni, Francesco, Barbieri). He was born in Cento, Ferrara, and received his nickname Guercino because of a pronounced squint. At a very early age he was adept with brushes, painting a figure of the Virgin on the side of his father's house before he was ten. He worked in the studio of a number of painters but to a high degree taught himself. His styles changed several times during his painting life, but they all retained a somewhat theatrical use of light and shade and also posing.

HALS, FRANS (c 1580–1666)

Born at Antwerp and attended the studio of Karel van Mander. **Frans was to become next to Rembrandt the greatest portrait-painter the Dutch school produced.** His manner had great virility and verve; examined closely it can be seen how broad much of his treatment was, often only two or three strokes being sufficient to give the impression of a finger, when viewed from a few yards away. Stories that have been circulated about him being a wife-beater and a drunkard seem to be scurrilous and without foundation; he may have frequented taverns to enjoy company but a close study of him and his works does not uphold the rumours above. His best-known work is probably 'The Laughing Cavalier' in the Wallace Collection, London, but it is just one in a series of lively portraits that he left for posterity. In the 18th and 19th centuries his prices seem ludicrously low. In 1774 'The Laughing Cavalier' was £15. In 1865 £2040. In 1855 the 'Lady in Black Dress' £2 15s. In 1885 'A man in black' plus a 'still life' was £5 5s, in 1913 the portrait alone went for £9000. In the 20th century the appreciation of the great painter increases; in 1960 'An unknown Cavalier' brought £182 000, 75 years before that it had gone for £15.

Frans's brother **Dirck** (1591–1656) was a pupil of Abraham Bloemaert, and painted 'conversation pieces', soldiers, cavaliers, ladies, making music, dancing and eating. Frans had five sons: Herman the elder was a 'conversation-piece' painter; Johannes who worked on genre figures; Frans the Younger painter of portraits and still life, he also copied some works of his father; Nicolas painter of peasant scenes and landscapes; and Reynier a mediocre artist working on genre, some of his works have been passed off under the name of his uncle, Dirck.

HAMILTON, HUGH DOUGLAS (1734–1806)

Born in Dublin and studied with James Mannin. He was to develop an individual manner with pastel portraits that became popular in both Dublin and London. Later on the advice of Flaxman he turned to oils, and became more ambitious with subjects in the vein of 'Cupid and Psyche in the Nuptial Bower' in the National Gallery, Dublin.

HAMMERSHØJ, VILHELM (1864–1916)

Danish artist who worked principally on interiors carried out with gentle colour schemes, and when people were included there was an air of sensitive atmosphere.

HANLY, PATRICK (1932–)

Born in Palmerston North in New Zealand, he started out as a hairdresser by day and studying painting at the local Technical College by night. In 1952 he was at the Canterbury School of Fine Arts, and later the Chelsea School of Art. He travelled to Spain, Holland and again to England and then he returned to New Zealand and settled in Auckland. His style has embraced many influences, the strong colour from Die Brücke, the near Abstract and figure work slightly reminiscent of Bacon down to simplified symbolism.

HARPIGNIES, HENRI (1819–1916)

French landscape-painter strongly influenced by the Barbizon school and especially Corot. Some of his best works are those painted in the Valley of the Loire.

'The Jolly Toper' by Frans Hals. (Courtesy Rijksmuseum, Amsterdam)

'Juan de Pareja' by Velázquez. (Courtesy Christie's)

Below: 'Annunciation' by Leonardo da Vinci. (Courtesy Uffizi, Florence)

'Tuscany I 1977' by Brett Whitely. Born in 1939 in Sydney, Australia. He studied at the Julian Ashton School (83 × 101·9 in (2109 × 2590 mm)). (Courtesy Fischer Fine Art Ltd)

'Le bosquet près de la rivière' F 12000, 1975, Hôtel Rameau, Versailles.

HARTLEY, JONATHAN SCOTT
(1845–1912)
American sculptor who was well known for his portrait busts of such as George Innes and John Gilbert.

HASSAM, CHILDE (1859–1935)
American painter who attended the Boston Art School and then later in Paris worked in the studio of Boulanger. Back in America again he worked on such subjects as 'The Strawberry Tea Set', 'Church at Gloucester' and 'The New York Window'. He also made a number of etchings and lithographs.
'Bridge at Posilippo, Naples' $29000, 1975, Sotheby Parke Bernet, New York.

HAYDON, BENJAMIN ROBERT
(1786–1846)
Born in Plymouth he studied at the Royal Academy Schools under Fuseli. He became a painter of historical subjects and at times satire. He was a strange character, liable to uncontrollable outbursts, fits of depression, imagined misfortunes. At last driven to despair took his own life. Two of his best compositions are 'Chairing the Member' and 'Punch on Mayday', both in the Tate Gallery.

HAYMAN, FRANCIS (1708–76)
Historical and portrait-painter, born at Exeter, studied with Robert Brown. Much of his fame rests on the decorations he carried out for the Vauxhall Gardens. **He was a founder-member of the Royal Academy** and its first Librarian.

HECKEL, ERICH (1883–1970)
Painter and engraver born at Döbelin, Germany. He met Kirchner and Schmidt-Rottluff and with them founded the Brücke. 'A lock-gate' DM 90000, 1975, Hauswedell & Nolte, Hamburg.

HEDA, WILLEM CLAESZ (1594–1680)
Born at Haarlem. He specialized in still life, and few could better him at showing, glass, metals, ceramics and various foods.

HEEM, JAN DAVIDSZ DE (1606–*c* 83)
Dutch flower-painter, the pupil of his father
David de Heem the Elder (*c* 1570–*c* 1632). His
compositions had great finish and were in much
demand by the flower-loving Dutch.

**HEEMSKERCK, MAERTEN JACOBZ
VAN** (1498–1574)
Born at Heemskerck near Haarlem, he studied
with Cornelisz Willemz and Jan Scorel. He was
influenced by the Italian, both by visits to Rome
and by examining the works of Masters. The
Fitzwilliam Museum, Cambridge, England, has
an unusual self-portrait in which Heemskerck
has placed a view of the Colosseum in the
background.

HELST, BARTHOLOMEUS VAN DER
(1613–70)
Born at Haarlem, he was a pupil of Nicolaes Elias.
Principally he was a portrait- and group-painter in
the manner of Frans Hals and Rembrandt.

HENRI, ROBERT (1865–1929)
He studied at the Pennsylvania Academy of Fine

Art, and later in Paris at the École des Beaux-Arts
and the Académie Julian; he was associated with
Bouguereau. He taught at the Art Students'
League in New York. **In his book *The Art Spirit*
he gave some extremely sound advice on
aesthetics and the finer points of picture-
making.**

HENRY, PAUL (1877–1958)
Born in Northern Ireland, he worked in
Whistler's studio in Paris; returning to Ireland he
settled in Connemara. No painter has succeeded
quite so impressively in capturing the atmosphere,
colour and texture of the west of Ireland; he hits
off the strange and beautiful light only to be found
out there. His wife Grace also painted, although
not attaining the quality of her husband; a
portrait by her of Paul is in the Ulster Museum.

HEPWORTH, DAME BARBARA
(1903–75)
Born in Wakefield, Yorkshire, she was trained at
the Leeds School of Art and the Royal College of
Art, London. She was married to Ben Nicholson

'The Boy and the Rainbow' by Robert Henri. (Courtesy The Detroit Institute of Arts)

'La Pêche miraculeuse' by Conrad Witz. (Courtesy Musée d'Art et d'Histoire, Geneva)

'Left-Left
We left our name
On the Road
On the Road
On the Famous Road
On the Famous Road
On the Famous Road
Of Fame'
by Jack B Yeats.
(Courtesy National
Gallery of Ireland)

until 1951. In 1939 she went to St Ives and there she lived and worked, and helped to encourage a local art group. In 1953 she won the second award in The Unknown Political Prisoner Competition and the Grand Prix at the 5th São Paulo Biennale in 1959. One of her outstanding bronzes is the 21 ft (6·40 m) high free-form abstraction 'Single Form' which stands on a granite plinth in the pool in front of the United Nations Building in New York; it stands in memory of Dag Hammarskjöld. Barbara in her work was deeply concerned with surface finish, whether of stone, wood or bronze and with this a never-ending exploration into form and rhythmic relationship with the space it stood in.

HERRERA, FRANCISCO DE, THE ELDER (1576–1656)

Born at Seville and studied under Luis Fernandez and worked in fresco; also etched and designed medals. **The latter occupation turned him to forgery and coining fake money.** On being discovered, he took refuge in the Jesuits' College; while in hiding he painted 'St Hermengild in Glory' which secured him a pardon from Philip IV. He was a violent and ill-tempered man to such an extent that all his children ran away, his son robbing him before going to Rome and his daughter becoming a nun.

HERRERA, FRANCISCO DE, THE YOUNGER (1622–85)

Born in Seville and went to Rome to escape his father. He studied architecture and perspective and painted numbers of still life that brought him considerable work; his showing of fish was so good that he was called in Italy 'il Spagnuolo degli pesci'. On his father's death he returned to Seville, and later moved to Madrid and was appointed as Painter to Charles II.

HESSELIUS, GUSTAVUS (1682–1755)

He is claimed to be the first portrait-painter in America; of Swedish descent it is likely he was assisted in a long series of portraits by his son John. There is a 'Last Supper' by him in a Maryland church and the Pennsylvania Historical Society has a portrait of him and his wife.

HICKS, EDWARD (1780–1849)

The most celebrated of the early American Primitive painters. He was a travelling Quaker preacher and a sign-painter. His most famous picture is 'The Peaceable Kingdom'; it was based on Isaiah 11, about the savage animals lying down peaceably with farm and domestic stock.

HIGHMORE, JOSEPH (1692–1780)

Born in London, the son of a coal-merchant, he was a pupil of Sir Godfrey Kneller. He worked

'George Clifford, 3rd Earl of Cumberland' by Nicholas Hilliard c 1590. This miniature measures 10⅛ × 7 in (257 × 177·8 mm) which is a large size for Hilliard; many of his exquisite examples are little more than 2 or 3 in (50 or 80 mm) in height. (Courtesy National Maritime Museum, Greenwich)

with portraits and 'conversation pieces' and became Sergeant-Painter to William III.

HILLIARD, NICHOLAS (c 1547–1619)

Born in Exeter, he worked with a goldsmith and then moved to London to set up as a miniature-painter. His style had an exquisite manner hardly ever surpassed; largely he worked on card, strained chicken skin or sometimes on the backs of playing-cards. He was a great favourite with Queen Elizabeth I, holding the positions of her Goldsmith, Carver and Portrait-Painter. His excellence moved the poet John Donne to make this illusion in his 'The Storm':

'A hand, or eye,
By Hilliard drawne, is worth an history
By a worse painter made.'

One of his most outstanding works is the delightful 'Portrait of a Young Man' in the Victoria and Albert Museum; a painting typifying the period with the costume, roses and the

'Wooded Landscape with Watermill' by Meindert Hobbema. (Courtesy The Minneapolis Institute of Arts)

expression of the sitter. Lawrence, his son, taught by father continued in a like but lesser manner, other **pupils included Isaac Oliver.**

HITCHENS, IVON (1893–)

An English painter who treats landscape as an inspiration for near-Abstract compositions with cool clear colours very broadly applied.

HOBBEMA, MEINDERT (1638–1709)

Born in Amsterdam and possibly a pupil of Jacob van Ruisdael. He was one of the outstanding landscape-painters of his era, although he did not give as much time to his talent as he might have as he took a post with the Excise when he was 30. The National Gallery, London, has his fine 'The Avenue Middelharnis' as well as eight others.

HÖCKERT, JOHAN FREDRIK (1826–66)

Swedish historical and genre-painter who attended the Stockholm Academy of Art and worked under Johan Boklund, later becoming a Professor there himself. His pictures include 'Queen

'A grand procession' by David Hockney. (Courtesy Cologne Art Fair and Thomas, Munich)

Christina Ordering the Execution of Monaldeschi' and 'Divine Service in a Lapland Chapel'.

HOCKNEY, DAVID (1937–)

Born at Bradford, Yorkshire, he studied at the Bradford School of Art and then at the Royal College of Art. Painter in a progressive style, often large canvases with well-diluted colour, in another mood clear almost clinical portraits. He has produced a great many etchings and **in 1961 received the Guinness Award for Prints at the The Graven Image Exhibition.**
'The sprinkler' £18000, 1975, Sotheby's, London.
'Henry' (Lithograph) L 280000, 1975, Finarte, Milan.

HODGKINS FAMILY

The father **William Mathew** (1833–98) was born in Liverpool; more or less self-taught he became a skilled water-colourist. Emigrating in 1859 he went first to Australia and then on to New Zealand, where he set up in Dunedin as a solicitor and amateur artist. 'After rain, Lake Wakatipu' in the Auckland City Art Gallery shows his considerable command and forceful manner with his medium. **Frances Mary** (1869–1947) was his second daughter. She studied at the Dunedin School of Art, then travelled widely in Europe and taught at the Académie Colarossi. After this went backwards and forwards between Europe and New Zealand until she settled in England.

HODLER, FERDINAND (1853–1918)

Born in Berne, he studied in Geneva under Barthélémy Menn and worked most of his life in Geneva. Early work with portraits and landscape was proficient but inclined to lack excitement. Then his style changed to a form of mystic symbolism, with compositions such as 'Towards the Infinite' in the Kunsthaus in Zürich; here he employs a simplification of anatomy and rhythmic arrangement of forms.
'Portrait de Mme Günzburger' Sfrs 180000, 1975, Auktionshaus am Newmarkt, Zürich.

HOFFLAND, THOMAS CHRISTOPHER (1777–1843)

Born in Nottingham, he had a time in the studio of John Rathbone, afterwards he taught a little himself and then went to London to make copies of paintings in the British Institution. Finding this made money and there were prospects in London he set up a studio there. In 1814 the governors of the British Institution awarded him 100 guineas for his best landscape 'A Storm off the Coast of Scarborough'.

HOGARTH, WILLIAM (1697–1764)

Born in Ship Court, Old Bailey, London, he

began his studies with a silversmith, Ellis Gamble, and after this it is likely that he attended Sir James Thornhill's academy in St Martin's Lane. He was principally an artist working on satire, levelling his brush at social evils and excesses. His two most famous series are 'Marriage à la Mode' and 'A Harlot's Progress'. The Marriage series was purchased at an auction held by the artist for £126. In another manner he painted the fine altarpiece for St Mary Redcliffe, Bristol. When he turned to portrait-painting he demonstrated a delightful free manner as seen in the 'Shrimp Girl' in the National Gallery, London. His graphic work and paintings were so heavily plagiarized that **Hogarth was largely behind the first Copyright Act passed by Parliament to protect an artist's original work from plagiarists and misuse without permission.**

HOLBEIN THE ELDER, HANS
(1465–1524)
Born in Augsburg he may have worked in the studio of Martin Schongauer. Principally a painter of religious subjects, his career latterly was followed by ill fortune.

HOLBEIN THE YOUNGER, HANS
(1497–1543)
Also born in Augsburg, his training was started by his father, but other details of tuition are a little confused. One of the most accomplished portrait-painters of his time, his sitters included the Duchess of Milan, in the National Gallery, London; Catherine Howard, in Toledo, Ohio; Anne of Cleves, in the Louvre; and the great wall-painting of Henry VIII and his mother and father with his third wife Jane Seymour, only the cartoon now survives as the painting was destroyed in 1698. His graphic work includes his 'Dance of Death', a series of 51 woodcuts.

In 1515 he journeyed to Basle where he found good employment and then became a naturalized Swiss citizen in 1520. In 1526 on a visit to England he met Sir Thomas More and stayed for two years painting the More family, and making a number of portrait-drawings. Back to Basle for a few years and then he returned to England where finally it appears the Plague caught up with him.

HOMER, WINSLOW (1836–1910)
Born in Boston, he studied at the National Academy of Design, New York. His career commenced as an illustrator. Then he began painting, in water-colour and oil, scenes from the Civil War. After a visit to England in 1881 he returned to America and settled on the Maine Coast and worked at fishing-boats and the sea. His most memorable picture is the 'Gulf Stream', in the Metropolitan Museum of Art, New York.

HONDECOETER, MELCHIOR D'
(1636–95)
His tuition included working with his father Gisbert Hondecoeter and later Jan Weenix. He is noted for his fine paintings of birds.

HONE, NATHANIEL (1718–84)
Born in Dublin and self-taught he gained some reputation for working in oils and painting miniatures. He came over to England and found success. In the National Portrait Gallery, London is his well-handled portrait of John Wesley. **A founder-member of the Royal Academy.** His painting 'The Conjuror' (now in the National Gallery, Dublin) was refused admission in 1775 as it was seen to be a satire on Sir Joshua Reynolds and to contain a nude figure intended to be Angelica Kauffmann; Hone denied this.

HONTHORST, GERRIT VAN (1590–1656)
A pupil of Abraham Bloemaert, he spent a number of years in Rome and on his return was to be one who more than most spread the influence of Caravaggio.

HOOCH, PIETER DE (1629–83)
Dutch painter associated with the courtyards and houses of Delft influenced by his contemporary Vermeer.

HOOGSTRATEN, SAMUEL VAN
(1627–78)
Born at Dordrecht, he was a pupil of Rembrandt and painted portraits, interiors and landscapes.

HOPPER, EDWARD (1882–1967)
An American painter who studied at the Chase School of Art, New York and later in Paris where the influence of Sisley, Pissarro and Renoir inspired him. On his return to America his landscape manner was characterized by a use of warm sunlight.

HOPPNER, JOHN (1758–1810)
Born in Whitechapel, London, of German parents, he studied at the Royal Academy Schools and later became Portrait-Painter to the Prince of Wales. A number of his pictures suffered through the use of asphaltum.

HOSKINS, JOHN (c 1595–1665)
A leading miniaturist, and appointed to Charles I in 1640, later his work was surpassed by that of his nephew and pupil Samuel Cooper. Hoskins made a number of miniature copies of Van Dyck's paintings.

HOUDON, JEAN-ANTOINE (1741–1828)
French sculptor who worked with Pigalle. His work includes the statue of George Washington in Richmond, Virginia; 'Benjamin Franklin', 'Thomas Jefferson' and 'Madame Victoire'.

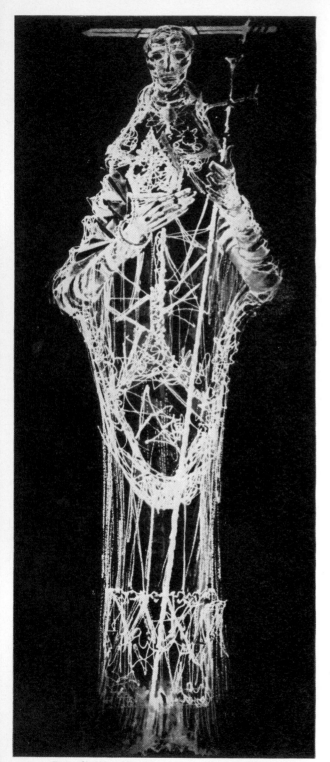

From the West Window at Coventry Cathedral – 'St Thomas Beckett (1118–70), Archbishop of Canterbury' by John Hutton. (Courtesy of the Artist)

HUDSON, THOMAS (1701–79)

Born in Devon, he studied with Jonathan Richardson. He had a period of success as a fashionable portrait-painter, although his faces and hands were inclined to lack animation, but his handling of clothes, fabrics and laces was excellent. **Hudson's most famous pupil was Sir Joshua Reynolds.**

HUGGINS, WILLIAM (1820–84)

Born in Liverpool he studied at the Mechanics' Institution and the Liverpool Academy. Later he became one of the best considered animal painters. He had a skill with foreign animals and was said to have followed Wombwells' Menagerie for a considerable period making studies.

HUNT, WILLIAM HOLMAN (1827–1910)

Born in London, he trained at the Royal Academy Schools and became one of the three principal members of the Pre-Raphaelite Brotherhood. Well-known paintings by him include 'The Hireling Shepherd' in Manchester, 'The Scapegoat' in the Lady Lever Gallery, and 'The Light of the World' (versions of this are in St Paul's Cathedral and Keble College, Oxford). These and other compositions by him show a quite extraordinary pursuit of detail with large pictures, a manner normally associated with miniatures.

HUTTON, JOHN (1906–)

Born in New Zealand, painter and glass-engraver. He has developed a highly individual manner in handling the glass, producing bold images with texture, clear delineation and power. Examples of engraved glass are in the West Doors of Guildford Cathedral, great glass screen for Coventry Cathedral; and the 'Dunkirk War Memorial'.

HUYSUM, JAN VAN (1682–1749)

Born in Amsterdam. After some instruction from his father Justus van Huysum, he became one of the most famous flower-painters. His are pictures noted for arrangement, colour and the inclusion of small details such as: drops of water on leaves and petals, flies, birds' nests with eggs and butterflies. 'Still Life with Flowers', £14700, 1975, Christie's, London.

INGRES, JEAN AUGUSTE DOMINIQUE (1780–1867)

Born at Montauban, France, he became a pupil of Jacques Louis David, and won the Prix de Rome in which city he worked for several years. Although he was labelled as a follower of Classicism, his manner was far removed from that of David, it harked back more to that of the High Renaissance and Raphael. His figure studies had modelling that suggested the finish of white marble rather than flesh, although they had at times a rare plasticity. Noteworthy is his 'La Grande Odalisque' and 'Madame Rivière', both in the Louvre. Historical mythological pictures included 'Dream of Ossian', 'Bonaparte as First Consul', 'Napoleon as Emperor' and 'Triumph of Romulus over Acron'.

INNES, GEORGE (1825–94)
Born in Newburgh, New York, he was largely self-taught. On trips to Europe he felt the influence of the Barbizon school and his final landscape style was soft, subtle and redolent with atmosphere. Pictures of merit include, 'Delaware Water Gap', 'The Lackawanna Valley' and 'Niagara Falls'.

ISRAELS, JOZEF (1824–1911)
After study in Amsterdam and the École des Beaux-Arts in Paris he turned to the study of fisher-folk and peasants, thus working in a manner similar to Millet.

IVANOV, SERGEI VASILYEVICH (1864–1910)
He attended the Moscow School of Art and the Academy of Arts, later teaching at the Stroganov School. A painter of historical subjects and genre who exhibited with Mir Iskusstva and was a member of the Peredvizhniki.

JAWLENSKY, ALEXEJ VON (1864–1941)
Born in Russia, he had a period of study with Repin, but after left for Germany and set up his studio in Munich in 1896. A close associate of Die Blaue Reiter, although he did not show with them. His art is a strange mixture of mysticism and strength with colour that gradually gave up its richness in exchange for warm and cool silvery greys. He did a series of Têtes Mystiques in 1917 which was followed by another series of Abstract Heads.
'Frau mit grünem Fächer' £46200, 1975, Christie's, London.

JOHN, AUGUSTUS EDWIN (1878–1961)
Born in Wales and trained at the Slade School, he became one of the outstanding British artists of his day. **As a draughtsman he had no equal**, his line-drawings of figures and heads were models of sensitivity and economy of means. His portrait of his wife 'The Smiling Woman' in the Tate Gallery, London, is a fine piece of brushwork in line with that of Frans Hals and Rubens. Gwen (1876–1939) his sister also went to the Slade School and later to Paris where she had lessons from Whistler. Her work was intimate and to a degree followed the colour and style of Whistler.

JOHN, CASPER (1930–)
Born in Allendale, South Carolina, he attended the University of South Carolina, having his one-man show at the Leo Castelli Gallery, New York. An experimenter with different media, from oils to bronze; with the latter he went for complete Realism to produce the likeness of two beer cans.

JOHNSON, CORNELIUS (1593–1661)
(also spelt Jonson and Janssens). Born in London, it is likely his parents from the Netherlands were refugees from the Duke of Alva's persecution. He became the leading portrait-painter prior to the coming of Van Dyck.

JONGKIND, JOHAN BARTHOLD (1819–91)
Born at Latrop, Holland, he was mostly self-taught and his style like that of Boudin presaged the work of the Impressionists; although unlike them he did not paint his oils out-of-doors, but worked them up from sketches in his studio. He lived in some squalor near Grenoble and like Van Gogh he went insane, and he ended his life in the asylum at Grenoble.
'Le Château de Nyon', £26000, 1975, Sotheby Parke Bernet.

JORDAENS, JACOB (1593–1678)
Born at Antwerp, he was a pupil of Adam van Noort. He was befriended by Rubens and worked on cartoons for tapestries for the King of Spain from the designs of Rubens. His own manner was strongly influenced by that of his friend; it was boisterous, and full of life and goings-on but lacked Rubens's colour, brushwork and taste. 'The King Drinks' (versions in Brussels and Vienna) is a painting typical of his style.

KALF, WILLEM (1619–93)
Born in Amsterdam and studied with Hendrik Plot. A still-life painter with particular skills for gold, silver and glass.

KANDINSKY, WASSILY (1866–1944)
Born in Moscow, he started out studying the law, but in 1896 he gave this up and turned to art. **He is one of the principal founders of Abstract Art.** In 1911 he was a founder of the Blaue Reiter group, and later taught at the Bauhaus. His principal writings setting out his theories are: *The Art of Spiritual Harmony* and *Point, Line and Surface*.
'Dicht', L 50000000, 1975, Finarte, Milan.

KANE, PAUL (1810–71)
Canadian painter of Indians, whom he studied while travelling on horseback. He studied in New York, also in France and Italy during a European trip. A number of his pictures are in the Royal Museum, Toronto.

KAUFFMANN, ANGELICA (1741–1807)
Swiss portrait and decorative painter, born at Coire, Grisons. At an early age she was a ready

student for painting, music and languages. She was taught by her father, a portrait-painter. In 1766 she came to England and got herself trapped into a clandestine marriage with the valet of the Swedish Count de Horn, who had pretended to her that he was the Count; this little escapade was closed by a payment of £300. She became friendly with Sir Joshua Reynolds and this started gossip. In 1768 **she was among the founder-members of the Royal Academy.** Her own work began quite a fashion in its somewhat pretty way and was much engraved, particularly by Bartolozzi. A portrait of Winckelmann is in the Kunsthaus, Zürich.

'Portrait of Thomas Jenkins and his niece Anna Maria' £6825, 1975, Christie's, London.

KEENE, CHARLES SAMUEL (1823–91)
English draughtsman, engraver, etcher and painter; he was illustrating for *Punch* for nearly 30 years, also produced the pictures for Charles Reade's *The Cloister and the Hearth.*

KEMENY, ZOLTAN (1907–65)
Born in Banica, Rumania, he trained at first as a cabinet-maker, then studied architecture, and designed fashion models, coming finally to sculpture. His style was to construct his works from scrap-metal, aiming for strange and weird textures and patterns that could suggest growth.

'Midi à quatorze heures' Sfrs 145000, 1975, Auktionshaus am Neumarkt, Zürich.

KENT, ROCKWELL (1882–)
Born in Tarrytown, New York, he studied with William Merritt Chase. American painter, etcher and illustrator. He was influenced by Robert Henri. Books illustrated include: Melville's *Moby Dick*, Chaucer's *Canterbury Tales* and Whitman's *Leaves of Grass.*

KIRCHNER, ERNST LUDWIG
(1880–1938)
Born in Aschaffenburg, he was at first self-taught and then attended the art school of Debschütz and Hermann Obrist; later he returned to his architectural studies which he had interrupted in Dresden. Influences on him included Munch, Van Gogh and Heckel, and emotionally he was affected by Nietzsche. As with others around him, his art was suppressed as degenerate by the Nazis and he committed suicide.

'Sangerin am Piano' DM 180000, 1975, Hauswedell & Nolte, Hamburg.

KLEE, PAUL (1879–1940)
Born near Berne, he studied in Munich, and practised painting and etching. He was associated with the Blaue Reiter group and taught at the Bauhaus. He returned to Switzerland when expelled by the Nazis in 1933. Klee's art is one of personal whimsy with a seasoning of mysticism. 'A Young Lady's Adventure' in the Tate Gallery, London, displays some of the influences on him, Picasso, Abstract, a version of Cubism and the colour theories of the German painters he had met. 'Gedanken in gelb' £73500, 1975, Christie's, London.

KNELLER, SIR GODFREY
(1646 or 1649–1723)
Born at Lübeck, he studied first with Bol, a pupil of Rembrandt, and he may then have had some contact with Rembrandt himself. He journeyed to Italy and there was influenced by Maratta. In 1674 he came to England, was made Principal Painter jointly with Riley and was knighted. Walpole mentions that Charles II to save himself sitting time was painted simultaneously by Kneller and Lely, and that Kneller was finished when Lely's was 'dead coloured' only (underpainted only). His most famous series was that of the Kitcat Club.

KOCH, ÖDON (1906–)
Born in Zürich, this sculptor started out learning tapestry, then teaching himself he turned to the third dimension. Stone is his favourite material, and his style vacillates between Abstract and the incorporation of Realist elements.

KOEKKOEK FAMILY
The father **Jan Hermann** (1778–1851) was born at Vere, Holland; he had a time in a tapestry factory, but his painting was largely self-taught. His river and sea scenes are admired. **Barend Cornelis** (1803–62) was the elder son, being born at Middelburg. He studied under his father, and later with Schelfhout and Van Os and in the Academy of Amsterdam. He published in 1841 *Souvenirs and Communications of a Landscape Painter.* Jan the younger son was also born in Middelburg and learnt from his father.

'The End of Winter' (Barend Cornelis Koekkoek) £30000, 1975, Sotheby Parke Bernet.

KOKOSCHKA, OSCAR (1886–)
Painter of Czech and Austrian parentage, he studied in Vienna and later taught at the Dresden Academy from 1919 and was influenced by the members of the Brücke. His style is a highly colourful and individual form of Expressionism. In 1947 he became a British citizen. His 'Polperro, Cornwall' shows his unusual viewpoint that he has also used with other scenes such as 'Thames Landscape'.

'Veronika', Sfrs 275000, 1975, Kornfeld und Klipstein, Berne.

KOONING, WILLEM DE (1904–)
Dutch painter, moved to America in 1926. He joined the New York Group of Abstract painters.

KRAFFT, ADAM (*c* 1460–*c* 1508)
German Late Gothic sculptor most of whose work
is in Nuremberg. His finest work is the 62 ft
(18·89 m) high tabernacle in the Church of
St Lawrence, there is also a crouching figure
holding a mallet supposed to be his self-portrait.

KRIEGHOFF, CORNELIUS (*c* 1812–72)
Born in Germany or Holland, he went to America
in about 1837; apparently he joined the US Army
and painted scenes in the Seminole War, and other
genre subjects connected with the French-
Canadian life.
'Tobogganing' £6600, 1975, Sotheby Parke Bernet.

KUNIYOSHI, YASUO (1893–1953)
Japanese-born painter who went to America
when he was 13 and attended the Los Angeles
School of Art and Design and later the Art
Students' League in New York. Influences on him
include Daumier, Delacroix and Pascin. He
developed a rich and imaginative use of colour.

KUPKA, FRANTISEK (1871–1957)
Czech-born painter, who studied in Prague and
Vienna, then came to Paris. He was associated
with the Abstraction-Création group.

LAM, WILFREDO (1902–   )
Born in Cuba and studied in Havana and Madrid.
He arrived in Paris in 1937 and contacted Picasso
and André Breton who lured him towards
Surrealism. He has been applauded as a leader of
African visual art; his compositions present the
savagery and primitive spirit of the jungle and
atmosphere which he experienced close to him.
'Totem', L 8 500 000, 1975, Finarte, Rome.

LANCRET, NICOLAS (1690–1743)
Born in Paris, he worked in the studio of Pierre
D'Ulin and Claude Gillot. He was an acquaintance of Watteau, whose manner of fêtes galantes
he was to imitate. He has a number of pictures in
both the Wallace Collection, London and the
Louvre.

LANDSEER, SIR EDWIN (1802–73)
Born in London, he received early tuition from his
father. **There are drawings that he did when
he was five years old**; and he was showing with
the Royal Academy when he was but 13. One of
Queen Victoria's favourite artists, he more or less
invented the idea of giving virtues and heroic
qualities to animals in paint. Thus a long series of
such pictures as: 'The Monarch of the Glen', 'The
Challenge', 'The Stag at Bay', 'Dignity and
Impudence', and 'The Old Shepherd's Chief
Mourner', in the Victoria and Albert Museum,
about which John Ruskin made some very
complimentary remarks. As a sculptor he modelled and had cast **the bronze lions round
Nelson's Column in Trafalgar Square.**
Although his values have slipped somewhat since
his day, they are now on the upward slant again.
'Queen Victoria on horseback' £9975, 1975,
Christie's, London.

LANFRANCO, GIOVANNI (1582–1647)
Born at Parma, his parents placed him as a page in
the service of Count Scotti at Piacenza. This
gentleman noticed the lad sketching in charcoal
and got him a place in the studio of Agostino
Caracci. He was influenced by Correggio. He
decorated a number of ceilings in the Illusionist
manner.

LARGILLIÈRE, NICOLAS DE
(1656–1746)
Born in Paris, he became a pupil of Antoine
Goubeau. In 1675 he visited England and worked
under Lely in the Royal Palaces. On returning to
Paris he specialized in portraits and was the main
rival to Rigaud.

LARIONOV, MIKHAIL (1881–1964)
Russian painter who studied in Moscow. He met
Gontcharova, and organized the first Knave of
Diamonds show. **In 1913 he published his
Manifesto on Rayonism, an important turn-
ing in the establishing of Abstract Art in
Russia.** Later he was to work on ballet décor for
Diaghilev in Paris.

LARSSON, CARL (1853–1919)
Swedish historical painter and illustrator, who for
the latter got himself the nickname the Swedish
Doré.

*'Sleighride' by the Canadian artist Krieghoff. Trappers
were amongst his favourite subjects. Today his paintings are
sought after by collectors. A few years ago, a woman at a
jumble sale bought what she thought was an old, painted
tray covered in thick brown 'coach' varnish. When cleaned
it turned out to be a scene of hunters in the white, frozen
wastes of the north, and was in fact a Krieghoff. For what
she had thought of as a piece of 'rubbish' and paid 1s 6d for
it, she received £2800. (Courtesy Christie's)*

LATHAM, JAMES (1646–1747)

Irish portrait-painter born in Tipperary, who studied in Antwerp and was admitted to the Guild of St Luke there about 1724. There is a fine portrait of George Berkeley by him in Trinity College, Dublin.

LA TOUR, GEORGES DE (1593–1652)

Born at Vic in Lorraine, he was considerably influenced by Caravaggio's use of light and shade. In fact for subtlety he surpassed this master. His treatments have considerable charm and atmosphere, worth seeing are 'St Joseph's Dream' in the Musée des Beaux-Arts, Nantes, and the 'Magdalene with the Lamp', in the Louvre. He attracted little attention until the early part of this century.

'A girl blowing on a brazier' £17850, 1975, Christie's, London.

LA TOUR, MAURICE QUENTIN DE (1704–88)

French pastel portrait master, appointed Painter to Louis XV in 1750.

LAURENCIN, MARIE (1885–1956)

Born in Paris, she attended the Lycée Lamartine. Later she was to meet Apollinaire and go to the Bateau-Lavoir. Her portraits and pictures of children have considerable charm, they are graceful and have an appealing atmosphere.

'Portrait de jeune fille' Sfrs 68 000, 1975, Auktionshaus am Neumarkt, Zürich.

'Tête de Jeune Fille' by Marie Laurencin. (Courtesy Christie's)

LAURENS, HENRI (1885–1954)

Born in Paris, from a working-class home in his early years he was engaged in heavy manual work. In 1911 he made a friendship with Georges Braque that lasted until Laurens's death. A leading Cubist sculptor, he worked in wood, stone, bronze and polychromed sheet-iron. An idea of his power can be seen with 'Mermaid' in the Musée d'Art Moderne, Paris.

LAWRENCE, SIR THOMAS (1769–1830)

Born at Bristol, and an infant prodigy, **by the age of ten he was a practising artist working at portraits in crayon.** He had a short time at the Royal Academy Schools. His career was a great success from the start, not only in England but also in Europe he was known as a fine hand with a portrait. **He became President of the Royal Academy in 1820.** He was only 22 when he produced the delightful and skilfully handled portrait of Miss Farren, in the Metropolitan Museum of Art, New York. The male portrait is exemplified by 'The Duke of Wellington' in the Wellington Museum, London, and 'George IV as Prince Regent' in the Vatican Gallery.

LE BROCQUY, LOUIS (1916–   )

Born in Dublin, this Irish painter is largely self-taught. He has been investigating the emergence of the head with influence from the early ages, and strong symbolism. He has done some noteworthy designs for textiles.

LE BRUN, CHARLES (1619–90)

French painter and the Director of the Arts under Louis XIV. A pupil of Vouet and for a time with Poussin in Italy. There are a number of decorations by him at Versailles.

LE CORBUSIER (1887–1965)

(Original name Charles-Edouard Jeanneret) Born in La Chaux-de-Fonds, Switzerland. Primarily one of the most important architects of this century, he also found time to paint. He founded L'Esprit Nouveau with Ozenfant in 1920, and brought in Purism, which was intended as a protective theory for Cubism.

LEECH, JOHN (1817–64)

English caricaturist and illustrator, he was with *Punch* for nearly 25 years and illustrations included Dickens's *Christmas Books* and Beckett's *Comic History of England*.

LÉGER, FERNAND (1881–1955)

French painter, who started out as an architectural designer, and then spent a short period at the École Nationale des Beaux-Arts; he set out on his career of an explorer into Cubism with his friend Delaunay. The components of his paintings whether human or otherwise have a solid massive quality and his colour is bright, strong and clean. **In 1924 he made the first abstract film entitled _Le Ballet Mécanique._** 'L'usine au Motif pour le moteur' F 250000, 1975, Palais Galliera, Paris.

LEGROS, ALPHONSE (1837–1911)
French artist best recalled for his fine etchings. He was a friend of Whistler who induced him into coming to England where he set up his studio, and he was Slade Professor at London University for 18 years from 1876.

LEIGHTON, LORD FREDERICK (1830–96)
Born at Scarborough, he travelled to many places in Europe and started his art training in Florence. His first success came with a large pastiche entitled 'Cimabue's Madonna carried in Procession', which was shown at the Royal Academy and purchased by Queen Victoria. He was elected **President of the Royal Academy in 1878.** With Alma-Tadema and Edward Poynter, Leighton set out to oust the tenets of the Pre-Raphaelites and to bring in a somewhat artificial Classicism, with heavy overtones of the Greek, and presented with a theatrical glamour and sentiment. **He was the first painter to be given a peerage.** Few painters' values have slumped quite as much in the period after their passing. 'The Sargonsian Bride' brought £2677 10s. in 1874, in 1960 it fetched £200.

LELY, SIR PETER (1618–80)
Born in Soest, Westphalia, his original name was Pieter Van der Faes. He was in the studio of Pieter Grebber. Lely came to London in 1641, and despite native opposition more or less swept the portrait field. **His studio in the end became practically a picture factory with lesser hands putting in the backgrounds, clothing and etceteras, while Lely would put in the faces and hands.** At his best he had considerable quality and his painting technique was sound. At the end he was seized with a fit of apoplexy while painting the Duchess of Somerset. He must have been a skilled diplomat, as he was in favour with Charles I, Oliver Cromwell and Charles II.

LEMOYNE, JEAN-BAPTISTE (1704–78)
French sculptor who studied with his father and with Le Lorrain. He was appointed Sculptor to Louis XV, but sadly his best works on a heroic scale were destroyed during the Revolution. Portrait busts of Voltaire and Mme de Pompadour give evidence of his quality; **pupils included Houdon, Pajou and Falconnet.**

LE NAIN BROTHERS
Born in Laon, they were **Antoine** (1588–1648), **Louis** (1593–1648) and **Mathieu** (1607–77), but little is known about their lives. Of the three Louis was probably the most talented, although with works attributed to them it is a difficult matter to find the truth, as it is known that at times all three brothers worked on the same picture. The National Gallery, London, has Louis's 'Family Portrait' which gives an idea of their subject-matter which was usually centred on peasant life, also it shows the influence of Caravaggio.

LEONARDO DA VINCI (1452–1519)
The incomparable one, he was probably the most talented of all the great host of artists of the High Renaissance. Apart from his genius as a painter **his notebooks reveal that he possessed invention backed by erudite reasoning on: architecture, botany, chemistry, geology, mathematics, mechanics, music, philosophy, physics, poetry, town-planning.**

He was born at Vinci, the illegitimate son of a Florentine notary and was brought up in his father's house. Later he was to work in the studio of Andrea Verrocchio, being trained in the elements of painting and sculpture; fellow pupils included Lorenzo di Credi and Botticelli. He came under the patronage of Lorenzo and was admitted to the Guild of St Luke in 1472. In 1482 he moved to Milan and entered the service of Lodovico Sforza. Towards the end of the century he was in Venice being used as a military engineer. Back to Florence in the first years of the 16th century. In 1508 to Milan again, and in 1517 he travelled to Amboise at the behest of François I and lived out his life under the kindly help of this monarch at Clos-Lucé, near Amboise, and in sight of the Royal Château. Of his paintings there are too few remaining, a number being lost. But those that we still have point clearly to his genius; in the handling of anatomy, with the subtle _sfumato_; in

'_A French Interior_' by Louis Le Nain. (_Courtesy National Gallery of Art, Washington, DC, Samuel H. Kress Collection_)

composition which was flawless but often complex as with the 'Virgin of the Rocks' of which there are two versions, one in the Louvre and the other in the National Gallery, London. There is extraordinary perception of character and thought as seen in the 'Mona Lisa' which is incidentally one of the most copied paintings in the world. A Venetian, Antonio Bin, has painted about 300 and on walls around Europe and America there are more than 200 further copies. She is also **the most reproduced picture, in black and white and colour, in the world.**

Perhaps in the 'Last Supper', painted between 1495 and 1498 in the Refectory of the Monastery of Santa Maria della Grazie, he achieved part of that ultimate he saw and was seeking to express. Sadly mutilated as it is from an unwise choice of technique, ignorant restoration, the action of time, humidity and concussion, the wall-painting still emanates an intense feeling of the consecration and power of Leonardo's thought.

In 1962 £804 361 was given for his 'The Virgin and Child with St John the Baptist and St Anne'. **Highest price recorded for a drawing.**

LÉPINE, STANISLAS (1835–92)
French painter, a pupil of Corot and an associate with Boudin, Jongkind and with them a forerunner of the Impressionists; principally a painter of the streets of Paris.

LE SUEUR, EUSTACHE (1616–55)
Born in Paris he studied with Vouet and became one of the leading French religious and mythological painters. Later influence came from Poussin. His 'Life of St Bruno' painted for the Charterhouse, Paris, is now in the Louvre and shows his style.

LE SUEUR, HUBERT (working 1610–43)
He worked during his early life in France, then came to England and did a considerable amount of mediocre work for Charles I; **the equestrian statue at Charing Cross is by him.**

LEVINE, JACK (1915–)
Born at Boston, Massachusetts, he worked under Denman Ross at Harvard. He developed into an Expressionist concentrating on social comment. 'Teresina' $6000, 1975, Sotheby Parke Bernet, New York.

LEWIS, WYNDHAM (1884–1957)
English painter and writer who was the founder of Vorticism, and publisher of *Blast*.

LHOTE, ANDRÉ (1885–1962)
Born in Bordeaux, besides painting he taught and wrote a number of thoughtful books including: *Treatise on Landscape*, *Treatise on the figure* and *Writings on Art*.

LICHENSTEIN, ROY (1923–)
American painter who takes everyday objects and things as a basis for pictures; such as giant blow-ups of parts from strip cartoons, or huge representations of hamburgers.

LIEBERMAN, MAX (1847–1935)
German painter who studied in Amsterdam where he was influenced by Israels, later to Paris and contact with the work of such as Millet. He became the leading German exponent of Impressionism. He was elected as President of the Berlin Sezession.

LIMBOURG BROTHERS
Pol, Hennequin and **Hermant** were all born in Flanders somewhere around the last years of the 13th century. In 1411 they were the Court Painters for the Duke of Berry for whom among other works they created the splendid illuminated manuscript 'Les Très Riches Heures du Duc de Berry', now in the Musée Condé, Chantilly. All three brothers were dead by 1416.

LINNELL, JOHN (1792–1882)
English painter, a friend and helpmate of William Blake, he studied with John Varley and also worked as a miniaturist.

LIOTARD, JEAN ÉTIENNE (1702–89)
Swiss artist accomplished in the use of pastel for portrait and genre. He travelled in Turkey dressed in Turkish dress for some years and retained it for the publicity value. He also worked in miniature.

LIPCHITZ, JACQUES (1891–1973)
Sculptor born in Druskieniki, Lithuania, Studying first at Vilno, in 1909 he went to Paris, then had a stint at the École des Beaux-Arts which was followed by the Académie Julian and the Académie Colarossi, and his work turned to Cubism with a lively force of movements. 'Chant des Voyelles' in the Kunsthaus, Zürich shows an attempt to adapt his style for readier comprehension by the viewer.

'La Danse' $28000, 1975, Sotheby Parke Bernet, Los Angeles.

LIPPI, FILIPPINO (c 1457–1504)
Son of Fra Filippo Lippi and likely to have been the pupil of Botticelli. Early frescoes are in the Brancacci Chapel of Santa Maria del Carmine, Florence, and another series is in the Strozzi Chapel, Santa Maria Novella.

LIPPI, FRA FILIPPO (c 1409–69)
Being left an orphan he was taken into a religious order when a child, although temperamentally an unsuitable candidate. He may have been a pupil of Masaccio. The first painting which bears a date is

the 'Tarquinia Madonna' in the National Gallery, Rome. He eloped with a nun and their son was Filippino above.

LISSITZKY, EL (1890–1941)
Russian typographic designer and poster and exhibition worker.

LOCHNER, STEFAN (active 1442–51)
Called Meister Stefan was the master of the Cologne school. He was born at Meersburg on Lake Constance. His finest work is the 'Madonna with Two Kneeling Kings' in Cologne Cathedral.

LONGHI, PIETRO (1702–85)
Born in Venice, he was instructed by Antonio Balestra and Giuseppe Maria Crespi. A painter of everyday scenes with often a touch of satire, these small pictures had great popularity and were much copied and imitated.
'The Display of the Elephant' £17850, 1975, Christie's, London.

LOPES, GREGÓRIO (c 1490–1551)
Portuguese painter appointed to the Courts of Manuel I and John III, he executed altar-paintings for the Convent of Christ, Tomar.

LORENZETTI BROTHERS
(active during the first half of the 14th century). They were **Pietro**, probably the elder and **Ambrogio**. Sienese painters who were considerably influenced by Giotto. In the Siena Town Hall are Ambrogio's frescoes 'Good and Bad Government', and Pietro's work can be seen in 'The Birth of the Virgin' in the Duomo Museum, Siena.

LOTTO, LORENZO (c 1480–1556)
Born in Venice, he probably worked under Vivarini, and he was influenced at first by Giovanni Bellini and later by Botticelli, Giorgione and Titian. He had a high perception when dealing with sitters, this appears in such as 'Andrea Odoni' in the Royal Collection, Hampton Court.

LOWRY, LAURENCE STEPHEN
(1887–1976)
English painter who found a visual poetry in the soot-stained townscapes of the north country peopled with little folk on matchstick legs.

LUCAS VAN LEYDEN (c 1494–1533)
Dutch portrait and religious subjects painter, he learnt from his father and Cornelius Engelbrechtsz. By the age of 15 he was renowned as an engraver, whose works today are highly valued, being second only to those of Dürer. In the Lakenkal Museum, Leyden is his impressive 'Last Judgement'.
'The Annunciation' (engraving) DM 1050, 1975, Karl u.Faber, Munich.

LUINI, BERNARDINO (c 1481–1532)
Born at Luino on Maggiore, he was a pupil of Stefano Scotto, but fell heavily under the influence of Leonardo, without being able to assimilate such power. The result was that what might have been successful work if he had been himself was deadened and failed to really succeed.

LURÇAT, JEAN (1892–1966)
French painter and a leading tapestry-designer. With the latter his colours are bright and strong and are combined with a sharp incisive style.

McCAHON, COLIN (1919–)
Born at Timaru, New Zealand. After arousing his imagination by visits to the galley in Dunedin and the Otago Art Society Exhibition, he finally got to the Dunedin School of Art in 1937 where part of his tuition came from Robert Field. His art that broke through has become bold, owing much to heavy strong colours and is often combined with literate statements with a religious tone. 'The King of Jews', 'Will He Save Him'.

MACKE, AUGUST (1887–1914)
German painter who studied at the Düsseldorf Academy and with Corinth. **A founder-member of the Blaue Reiter.** He was killed in action in 1914.

MACLISE, DANIEL (1806–70)
Born in Cork, he became one of the outstanding historical painters of his time. In the House of

'Woman with a Beard' by L S Lowry. (Courtesy of M Bloom Esq.)

Lords there are two frescoes by him, 'The Meeting of Blücher and Wellington' and 'The Death of Nelson'.

MAES, NICHOLAES (1634–93)
Born in Dordrecht, he was a pupil of Rembrandt. His portraits have an elegant quality as do his genre and 'conversation pieces'. Well-handled light and shade.

MAGRITTE, RENÉ (1898–1967)
Belgian Surrealist who studied in Brussels, and then went to Paris and came into the main stream of the movement with the inspiration of André Breton. His invention is varied and often relies on unrelated objects to produce shock. Later he returned to Belgium and **with Delvaux formed a Belgian group of Surrealists.**
'L'œil de la Montagne' Sfrs 154000, 1975, Galerie Motte, Geneva.

MAILLOL, ARISTIDE (1861–1944)
Born in Banyuls-sur-Mer. He began studying to be a painter and was under the spell of Puvis de Chavannes, Gauguin and Cézanne. Next he turned to tapestry and did not come to sculpture until about 1907. His style is somewhere between the vigour of Rodin and the classical. He worked primarily in bronze, the modelled clay allowing him a freedom, also providing a means towards the sensuous surface finish he sought. There is a

'Not to be reproduced' (1939). Portrait of Edward James, the reclusive patron of surrealist art, by René Magritte (1898). (Courtesy Boymans-van Beuningen Museum)

Torso by him in the Tate Gallery, London, which clearly shows his forceful physical treatment.
'Femme debout se coiffant' (bronze) £11200, 1975, Sotheby Parke Bernet.

MALEVICH, KASIMIR (1878–1935)
Russian painter who **dissatisfied with Cubism invented Suprematism. He produced the ultimate in 'non' pictures with the squares 'Black Square' and 'White on White'.**

MANDER, KAREL VAN (1548–1606)
Born at Meulebeke in Flanders, he worked with Lucas de Heere, a poet and painter of Ghent, then he went to the historical painter Pieter Vlerick. His resulting work was a blend of Mannerism from Italy and the Low Countries.

MANESSIER, ALFRED (1911–)
French painter, who has also worked in stained glass and tapestry design.

MANET, ÉDOUARD (1832–83)
Born in Paris he was a pupil of Couture's for six years, then he travelled extensively, and became strongly influenced by Velázquez and Tintoretto. Settling in Paris he painted joyously free from sterile doctrines and rulings. His choice of subjects often shocked. There was the highly controversial 'Le Déjeuner sur l'herbe' in the Louvre, 'Absinthe Drinker' and the new-look-type composition such as 'The Bar at the Folies-Bergère', Courtauld Collection, London, also the pert challenging nude 'Olympia' in the Louvre. In his own time the story of non-appreciation repeats itself. 'Rue Mosnier aux drapeaux' goes for £480 in 1898, in 1958 it pulls in £113000.

MANTEGNA, ANDREA (c 1431–1506)
Born at Vicenza, he started studying very early with Squarcione who adopted him, and was **received into the Guild of Paduan artists at the age of ten.** From his master he assimilated a classic influence, and he was intrigued by the possibility and uses of perspective and the foreshortening of figures, as can be noted in his 'Agony in the Garden', in the National Gallery, London. Mantegna, who in his figure and clothes painting may resemble Botticelli, was a considerable dramatist. Besides being a painter he was a skilled engraver.

MANZÙ, GIACOMO (1908–)
Born at Bergamo, he studied sculpture at Cicognini Academy while doing his military service. There appears some influence from Donatello in his use of subdued relief. He has completed doors for St Peter's in Rome and also Salzburg Cathedral.
'Passo di Danza' £29400, 1975, Christie's, London.

MARC, FRANZ (1880–1916)
German Expressionist painter with the Blaue
Reiter, a peace-loving, nature-probing mystic. He
developed a pleasing manner with animal
painting; horses and deer became part of rhythmic
patterns in his pictures. He was killed in action at
Verdun.

MARIN, JOHN (1870–1953)
American free-style water-colourist, he studied at
the Pennsylvania Academy of Fine Arts and then
the Art Students' League in New York. He was
considerably aided by the generous patronage of
Alfred Stieglitz.

MARINI, MARINO (1901–   )
Born at Pistoia, he attended the Academy at
Florence under Domenico Trentacoste, later there
were further studies in sculpture in Paris. His style
is one of stark simplicity with an undertone of
emotional turbulence. He works primarily in
bronze and wood, the latter sometimes being
polychromed.
'Horseman' (bronze) in the Walker Art Center,
Minneapolis, is a theme he has returned to a
number of times.
'Cavaliere' (bronze) £30000, 1975, Sotheby Parke
Bernet.

MARIS BROTHERS
Jacob (1837–99) was the eldest and best known,
he studied at the Antwerp Academy and with
Ernest Hebert in Paris; his subjects were domestic
interiors and intimate landscapes. **Matthias**
(**Matthijs**) (1839–1917) was a genre-painter who
worked most of his time in London. The youngest
was **Willem** (1844–1910) who painted Dutch
landscapes often with cattle included.

MARQUET, ALBERT (1875–1947)
French painter, who was a pupil of Gustave
Moreau. Later he went along with Fauvism.

MARSH, REGINALD (1898–1954)
American painter of city scenes, often with
striking compositions at night with the garish
street lights. He was a pupil of K H Miller and he
also etched and did illustrations for the *New Yorker*
and *Vanity Fair*.

MARTIN, JOHN (1789–1854)
Born near Hexham, he worked for a time
decorating china and painting heraldic devices on
coaches. As he started to paint seriously he
produced a number of extremely imaginative
landscapes generally on a biblical or a mythological theme, such as the 'Destruction of
Herculaneum' in Manchester and 'The Day of his
Wrath' in the Tate Gallery, London.

MARTINI, SIMONE (*c* 1284–1344)
Born at Siena, he was a pupil of Duccio. A painter
who put down his subjects with a simple pure
sincerity. In 1315 he produced a 'Maestà' for the
Town Hall at Siena. In 1317 he was called to
Naples, then a French kingdom, to paint for
Robert Anjou 'St Louis of Toulouse Crowning
the King', now in the National Gallery, Naples.
Later his figures, especially the faces, took on a
sense of emotive drive.

MASACCIO (1401–28).
(properly Tommaso di Giovanni). Born at San
Giovanni Valdarro about 25 miles (40 km) from
Florence, his nickname being interpreted as
clumsy Tom or hulking Tom. Little is known of
his early years, the first clear mention being in
1422 when he entered the Guild in Florence.
Masaccio was one of those precious élite, sadly to
go so soon, but he had time to do **the wonderful
frescoes in the Brancacci Chapel of Santa
Maria del Carmine, Florence**; one of the most
perfect scenes here being the Expulsion from the
Garden (Genesis), a powerful emotion-filled
passage of pure Expressionism.

MASOLINO (*c* 1383–*c* 1432)
A pupil possibly with Masaccio; also he may have
been with Ghiberti. His work is close in style if not
quality to Masaccio's.

MASSON, ANDRÉ (1896–   )
A French Surrealist, he parted company with
André Breton and started a splinter movement
that was to have a considerable influence on
Abstract Expressionism in America.

MATHIEU, GEORGES (1921–   )
A French Abstract lyrical and near-figurative
painter, generally using a dark background with
sharp symbolic marks in clean bright colour.
'Noire d'Arcy' L 13 000 000, 1975, Finarte, Milan.

MATISSE, HENRI (1869–1954)
He worked at first at the École des Beaux-Arts and
then in the studios of Bouguereau and Gustave
Moreau. During his long life, with the possible
exception of Picasso, he probably tried his hand at
more techniques than anyone else in the 20th
century: he painted, modelled, worked in his own
personal way with collage, illustrated books, and
then eventually entered upon his last major work
when he took on the decoration of the interior of
the Chapel of the Dominican nuns at Vence. **The
leaders of Les Fauves**, his colours at that period
were as bright and powerful as any of those of the
others.
   He had many moods; 'The Painter and his
Model' in the Musée d'Art Moderne in Paris; 'Joie
de Vivre', Barres Foundation, Philadelphia; and

'Portrait of Madame Matisse' in Leningrad. 'Danseuse au palmier' $230000, 1975, Sothby Parke Bernet, New York.

MATSYS, QUENTIN (1464/5–1530)
(also Massys or Metsys). Born in Louvain, he became a Master of the Antwerp Guild in 1491.

'Expulsion from the Garden' from "Paradise" by Masaccio in The Church of Santa Maria del Carmine, Florence.

On one hand he was the painter of caricature faces he saw around him, similar to the drawings of such subjects by Leonardo; and on the other a deeply conscientious painter of such as the St Anne altarpiece at the Musée Royale des Beaux-Arts in Brussels.

MATTA, ECHAURREN (1912–)
Born in Chile, he studied architecture with Le Corbusier, but then turned to a violent form of demoniac Surrealism. He has stayed away from Chile except for one short return trip in 1941 when he painted 'Listen to Living'.
'Les chants de la rétine' L 20000000, 1975, Finarte, Rome.

MAUVE, ANTON (1838–88)
Born at Zaandam, he was for a time a pupil of Van Os from which study he gained little; he absorbed more from his association with Josef Israels and Willem Maris. He painted in soft tones similar in manner to Daubigny and Millet.

MAZO, MARTINEZ JUAN BAUTISTA DEL (c 1612–67)
Spanish painter who was the pupil and son-in-law of Velázquez and succeeded him as Court Painter in 1661.

MEISSONIER, JEAN-LOUIS-ERNEST (1815–91)
French painter, mainly self-taught, but with a time in the studio of Léon Cogniet. He specialized in huge canvases of Napoleon's campaigns.

MELOZZO DA FORLI (1438–94)
Italian painter who is **credited with the invention of extreme foreshortening.** Most of the remaining work credited to him is fragmentary.

MEMLINC, HANS (c 1430–94)
Born at Seligenstadt, near Frankfurt am Main and it is probable he was a pupil of Roger van der Weyden. He settled in Bruges and must have been comfortably successful because in 1480 town records show he was among the 247 burgesses who advanced money to the town for expenses of the war between Maximilian and the King of France. His manner is northern, cool, unemotional in contrast to the influence that was beginning to flow from Italy. Fine works include: 'The Mystical Marriage of St Catherine' and 'The Shrine of St Ursula', both in the Hôpital Saint-Jean, Bruges; his portrait work is shown by 'Tommaso Portinari' and 'Wife of Tommaso Portinari', both in the Metropolitan Museum of Art, New York. In the first half of the 19th century his values were low, 1830 only £74 12s for 'The Death of the Virgin', but now they have risen sharply.

'Holy Family and Donors', an altarpiece in 1955 brought £88000.

MENGS, ANTON RAPHAEL (1728–79)

Born in Aussig, Bohemia, the son of Ismael Mengs, a Dane, the Dresden Court Painter. The father brought his son up in a rigorous manner seeing to his studies. Anton blossomed with talent, even if it may have lacked great originality. He worked in pastel, oil and fresco, and was perhaps most successful with the last. He became director of the Art Academy at the Vatican.

MENZEL, ADOLF VON (1815–1905)

German painter best known as the recorder of events, especially battles and campaigns in the life of Frederick the Great.

MERYON, CHARLES (1821–68)

Born in Paris, after a short time at sea he started to paint but discovered he was colour blind. He turned to etching and made some 100 plates of views of Paris of a very high standard. He died in an insane asylum.

METSU, GABRIEL (1629–67)

Born at Leyden and studied with Dou, he painted charming small-scale genre and 'conversation pieces', with great attention to detail, textures and light.

'Bearded Man in a Brown Coat' by Gabriel Metsu. (Courtesy Christie's)

METZINGER, JEAN (1883–1956)

Born in Nantes, he studied in Paris and was associated with Gleizes in bringing out **Du Cubisme the first book on the movement.**

MEUNIER, CONSTANTIN (1831–1905)

Born in Brussels, he studied with his brother Jean Baptiste. At first he expressed himself with paint, subject-matter was religious then Impressionist. Later he turned to sculpture, working figures on a large scale as 'The Docker' in the Musée des Beaux-Arts, Antwerp.

MICHEL, GEORGES (1763–1843)

Born in Paris he was in the studio of Leduc. He painted small views of the country close to Paris, and is also recalled as a restorer used by the Louvre for pictures from the Dutch school.

MICHELANGELO, BUONARROTI (1475–1564)

Born in Castel Caprese, a small town on the outskirts of Florence. He was apprenticed to Ghirlandaio, and later came under the golden patronage of Lorenzo the Magnificent. Few of the world's sublime geniuses have had to journey quite such a rough road as Michelangelo. His life is a record of bitter attacks, jealousy, misunderstanding even of relatives and friends. His parents nagged for money he often had not got. When Lorenzo died, Pietro continued the patronage, but in many ways wasting the talents, ordering him to make a statue in snow.

As with Botticelli and others he fell under the spell of Savonarola. He told Pietro that the House of Medici would fall, and for his pains was driven out. He went to Venice at the moment when Savonarola came to power in Florence and set up a republic and publically called for no allegiance other than to Christ Jesus, even the Papal authority was defied. Less than four years later Savonarola was to be fired at the stake. Michelangelo returned to Florence and carved the great statue 'David' from a single block of white marble, which now stands in the Accadèmia, Florence. To stand and examine this work for a period of time shows a facet of perfection. Pope Julius called him to Rome to commission a personal mausoleum. He came across Bramante, the architect, who filled with envy, whispered jealous words to the Pope, that the mausoleum could be an omen, and Julius abandoned the project. Michelangelo was turned out, unpaid for the work done, to find his way back to Florence.

Sent for again, this time to decorate the Sistine Chapel – a work Bramante had felt Michelangelo would not be able to carry out successfully, and he made what he intended as a harmful suggestion to the Pope, that Michelangelo should do it.

Michelangelo himself felt his greatest talent was with sculpture and begged to be allowed to withdraw, but the Pope refused him his request. So, for four years he worked in secrecy behind locked doors creating in fresco on the ceiling of the chapel what must surely be **the greatest single achievement in the arts of all time.** The main panels showed The Creation of the World; God Creating the Luminaries; God Blessing the Earth; The Fall of Man; The Creation of Adam; The Creation of Eve; The Temptation and Fall; The Sacrifice of Noah; The Deluge and The Drunkenness of Noah. The whole is a technical triumph against the difficulties of working from a scaffold and looking upwards to the moist plaster on which he must paint, probably often with the lime-tainted water from the colours falling down on to him.

Thus the great one went on. One of his very few oils 'The Entombment' is in the National Gallery, London; in Notre-Dame Cathedral, Bruges is a charming small Madonna redolent with love and care; in Rome stands his St Peter's, a triumph in yet another field. He was only 37 when he finished the Sistine, and the immense effort had undermined his health, yet for another half century he was to continue to pour out his genius lavish with his generosity in providing so much, while he himself lived a life of ascetic quality, content perhaps with the beauty and wonder he created. Piece after piece came into being: 'The Pietà' in St Peter's, Rome (subjected not long ago to an assault from a lunatic's hammer), the Medici Tomb; he even found time to write some exquisite sonnets. He died in the making of the 'Rondanini Pietà' in the Castello, Milan, worn out in the service of his genius.

MIGNARD, PIERRE (1612–95)
Born at Troyes, he studied with Vouet and succeeded Le Brun as Court Painter in 1690.

MILLAIS, SIR JOHN EVERETT (1829–96)
Born in Southampton, **an infant prodigy** he was admitted to the Royal Academy Schools when he was eleven. **With Holman Hunt and Rossetti he founded the Pre-Raphaelite Brotherhood in 1848.** He had a strong friendship with John Ruskin which came to a sharp end when Millais married Ruskin's former wife. He became **President of the Royal Academy in 1885.** Notable pictures include the 'Death of Ophelia', 'Christ in the House of His Parents', 'The Boyhood of Sir Walter Raleigh' and 'The North-West Passage'.

MILLES, CARL (1875–1955)
Swedish sculptor who studied in Paris and was considerably influenced by Rodin. He has numerous works in Stockholm on buildings and in parks. He went to America in 1929 and became a citizen in 1942.

MILLET, JEAN-FRANÇOIS (1814–75)
Born near Cherbourg, the son of a peasant, he studied first with a local artist then went to Paris and the studio of Delaroche. In 1849 he moved to Barbizon and stayed there painting for the rest of his life. He is essentially the honest painter of those scenes from which his family had come; the workers in the fields and country craftsmen portrayed in a sincere and unpretentious manner. His best-known scene is probably 'The Angelus' in the Louvre.

MIRÓ, JOAN (1893–   )
Born in Catalonia, he attended the Barcelona School of Fine Art. Affected by diverse influences that included Cézanne, Gaudí's architecture and Les Fauves, he turned to Surrealism and in 1925 took part in the 1st Surrealist Exhibition. His style has become more Abstract as he has progressed.
'Femme dans la Nuit' (oil on leather) L 35 000 000, 1975, Finarte, Milan.

MODERSOHN-BECKER, PAULA (1876–1907)
German painter, a forerunner of Expressionism, she portrayed the flaxen-haired children of Worpswede, with a rich, warm, dark palette.

MODIGLIANI, AMEDEO (1884–1920)
Born at Leghorn, he studied in Florence and Venice arriving in Paris in 1906. Affected by Cézanne's approach to Cubism, Modigliani distorted to his own ends, producing his strange, long, oval faces on overlong necks; the expression is generally one of melancholy accentuated by the heavy warm colours he was wont to use. A good-looking and amorous one he was heavily addicted to drink and drugs. He said 'I am going to drink myself dead' and he did just that.
'Le Garçon en culottes courtes' £189000, 1975, Christie's, London.

MOHOLY-NAGY, LASLO (1895–1946)
Hungarian designer, painter, photographer and sculptor. He taught in the metal workshop of the Bauhaus, and was a member of the Abstraction-Création group in Paris.

MONDRIAN, PIET (1872–1944)
Dutch painter who went to Paris in 1911 and being caught up with Cubism and other current vogues, left his naturalistic landscape-painting for an exploration of pure geometric abstraction. **He has exerted considerable influence in this field.**

MONET, CLAUDE (1840–1926)

Born in Paris he and his family moved to Le Havre; there he had some training, and met Boudin who encouraged him to paint nature in the open air. In Paris in 1859 at the Atelier Suisse he met Pissarro and Cézanne. For two years he did 'military service in Algeria and then was in the studio of Gleyre with Sisley, Renoir and Bazille.

Monet was the leading member of the Impressionists and stuck to their tenets more faithfully than the others. Searching and patient to achieve he would return again and again to a subject to capture the effects under varying lights. Thus his celebrated series of Haystacks which he showed at Durand-Ruel's in 1891, these 15 canvases painted at various hours of the day. His works were numerous and include such as 'Argenteuil Bridge' in the New State Gallery, Munich, 'Rouen Cathedral' in Rouen Museum, 'Old Fort at Antibes' in the Museum of Fine Arts, Boston, 'Vétheuil in Summer' in the Metropolitan Museum of Art, New York and 'The Studio-Boat' in the Rijksmuseum Kröller-Muller, Otterlo.

'La cathédrale de Rouen, la tour d'Albare le matin' £210000, 1975, Sotheby Parke Bernet.

MONTANEZ, JUAN MARTINEZ (1568–1649)

Spanish sculptor, chiefly religious subjects often of extreme Realism, notable examples in Seville.

MONTICELLI, ADOLPHE (1824–86)

French painter from Marseilles, he studied in Paris. His work was mostly portraits and still lifes at first, then he broadened into genre often associated with circuses and fairs. His painting method produced a pleasant soft slightly blurred effect.

MOORE, HENRY (1898–   )

Born in Castleford, Yorkshire; after training as a teacher, he joined the army in 1917 and was gassed at Cambrai and invalided home. He studied at the Leeds College of Art and later at the Royal College of Art, London. From 1926 he taught at the Chelsea School of Art until 1939; then, during the Second World War he was an Official War Artist and among other work did his well-known series of air-raid shelter drawings. He works in stone, lead, bronze, wood and plaster, and although a member of Unit One with Hepworth and Nicholson his subject-matter is not entirely Abstract. Inspiration comes with reclining forms, heads, mother and child groups. The resulting forms possess a sense of movement and at times as with such as the 'King and Queen' in the Open Air Museums, Middelheim, Antwerp, a feeling for the supernatural. His works will be found in nearly every major gallery or museum.

'Falling Warrior' $65000, 1975, Sotheby Parke-Bernet, Los Angeles. In 1972 the **highest figure for a living sculptor** was reached when £104000 was bid for his 'Reclining Figure' at Sotheby Parke Bernet, New York.

MOR, ANTHONIS (c 1519–75)

Portrait-painter from Utrecht who was in the studio of Jan Scorel, he became Court Painter to the Spanish Netherlands. His manner was influenced by Titian.

MORALES, LUIS DE (1509–86)

Spanish painter who worked almost entirely on religious subjects and infused his own passionate feeling into his paintings.

MORANDI, GIORGIO (1890–1964)

Italian painter, largely concerned with still life, subjects often just a few bottles and pots, who was associated with the Metaphysical painters and the Novecento group.

'Still Life' L 20000000, 1975, Finarte, Rome.

MOREAU, GUSTAVE (1826–98)

Frenchman who studied with François Picot and was influenced by Chasseriau. His work was often based on biblical or mythological themes.

MORISOT, BERTHE (1841–95)

Born in Bourges, she studied with Chocarne, then Guichard and lastly Corot. In 1868 she met Manet and in 1874 married his younger brother Eugène. She showed with all but one of the Impressionist exhibitions. Her sensitive colour-handling was to influence her brother-in-law.

'Emma Rodeau dans un Jardin' £21025, 1969, Christie's, Tokyo.

MORLAND, GEORGE (1763–1804)

Born in London, he studied with his father, Henry Robert, and was showing with the Royal Academy by the time he was ten. He painted English rustic scenes somewhat in the manner of Brouwer, with more emphasis on animals, particularly horses. Of somewhat intemperate habits and careless with money, the latter part of his life was spent on the run from his creditors and trying to paint enough pictures to buy them off; much of this output was, needless to say, 'skimped' and of doubtful quality. He expired in a sponging-house in Eyre Street Hill.

MORONI, GIOVANNI BATTISTA (c 1525–78)

Italian painter who studied with Moretto and was influenced by Lotto. He was a good hand at portraits in a manner somewhat similar to Holbein with a leavening of the Venetian school, note his 'Portrait of a Tailor' in the National Gallery, London.

MORRIS, WILLIAM (1834–96)

English craftsman, designer, writer and Socialist. Influenced by John Ruskin, he worked to combat the soul-destroying action of industrialism, ideas similar in many ways to those Gropius was to activate in the Bauhaus in the 1920s.

MORSE, SAMUEL (1791–1872)

Born in Charlestown, Massachusetts. After leaving Yale he trained under Washington Allston and chose subjects from mythology, but earned his living with portraits. He was the **first President of the National Academy**. He introduced Daguerre's process of photography to America, and also invented the electric telegraph and Morse code.

MOSER, MARY (about 1740–1819)

English flower-painter, she was **one of the founder-members of the Royal Academy.**

MOSES, GRANDMA (1860–1961)

(Anna Mary Robertson). American Primitive painter who had considerable success in the United States and even rose to an exhibition in Paris in 1951.
'Cambridge Valley' $9250, 1975, Sotheby Parke Bernet, New York.

MULREADY, WILLIAM (1786–1863)

Born at Ennis, County Clare, Ireland, the son of a leather breeches-maker. The family moved to Dublin and then London, where William had some tuition from Graham and the sculptor Banks. His work was in the manner of small rather sentimental genre and conversation pieces.

MULTSCHER, HANS (c 1400–c 1465)

German sculptor and painter working in and around Ulm. There are the figures at St George and St John the Evangelist on Ulm Cathedral.

MUNCH, EDVARD (1863–1944)

Norwegian painter and one of the foremost Expressionists, he was born at Löten and after studying in Oslo he spent almost 20 years travelling abroad, in Paris, the south of France, Italy and Germany. Munch is fascinating, for blended into his painting are root influences from the ancient Nordic world, and forces of the supernatural. In a peaceful mood, there is 'Girls on the Bridge' in the National Gallery, Oslo, or in shattering contrast 'The Cry' in the same gallery. He produced a large number of prints, many of them modelled on themes that he had painted. In Oslo is the Munch Museum, which provides a good opportunity to study an exceptional artist.
'Jungen am Meer $61 000, 1975, Sotheby Parke Bernet, New York.

MUNNINGS, SIR ALFRED (1878–1959)

English horse-painter with a broad and skilled manner and a great feel for a horse. **President of the Royal Academy** and an outspoken one against the excesses of Modern Art.

MÜNTER, GABRIELE (1877–1962)

German painter and engraver, in 1909 she was one of the two women members of the New Artists' Federation, she also showed with the Blaue Reiter in Munich. For a time she was married to Kandinsky. Her style can be noted with 'Pensive Woman' in the Städtische Galerie in Lenbachaus, Munich.

MURILLO, BARTOLOMÉ ESTEBÁN (1617–82)

Born in Seville, he was orphaned when quite young and raised by an uncle, who put him to study with a relative Juan del Castillo. His work evolved as religious subjects with a somewhat sentimental undertone, and pictures of street

'Racehorses' by Sir Alfred Munnings. (Courtesy Christie's)

urchins and fruit-sellers. Among his better paintings are 'Beggar Boys Throwing Dice' in the Alte Pinakothek, Munich, and 'The Ragged Boy' in the Louvre. His death was apparently brought on by a fall from a scaffold while painting a composition for the Capuchins in Cadiz.

MYTENS, DANIEL (*c* 1590–before 1648)
Born at The Hague and trained there he came to England and was Court Painter to Charles I. He had considerable quality, as may be noted with 'The First Duke of Hamilton', in the National Gallery of Scotland, Edinburgh.

NAIRN, JAMES McLACHLAN (1859–1904)
Born in Scotland and attending the Glasgow School of Art, he arrived at Dunedin, New Zealand in 1890. Early on he and his followers aroused the bile of newspaper critics who spoke of their 'chromatic lunacy'. Yet with Nairn it was his colour that was his quality, either with 'Winter Morning, Wellington Harbour' in the National Gallery, Wellington or the well-controlled 'Wharf at Kaikoura' in the Auckland City Gallery.

NAMATJIRA, ALBERT (1902–59)
Born at Hermannsburg, Australia, he was a pure-blooded Aboriginal of the Arunta tribe in Central Australia. He learned the use of water-colour from Rex Battarbee, and in 1938 had a sell-out with an exhibition in the Fine Arts Society Gallery, Melbourne.

NANNI DI BANCO (*c* 1385–1421)
Born in Florence the sculptor was trained by his father. A number of his works were specially designed to go with the Duomo and Or San Michele, projects on which he worked at the same time as Donatello; the figures harking back to the early Romans and Greek with vigorous lifelike faces and informal and natural drapery.

NASH, PAUL (1889–1946)
English painter, a Slade School student, who later dabbled with Surrealism; he was a member of Unit One. An Official War Artist during the Second World War he produced a number of moving paintings of aircraft, such as 'Dead Sea' in the Tate Gallery, London, pictures with sym-bolism using a limited palette and a sensitive mysticism akin to some of the works of Chirico.

NATTIER, JEAN-MARC (1685–1766)
Born in Paris and instructed by his father and later at the Académie, he distinguished himself as a portrait and historical painter.

NESTEROV, MIKHAIL VASILYEVICH (1862–1942)
Born in Moscow he studied at the Moscow School of Art and then the Academy of Art. At first a painter of religious subjects and landscape, latterly he turned to portraits and genre. Honoured Art Worker of the R.S.F.S.R.

NEVINSON, CHRISTOPHER RICHARD WYNNE (1889–1946)
English painter involved with Vorticism, and a personal Cubism, especially with paintings of the First World War.

NICHOLSON, BEN (1894–)
The son of Sir William Nicholson. After a single term at the Slade School he left and journeyed to America, Spain, Italy and France. His manner is influenced by Cubism and Piet Mondrian. Although coming close to pure Abstract, a number of figurative shapes come into his compositions. A member of Unit One.
'August 24–52' $32000, 1975, Sotheby Parke Bernet, New York.

NICHOLSON, SIR WILLIAM (1872–1949)
He studied at the Académie Julian in Paris. Apart from portraits, landscape and still lifes, he was an advanced **designer of posters working with James Pryde as the Beggarstaff Brothers.**

NOGUCHI, ISAMU (1904–)
Born in Los Angeles, his parents were the Japanese poet Yone Noguchi and the novelist Leonie Gilmore. With a Guggenheim Fellowship he was able to have a period in Paris and with Brancusi, he also met Giacometti. He went to Peking to study drawing and then to Japan with the potter Kyoto. His style is Abstract with added symbolistic passages, with attention to outlines and surfaces.

NOLAN, SIDNEY (1917–)
Born in Melbourne, studies were intermittent at technical schools and by absorbing manners and techniques in the National Gallery there. Influences on him have come from such contrast-ing sources as Paul Klee and Aboriginal Art. His paintings built round Ned Kelly the bushranger caused wide comment and admiration. One of the most stimulating painters to emerge from Australia.

NOLDE, EMIL (1867–1956)
The founder of the important school of German Expressionism, born in Nolde, North Schleswig, his real name was Hansen. He attended a Flensberg wood-carving school, and later Hoelzel. He founded the Neue Sezession in Berlin and was associated with the Brücke and the Blaue Reiter, but despite this he was a solitary one, drawing on folk-tales and religious motifs for his paintings. Colour at the start was muted and then as his confidence and knowledge grew, colour flamed, glowed and was vibrant with contrasts in

his oils, water-colours and gouache. Few painters have achieved the sheer enjoyment his colour brings to the viewer. Ordered to stop painting in 1941 by the Nazis he produced his 'Ungematte Bilder', hundreds of small water-colour studies, some of which he worked up into full paintings after the war.

'Die Jüdin' DM 55000, 1975, Hauswedell & Nolte, Hamburg.

NOLLEKENS, JOSEPH (1737–1823)
Leading English sculptor of his time, he spent the years between 1759 and 1770 in Rome, learning, copying and restoring damaged works. When he came back to England he rapidly made a name and was elected as an Academician in 1772. Sitters included Pitt, Garrick, Fox, Sterne and Dr Johnson. His style was individual and happily free from vogues and influences that could have undermined the strength.

NORTHCOTE, JAMES (1746–1831)
A pupil of Sir Joshua Reynolds, an aspirant historical painter and one who produced a large number of portraits, that tended to ape his master's manner but appeared weak beside him.

O'CONNOR, JAMES (1792–1841)
Irish landscape-painter born in Dublin, he was brought up as an engraver by his father whose profession it was. In 1813 he set out with Francis Danby and George Petrie on an ill-starred trip to London, funds ran out and he had to return to Dublin. In 1822 he returned to England, which became his home. In the National Gallery of Ireland, in Dublin, is his 'The Poachers' which displays a well-handled atmosphere with a brave use of cloud pattern.

'Wooded landscapes' (a pair) £1050, 1975, Sotheby Parke Bernet.

OLIVER AND SON
Both miniaturists, the father was **Isaac** (1564–1617) of French Huguenot parentage, who studied with Hilliard, and then became his rival. His sitters included Elizabeth I and Mary Queen of Scots. The son was **Peter** (c 1594–1647) who studied with his father. His colour was richer than his father's. He was employed by Charles I to make miniature copies of works by such as Titian, Raphael, Correggio and Holbein in his collection; reputedly the King liked to carry these copies with him on his travels.

Highest price for a portrait miniature, £65 100 at Christie's in 1971 for 'Frances Howard, Countess of Essex and Somerset' by Isaac Oliver.

OPIE, JOHN (1761–1807)
Born at St Agnes, Cornwall, he was set up in London by John Wolcot (satirist Peter Pindar) as the self-taught Cornish Wonder. His early works which dealt with light and shade effects and peasants, children and genre had quality, but this was lost as he let himself get caught up in the portrait race.

ORCAGNA, ANDREA (c 1308–68)
Working in Florence, he was painter, sculptor and architect. The Loggia dei Lanzi is his, so is the 'Burial and Assumption of the Virgin', signed and dated 1359, a well-conceived and handled relief, of the Tabernacle in Or San Michele. The only known painting certainly by him is the Strozzi altarpiece 'Christ, the Virgin and the Saints' in Santa Maria Novella.

OROZCO, JOSÉ CLEMENTE (1883–1949)
Mexican painter, who trained first as an architect. He started to paint in water-colour, then found fresco gave him the means for the expression he sought. His 'Cortes and Malinche', an early work, in the National Preparatory School of San Ildefonso, is a monumental display of the two figures; at Guadalajara, in the Instituto Cabañas, in the Salon des Actos, in the cupola there must be one of the strongest coloured frescoes ever. It shows Man of Flames and is a shock composition with two gesticulating figures in dark warm greys, and above them lurid in rich reds and dark madders is a figure being devoured by flames.

ORPEN, SIR WILLIAM NEWENHAM MONTAGUE (1878–1931)
Born in Dublin, he attended first the Metropolitan School before going to the Slade School. He was a member of the New English Art Club and an official artist in the First World War; his most famous work from this period was 'The Signing of the Peace Treaty, at Versailles' now in the Imperial War Museum, London. In the Manchester City Art Gallery of interest is his 'Homage to Manet' which includes portraits of George Moore, Sickert and Wilson Steer at a table in front of a representation of Manet's portrait of Eva Gonzales.

OSBORNE, WALTER FREDERICK (1859–1903)
Irish painter, who studied at the Royal Hibernian Academy, Dublin, later in Antwerp with Verlat; he led the so-called Antwerp School of the 1880s. Walter was happiest with small street scenes such as 'Saint Patrick's Close, Dublin', in the National Gallery of Ireland.

OSTADE, ADRIAEN VAN (1610–85)
Born at Haarlem and was probably in the studio of Frans Hals or Salomon van Ruysdael. His *métier* was with small low-life genre-paintings.

A Caprice, 47×56 in (1194×1422 mm), by Giovanni Paolo Panini. (Courtesy Sotheby Parke Bernet)

OUDRY, JEAN-BAPTISTE (1686–1755)
Born in Paris, he became a pupil of Largillière. Besides painting, he designed tapestries and worked as an illustrator. He was Director of the Beauvais tapestry works. But his best manner was with animals, in particular hounds and wild ones. The backgrounds may have been a little weak, but his anatomy and treatment of coats was well handled.
'Chien flairant du Gibier' F80000, 1975, Palais Galliera, Paris.

OVERBECK, JOHANN FRIEDRICH
(1789–1869)
The leader of the German religious-inspired Nazarenes. He concentrated on historical and religious subjects.

OZENFANT, AMÉDÉE (1886–1966)
Born at Saint-Quentin, he was in the studio of Dunoyer de Segonzac and had been one of the main theorists of a number of modern trends.

PACHECO, FRANCISCO (1564–1654)
Spanish painter and writer, **best recalled as being the master of Velázquez**.

PACHER, MICHAEL (*c* 1453–98)
A native of Bruneck in the Tyrol, he was a painter and wood-carver, his best-known work is the St Wolfgang altarpiece in the Salzkammergut.

PAJOU, AUGUSTIN (1730–1809)
French sculptor, a pupil of Lemoyne, he also studied in Rome. He decorated the Opera House at Versailles and also executed numerous excellent

portrait busts, including one of his teacher and one of Descartes.

PALMA VECCHIO (c 1480–1528)
Born in Serimalta, near Bergamo, a pupil of Giovanni Bellini, he was influenced by Titian and Giorgione. One of his finest works is the 'St Barbara and other Saints' in Santa Maria Formosa, Venice.

PALMER, SAMUEL (1805–81)
English painter primarily working in water-colour; a prodigy he was showing with the Royal Academy from the age of 15. He met William Blake and became affected by his mysticism, a new feeling came into his work, a symbolism was evident.
'Near Sevenoaks' (water-colour) £5100, 1975, Sotheby Parke Bernet.

PANINI, GIOVANNI PAOLO (c 1692–1764)
Born at Piacenza, he went to Rome where he studied with Locatelli, and was impressed with Salvator Rosa. He produced numbers of land-scapes with delicately treated ruins that had considerable popularity with tourists.

PANKOK, OTTO (1893–1966)
German graphic artist, draughtsman and painter. His prints have considerable strength, and exploit the possibilities of large areas of black and white combined with a free treatment in the cutting.

PAOLOZZI, EDUARDO (1924–)
Born in Edinburgh, the sculptor attended the Slade School and later worked in Paris. He is concerned with materials, not to be worked necessarily by hand, but rather to be built into a fresh image; some present a strange impression of being patched-up forms, with the conglomeration of things that he uses.

PARMIGIANINO, FRANCESCO (1503–40)
Born in Parma, he had an advanced talent, as by the age of 19 he was decorating the south transept of the cathedral there. Influenced by Correggio he developed an individual manner with slightly elongated figures. He was captured in the Sack of Rome in 1527 but managed to escape to Bologna, from thence to Verona and Venice, and finally back to Parma, where he spent the last years of his life working on frescoes in Santa Maria della Steccata and getting into difficulties with the overseer, who in 1539 had him imprisoned for failing to fulfil his contract. The following year he died.

PATENIER, JOACHIM (c 1485–1524)
Born at Dinant or at Bouvignes on the Meuse, he became a member of the Antwerp Guild in 1515. He used a distinctive manner with landscape, often with large strange rock shapes in the background.

PEALE FAMILY
The father, an American portrait-painter and miniaturist, was **Charles Wilson** (1741–1827), who studied with Gustavus Hesselius, John Singleton Copley and Benjamin West. He was **one of the founders of the Pennsylvania Academy of Fine Arts**. Charles Wilson's brother **James** (1749–1831) was a qualified portrait-painter and twice had George Washington sit for him. Charles Wilson had eleven children, all of whom he named after famous painters. Some were to achieve recognition including: Rubens, a painter of still life, Raphael, a miniaturist, and Rembrandt, a painter of portraits and historical subjects, including 'The Court of Death' a large mural in the Detroit Institute of Arts.

PECHSTEIN, MAX (1881–1955)
A member of the Brücke group for a short time, he began life as a house-painter and then studied at the Dresden Academy. He was strongly in-fluenced by Matisse with the handling of colour.

PERMEKE, CONSTANT (1886–1952)
Belgian painter who studied in Bruges and Ghent. He was wounded in the First World War and was in hospital in England where he began seriously to paint. His figures are thick and powerful, modelled on labourers in the landscape of Belgium; colour is strong and warm.

PEROV, VASILY GRIGORYEVICH (1833–82)
The natural son of Baron Kruedener, he studied at Arzamas under Stupin and later at the Moscow School. Painter of genre, historical subjects and portraits, among his sitters were Turgenev and Dostoievsky.

PERUGINO (c 1445–1523)
He worked in the studio of Verrocchio where he would have met Leonardo. One of his most important works was a fresco in the Sistine Chapel, Rome, 'Christ giving the Keys to St Peter'. He was influenced by Signorelli and also to a degree by Memlinc.

PEVSNER, ANTOINE (1886–1962)
Born in Orel, Russia, he attended the Art School in Kiev. Further study in Paris brought him into contact with Archipenko and Modigliani. On his return to Moscow in 1917 he was appointed to teach at the Moscow Academy, where Kandinsky, Malevich and Tatlin also taught. With his brother Naum Gabo he was co-signatory of the Realist

'Birth of Christ' by Perugino. (Courtesy The Art Institute of Chicago)

Manifesto. His work is concerned with spatial relationship, not masses, but with convoluting forms that are free in their movement and not bound to a base other than as a support.

PICABIA, FRANCIS (1878–1953)
French painter and associate of Marcel Duchamp. He was largely responsible for the spreading of the Dada style to America. He also used collage and construction as methods of expression.

PICASSO, PABLO (1881–1973)
Born in Malaga, the son of an art teacher, he was encouraged from his early years and attended the Madrid School of Fine Arts when he was 16. **No single figure has influenced the art scene of the 20th century more, enlivened it, turned it around than Picasso.** Few artists also have worked in so many ways, oils, water-colour, gouache, pen and ink, etching, sugar lift, aquatint, collage, ceramics and modelling for bronze. In his long life he must have touched on nearly every art movement and theory prevalent around him.

His work progressed from the strictly academic of his teens to his Blue period, studies of outcasts, colour cool and melancholy. His Rose period showed some joy and warmer colour as he tackled wandering actors and those in circuses. After the First World War much of his work became more violent, culminating in the large 'Guernica', a bitter pungent comment on the Spanish Civil War. Examples of his work in one or more media will be found in most museums and galleries of Modern Art.

'Self-portrait' £282 500, 1975, Christie's, London.
'Femme Vase' (bronze) $16 000, 1975, Sotheby Parke Bernet, New York.
'Femme Assise' (water-colour and gouache) DM 92 000, 1975, Hauswedell & Nolte, Hamburg.
'Nature Morte à la guitare et clarinette' (pencil, charcoal and wash) £15 700, 1975, Christie's, London.
'Minotauromachie' (etching No 12/50, fifth state) Sfrs 265 000, 1975, Kornfeld und Klipstein, Berne.

PIERO DI COSIMO (c 1462–c 1521)
Italian painter, a pupil of Cosimo Rosselli and may have worked with him on his frescoes in the Sistine Chapel. He was influenced by Leonardo and for much of his time painted rather obscure subjects based on allegories and mythology. The National Gallery, London has two of his pictures 'The Death of Procris' and 'Fight between the Lapiths and the Centaurs'. He also worked on designs for festivals, masques and processions, such as the Trionfi, included here was his fearful 'Triumph of Death' done in 1511.

PINTURICCHIO, BERNARDINO (c 1454–1513)
One who was probably working with Perugino in

the Sistine Chapel as there is some evidence of his influence. Most important works were the frescoes in the Borgia Chambers in the Vatican.

PIPER, JOHN (1903–)
Born at Epsom, England, and attended the Royal College of Art and the Slade School, London. In the 1930s he came across Braque, Brancusi, Léger and Nicholson. Yet he has remained strongly individual, with his dramatic but sensitive handling of landscape. A more recent adventure has been into stained glass with great success, the windows of Basil Spence's Coventry Cathedral are an example.

PIPPIN, HORACE (1888–1946)
A self-taught black American painter who was neglected until being 'discovered' in the 1930s. He has a pure forceful Primitive style. One celebrated picture is his 'John Brown goes to his Hanging' in the Pennsylvania Academy of Fine Arts.

PIRANESI, GIOVANNI BATTISTA (1720–78)
Born in Venice he was the pupil of Valeriani. He became an architect and engraver and left a great number of plates of Roman antiquities, prisons, ruins and architectural features. These were handled in a distinctive manner, with dramatic use of light and shade. A set of 14 etchings, first state, before signing or numbering fetched £19000 in 1975 at Sotheby Parke Bernet.

PISANELLO, ANTONIO (1395–c 1450)
Born perhaps in Pisa, but if so, he would have moved at an early age to Verona. He may have studied with Gentile da Fabriano. The earliest work still existing by him is the 'Annunciation' in Verona, the known pictures all show considerable observation and an attention to detail, epitomized by the attractive 'Vision of St Anthony and St George' in the National Gallery, London, which contains a glorious costume and hat for St George.

PISANO, ANDREA (c 1290–1348)
Italian goldsmith, sculptor and architect who between 1330 and 1336 completed the first pair of doors for the Baptistery in Florence. He also completed the top two storeys of Giotto's Campanile.

PISSARRO, CAMILLE (1830–1903)
Born at Saint-Thomas in the West Indies, working as a clerk in his father's store; he ran away with a Danish painter and eventually arrived in France in 1855. He met Corot who was somewhat critical of the new adventures the Impressionists were setting out on. Pissarro experimented with Pointillism, a method that took to the limit the Impressionists' theory of divided colour. He soon

tired of the discipline of this and returned to the more moderate manner adopted by Monet. With pictures such as 'Market Gardens with Trees in Blossom', in the Louvre, he achieves a marvellous sense of reality and atmosphere. A prodigious worker, Pissarro produced a great number of pictures and prints, using: oils, gouache, pastel, drawing in pencil and pen and ink, etching and lithography.
'Soleil, après-midi, rue de l'Épicerie à Rouen' £120 000, 1975, Sotheby Parke Bernet.

PLASTOV, ARKADY ALEXANDROVICH (1893–)
Born in Moscow, he studied under Mashkov in Moscow and attended the Stroganov School sculpture classes. He lives and works in Moscow and the village of Prislonikhe. His subjects move round the glory of work as with 'Collective Farm Threshing' in the Russian Museum of Art, Kiev.

PO, LENA
Russian blind sculptress, who has achieved in a quite remarkable way the ability to model, as with her 'The Sprite', which captures the sense of movement and anatomy of a child.

POLLAIUOLO, ANTONIO DEL (c 1432–98)
Italian painter, sculptor, engraver and goldsmith, who with his brother Piero del (c 1441–96) had a highly successful workshop in Florence. Antonio is taken as the greater talent of the two; in the National Gallery, London, is 'The Martyrdom of St Sebastian' which was probably worked on by both of them. Anatomy was one of their most earnest studies and Vasari says that Antonio was **the first to dissect dead bodies to further his knowledge.**

POLLOCK, JACKSON (1912–56)
American painter who attended the Los Angeles Art School and later worked under Benton in New York. His work progressed from Realism to Surrealism and at last to Action painting.

PONTORMO, JACOPO (1494–1556)
Born at Pontormo near Empoli, he was in the studio of Andrea del Sarto, and may have had some tuition from Leonardo. He was a **leading figure with the coming of Mannerism**. He decorated the Medici Villa at Poggio a Caiano, after this came the frescoes of the Passion at Certosa not far from Florence. His finest work is the Deposition altarpiece in Santa Felicità, Florence. **Of his pupils Bronzino is the most famous.**

PORTINARI, CÂNDIDO (1903–62)
Brazilian painter known best for his huge murals,

although he did produce a number of small paintings usually based on peasant activities.

POSADA, JOSÉ GUADALUPE (1851–1913)
Mexican engraver, born at Aguascalientes of peasant background. He is reputed to have made up to 15000 prints for use with broadsheets, pamphlets, many with savage satire, and macabre cruel images.

POTTER, PAULUS (1625–54)
One of the leading Dutch animal painters, he studied with Camphuysen and Pieter de Weth. Best-known painting is 'The Bull' at The Hague. 'Une groupe de chiens dans un intérieur' F 200000, 1969, Palais Galliera, Paris.

POUSSIN, NICOLAS (1594–1665)
Born at Villiers, Normandy, he went to Paris, but little is known of him there until he appears in 1621 helping Champaigne at the Luxembourg Palace. Three years later he went to Rome and was in the studio of Domenichino. In 1640 he returned to France to work for Cardinal Richelieu, but finding himself in discordant competition with other painters of the Court, he went back to Rome after just over a year and remained there. His paintings have an intense feeling of harmony sometimes with a predominance of a rich warm blue; subject-matter often came from mythology, as with 'Bacchanal' in the National Gallery, London and 'Triumph of Pan' in the Louvre. His values are difficult to assess as today it would be an occurrence for an important work by him to come on the market.

In 1951 his 'Holy Family' did make £25000, when in 1810 it made just £640 10s.

POWERS, HIRAM (1805–73)
American sculptor who studied abroad and eventually settled in Florence. While still in America he did some work in wax modelling at the Cincinatti Museum and his portrait busts included Daniel Webster and Chief Justice John Marshall.

PRIMATICCIO, FRANCESCO (1504–70)
From Bologna, he was the leader of the First School of Fontainebleau – doing many things – painter, architect or sculptor and interior decorator. He worked in the studio of Giulio Romano.

PRUD'HON, PIERRE-PAUL (1758–1823)
Born at Cluny, the 13th child of a stone-mason, he was sent to Dijon Academy by the Bishop of Mâcon. In Italy he was impressed by the work of Correggio and Leonardo, both of which helped to mould his own style.

PUGET, PIERRE (1620–94)
French sculptor who was in the studio of Cortona, and who had a considerable command of modelling anatomy. He was not in great favour with Louis XIV or Colbert. In his early days he worked at Marseilles on the construction and decoration of galleys. Outstanding group is the life-size white marble 'Milo of Crotona'.

Aboriginal Chief sculptured by William Ricketts. Open Air Art Gallery, Central Australia. (Courtesy Australian News and Information Bureau, Canberra)

Show sculpture park. St Louis, Missouri. (Photo by the Author)

QUERCIA, JACOPO DELLA (1374–1438)
Italian sculptor, one of the competitors for the Baptistery doors, in Florence, which went to Ghiberti. Leading works include the Gaia Fountain in Siena and the doors of San Petronio, Bologna.

RAEBURN, SIR HENRY (1756–1823)
Born in a small village now absorbed into Edinburgh, he was left an orphan when still young and was apprenticed to a jeweller. He was to become **Scotland's leading portrait-painter, very successful, completing close on 1000 portraits**, also some miniatures. He had a lively broad manner and caught at times vivid likenesses as with 'Mrs Campbell' in the National Gallery of Scotland and 'Mrs Lauzun' in the National Gallery, London. He was the **first President of the Royal Scottish Academy**.

RAMSAY, ALLAN (1713–84)
Born in Edinburgh, he was the pupil in London of Hans Hysing and later in Rome under Francesco Fernandi Imperiali and in Naples with Solimena. On his return he set up his studio in London and had every success. His portrait of George III displays quality not only in the posing but in the handling of the robes.

RAPHAEL (1483–1520)
(properly Raeffaello Sanzio). Born at Urbino, the son of the painter Giovanni Santi, by the time he was 17 he was working with Perugino. Often he is dismissed as a painter of Madonnas; actually he was in the line of Duccio and Francesca, composing his pictures with sincerity and intense purity of design as with his 'Crucifixion' in the National Gallery, London, where the feeling of the terrible moment is banished, and the setting is for the coming victory of the Resurrection. From Florence he was summoned by Pope Julius II to Rome, where among other commissions he received was that for frescoes for the *Stanze* in the Vatican. He chose the subjects of Theology, Philosophy, Law and Poetry. Perhaps that on Philosophy, known as the 'School of Athens', is the greatest – an immense composition with many figures, beautifully balanced, without stiffness, full of life. As far back as 1901 the Colonna altarpiece brought £100000, and in 1931 the 'Alba Madonna' fetched £240800.

RAUSCHENBERG, ROBERT (1925–)
Born in Port Arthur, Texas, he studied at the Kansas City Art Institute, later in Paris with the Académie Julian. A mixed media worker, using collage and often high relief with objects secured to the picture plain.

RAY, MAN (1890–)
Born in Philadelphia, he has experimented with paint, photography and films; he was associated with Moholy-Nagy in exploring the possibilities of an art form using enlarging-paper, but with a hand-produced negative rather than a camera-made one.

REDON, ODILON (1840–1916)
Born in Bordeaux he developed along two quite separate lines. The painter of charming flower pieces, gently treated with a near-Impressionist manner, and the producer of weird human heads, phantoms and objects from nightmares, heavy with symbolism.
'Tête Astrale' F 500000, 1975, Hôtel George V, Paris.

REMBRANDT VAN RIJN (1606–69)
He was born in Leyden, the son of a miller. After a short spell at Leyden University he entered the studio of Van Swanenburgh and then went to Pieter Lastman; both these artists had made the Italian Tour, and so would have been able to pass on influences, such as Caravaggio. Yet Rembrandt rebelled against the commercialized studio and left to battle alone for what he must have known was in his reach.

His subjects covered everything: portraits, among the finest ever painted, large groups of figures, still life, landscape, interiors, religious subjects; not only did he paint and draw but he also produced some of the most wonderful etchings ever made. Perhaps the greatest facet of his genius was his sheer magic with the handling of light; with 'The Adoration of the Shepherds' in the National Gallery, London, there is a perfect lesson in the indication of the highlights and the skilful tracing of details into the shadows.

Rembrandt was married twice; first to Saskia van Uilenburgh, second to Hendrickje Stoffels. He did taste worldly success; he had a fine house, a collection of fine paintings. Yet he would not compromise when commissioned. This was the case with the great canvas of 'The Night Watch' in the Reichsmuseum, Amsterdam (which was viciously and stupidly slashed with a knife in 1976). Instead of painting a series of easily recognized portraits, he did the picture in the way he saw it as a dramatic arrangement of figures. The patron was furious and rejected the painting and in a short space of time fashion turned against him. As his circumstances narrowed, so came sadness; his mother died, so did Saskia. He married Hendrickje, his servant-girl, to find some help, affection, care for the evening years, an act that turned the snob establishment against him more than ever.

Self-portrait, Rembrandt. (Courtesy Wallraf-Richartz Museum, Cologne)

'Portrait of a boy' (probably his son Titus) by Rembrandt. (Courtesy Christie's)

He was alone, wrapped in his visions of creative genius. From this time came many of his landscapes, portraits of humble folk and more and more of the deeply revealing self-portraits, a series of which he had kept going since his student days. Hendrickje worn down by privation and work died, and then his beloved son Titus. Deeper into mortal gloom he appeared to be pulled and the harshness of the world can be read in the face of that last self-portrait; yet the flame of genius could never be extinguished, try as they might. The Wester Kerk, Amsterdam, records that there was put to rest in a pauper's grave on 'Tuesday, October 8, 1669, Rembrandt van Rijn, painter'.

In the 18th century his value was still not seen, Louis XVI in 1777 bought 'Supper at Emmaus', now in the Louvre, for £442. Today he stands where he should be, with that small host of the very great.

'Saskia as Minerva' in 1965 in London £125000, in 1975 in Paris at the Palais Galliera F 1 300000.

'Jacob bénissant les fils de Joseph' (pen and ink) Fl 40000, 1975, Sotheby Mak van Waay, Amsterdam.

'The Agony in the Garden' (etching and dry-point) £4400, 1975, Sotheby's, London.

REMINGTON, FREDERICK (1861–1909)
American sculptor, painter and illustrator, who attended the Yale Art School and the Art Students' League, New York. He was with the US Army in battles against the Indians, and it is from this experience he has drawn many of his subjects. 'Bronco Buster' ($31\frac{1}{2}$ in (800 mm) bronze) $45000, 1975, Adam A. Weschler & Son, Washington, DC.

RENI, GUIDO (1575–1642)
Italian painter from Bologna, he was a pupil with Albani and Domenichino of Calvaert, and was strongly influenced by Caravaggio and the Caraccis. His work was heavily criticized by Ruskin, perhaps somewhat unkindly.

RENOIR, PIERRE AUGUSTE (1841–1919)
Born in Limoges, his artistic life began with painting fans for missionaries destined for hot countries, and decorating china. With saved money he managed to get to the Atelier Gleyre where he met Sisley, Monet and Bazille. For a time he worked as a Divisionist, then passed through various influences including Ingres and Courbet. He travelled to Italy, Venice in particular exciting his sense of colour. From this

time dates the beginning of the flowering of his
style, full-modelled figures, warm with sunlit
colours. A landscape such as 'The Path up through
the Field', in the Louvre, displays Renoir's skill
with light, it appears to permeate to every part of
the composition; or with 'Ball at the Moulin de la
Galette', also in the Louvre, how well he indicates
the dappled sunlight. In this way Renoir was a
quiet revolutionary.
'Jeune Fille au bouquet de Tulipes' £202000,
1975, Sotheby Parke Bernet.

REPIN, ILYA (1844–1930)
Born in Chuguyev in the Ukraine, he studied first
under Bunakov and worked with icon-painters,
then he went to the Society for the
Encouragement of Art and lastly to the Academy
of Arts. A portrait-painter, historical and genre
subjects, his work is considered a model for the
Socialist Realist school in Russia. He was a leading
figure with the Wanderers group.
'A casualty of War' $2300, 1975, Sotheby Parke
Bernet, New York.

REYNOLDS, SIR JOSHUA (1723–92)
Born in Plympton, after some parental objection
he was sent to the studio of Thomas Hudson.
From an early age he felt he had a mission in life to
introduce Italian principles with regard to
painting to the English school; so he set off for
Italy, all the way round by sea, helped by his friend
Captain Keppel. Once there he roamed at will,
absorbing influence, making notes and collecting a
useful folder of master drawings. When he
returned to London he began the practice of
portrait-painting and had great success. The high
and the mighty flocked to his studio in Leicester
Fields.

When the Royal Academy was founded in 1768
he was elected its first President, a position he
held until 1790. With Dr Johnson he ran The Club
in Gerrard Street, where weekly dining and
discussion took place; other members included
Garrick, Goldsmith and Burke.

Reynolds often accused by carping critics of
borrowing, was actually one of the most original
artists in the way he posed his sitters, even causing
Gainsborough to exclaim one day: 'Damn him,
how various he is.'

RIBALTA, FRANCISCO (1565–1628)
Spanish painter who may have been in the studio
of Navarrete, and was strongly influenced by
Caravaggio, whose principles of chiaroscuro he
was the first to introduce into religious painting in
Spain.

RIBERA, JUSEPE (c 1590–1652)
Born in San Felipe, near Valencia, his father
intended he should follow the law, but the young

*'Stampede from Lightning' by Frederick Remington (1861–
1909). Born in New York, he trained at Yale Art School,
and fought with the US troops against the Indians.
(Courtesy Christie's)*

man rebelled and journeyed through to Italy, He,
like many another, came under the spell of
Caravaggio, and this led to his manner for using
the artificial dramatic light to intensify the
emotions in pictures of martyrdom and saints.

RICHARDS, CERI (1903–)
Born at Dunvant near Swansea, he was trained at
Swansea and at the Royal College of Art. He
experimented with Surrealism and then turned to

*Head of St Philip, one of a group of Apostles by Tilman
Riemenschneider. (Courtesy Mainfränkisches Museum,
Würzburg)*

Centre National d'Art et de Culture Georges Pompidou in Paris. (Courtesy French Government Tourist Office)

Italian Art 17th and 18th century. Room 238 in the Hermitage, Leningrad. (Courtesy Novosti Press Agency)

Abstraction. In the Chapel of the Metropolitan Cathedral, Liverpool, is a pleasing altarpiece in colours and forms that echo the stained glass at the top of the great cone.

RIEMENSCHNEIDER, TILMAN
(*c* 1460–1531)
The German sculptor was born in Osterode, Harz; little is known of his early life. Apparently he came to Würzburg in 1483 and remained working there. He was one of the most skilled carvers of wood and stone, being able to impart to material a considerable sympathy and emotion. The quality of his talent can be judged from the altarpiece at Creglingen, in limewood, uncoloured or stained; or the moving purity of the white sandstone figure of Eve now in the Mainfränkisches Museum, Würzburg, commissioned by the Burgomaster and councillors on 5 May 1491.

RIGAUD, HYACINTHE (1659–1743)
French Official Portrait-Painter to Louis XIV and Louis XV, he, with Largillière, responded to the lavish pomp of the Court, presenting flattering likenesses in the rich robes and settings of their patrons.

RILEY, JOHN (1646–91)
Born in London, he was in the studio of Isaac Fuller and later Gerard Zoust. One of the most skilled of English portrait-painters, he was somewhat put in the shade by the workshops and mass production of Lely and Kneller. He was a reserved and rather shy person, who must have been severely jolted when painting Charles II to hear the monarch exclaim 'Is this like me? Then, odd's fish, I'm an ugly fellow.' His greater quality lies with pictures of the less exalted such as 'Scullion' at Christ Church, Oxford, and 'Bridget Holmes, Housemaid' in the Royal Collection.

RIOPELLE, JEAN-PAUL (1923–)
Born in Montreal the French-Canadian painter has lived in Paris since 1948. He has felt his way through the maelstrom of theory and style that has been the scene, starting with Surrealism and moved with this still underlying his work to a greater sense of freedom and expression.

RIVERA, DIEGO (1886–1957)
The Mexican painter who was working in Paris during the genesis of Cubism, one who knew most of the leading figures associated with the movement; despite the strong influences he developed his own manner which was to emerge as an immensely powerful imagery, culled in part from the ancient Aztec and Mayan with perhaps the freedom discovered by Gauguin. In the corridors of the Palacio Nacional, Mexico City is his 'The Market Place of Tenochtitlan-México'.

Kröller-Müller sculpture park, Holland. (Photo by the Author)

Carried out in fresco, it is a quite remarkable presentation of a very confused scene, with hundreds of figures and panoramic views, yet all clearly stated; in contrast is the arresting display of massive anatomy in another fresco 'Man Masters the Elements' in the Salon de Actos of the National Agricultural School at Chapingo, Mexico.

RIVERS, LARRY (1923–)
Born in New York, he studied first at the Juillard School of Music and then the Hans Hoffman School of Fine Arts, New York. His portrait manner has an influence of the broken image from Francis Bacon, other subjects are often related to Pop Art symbolism.
'Summer' $13 000, 1975, Sotheby Parke Bernet, New York.

ROBERT, HUBERT (1733–1808)
French painter who specialized in landscapes with romantic ruins. In Rome he was a friend of Panini and Piranesi. Later he became Keeper of Louis XVI's paintings and then was **one of the early Curators of the Louvre**.
'Le Temple Antique' F 80 000, 1975, Palais Galliera, Paris.

ROBERTS, TOM (1856–1931)
Born in Dorchester, England and died at Kallista,

'Les Bourgeois de Calais' – Musée Rodin. (Courtesy Commissariat Général au Tourisme)

Victoria, Australia. He studied at the Collingwood branch of the Mechanics' School of Design and the Melbourne National Gallery School under Thomas Clark. **He was the founder of the artists' camps at Box Hill and Heidelberg.**

RODIN, AUGUSTE (1840–1917)
Born in Paris, he worked in the studio of Horace Lecoq de Boisbaudran. He failed his entrance examination three times for the École des Beaux-Arts, and then while working as an assistant to a decorative mason attended night classes with Barye. After this he was at the Sèvres factory. In 1864 he submitted 'The Man with a Broken Nose' to the Salon and it was rejected. He went to Italy in 1874 and was greatly moved by the intense spirituality he saw in Michelangelo's work, also by his tremendous command over anatomy. On his return such was his mastery that he was accused with life-size figures of actually casting them from the model. He dispersed his critics by producing the over-life-size figure of St John the Baptist, now in the Tate Gallery, London. One of his finest portraits in bronze is that of Balzac, which at the time drew scorn even to the point that the Société des Gens de Lettres who had commissioned it, refused to honour their commitments. Today his

genius has been at last recognized and there are numbers of his best works to be seen either in the Rodin Museum in the Hôtel Biron, Paris or the Rodin Museum at Meudon.
'La Cathédrale' (bronze 25 in (635 mm)) £5775, 1975, Christie's, London.
'L'Amour et Psyché' (bronze 8 in (203 mm)) DM 33000, 1975, Hauswedell & Nolte, Hamburg.

ROMNEY, GEORGE (1734–1802)
He was born at Walton-le-Furness, Lancashire, and studied with a weird character in Kendal, Christopher Steele, who neglected the tuition and more or less used Romney as a studio drudge. After this he led a strange life, nearly illiterate, he married and then abandoned his wife in Kendal and set up a studio in London where he found some considerable success. He is most famous for the series of portraits of Emma Hamilton in various guises. He travelled to Italy and broadened his style thereby; probably he is best seen in his sketches, which are lively and forceful. As success waned he left London and returned to his poor wife.
'Portrait of Mrs Roger Smith' £17850, 1975, Christie's, London.

ROSA, SALVATOR (1615–73)
Italian Romantic painter who probably studied with Ribera. He produced many pictures with ragged exciting landscapes as the backgrounds to battles, bandits, saints and peasant labourers. He had a great vogue with English collectors. His values have moved up and down; high in the 18th century, down sometimes to a few pounds in the 1920s and now reaching up into the thousands of pounds.

ROSSELLINO BROTHERS
Florentine sculptors, **Bernardo** was the elder (1409–64) and **Antonio** (1427–79). Bernardo was also an architect and the more talented, among his best work is the tomb of the Humanist and Florentine Chancellor Leonardo Bruni in Santa Croce, Florence. Antonio also was a producer of tombs and can be judged by that of the Cardinal-Prince of Portugal that is in San Miniato, Florence.

ROSSETTI, DANTE GABRIEL (1828–82)
He was the third principal member of the Pre-Raphaelite Brotherhood, with Holman Hunt and Millais. With his work he set out to capture an illusive idealism redolent with poetic religion.

ROSSO, GIOVANNI BATTISTA (1494–1540)
(also known as Rosso Fiorentino). Italian painter who was **one of the founders of Mannerism**. He was working in Florence up till 1523 and then

went to Rome; surviving the Sack of 1527 he was asked to go to France and became one of the leading lights of the First School of Fontainebleau, working particularly on a series of allegorical pictures based on the life-style of François I.

ROTHKO, MARK (1903–70)
Born in Dvinsk, Russia, he went to America and studied at Yale and the Art Students' League, New York. His manner is a form of Soft-edged Abstract, often working with a limited palette. 'Green, blue and green on blue' £17850, 1975, Christie's, London.

ROUAULT, GEORGES (1871–1958)
He started out as an apprentice to a stained-glass painter and helped to restore ancient windows. Next he went to the École des Beaux-Arts to study under Gustave Moreau. Apart from participation in the first show of Les Fauves he kept away from the mass of theories and groups. The influence of Matisse is strong in Rouault's 'Three Judges' in the Tate Gallery, London. His choice of subject ranged from religious and spiritually moving paintings to the mundane and voluptuous nudes as 'At the Mirror' in the Musée d'Art Moderne, Paris. He also worked on sets for the Ballets Russes and did some engraving. 'Le Lutteur' (oil and gouache) £26250, 1975, Christie's, London.

ROUBILIAC, LOUIS-FRANÇOIS (1695–1762)
French sculptor who worked with Nicolas Coustou and then settled in England. Fine works include: 'Monument to the Duke of Argyle' in Westminster Abbey, 'Sir Isaac Newton' at Trinity College, Cambridge and 'George Frederic Handel'.

ROUSSEAU, HENRI (1844–1910)
(called Le Douanier). Born at Laval in Mayenne, at 18 he was playing a saxophone in the 52nd Infantry Band. The year 1871 saw him in action in the Franco-Prussian War and after his discharge he entered the Customs service – hence his nickname. He began to paint, untaught, a pure Primitive who just answered an urge within him. His style which could have influenced many others since his time, was colourful and had at times a pleasant mystical quality as with 'The Snake Charmer' in the Louvre, or 'Sleeping Gipsy' in the Museum of Modern Art, New York. 'La Promenade aux Buttes-Chaumont' £40000, 1975, Sotheby Parke Bernet.

ROUSSEAU, THÉODORE (1812–67)
French landscape-painter associated with the Barbizon school. His scenes are influenced by the Dutch landscape-artists of the 17th century;

'War' by Douanier Henri Rousseau. (Courtesy the Louvre, Jeu de Paume, Paris)

the scenes were often with late afternoon or evening light, and were filled with a convincing atmosphere.

ROWLANDSON, THOMAS (1756–1827)
The celebrated caricaturist, satirist and etcher was born in the Old Jewry, London. He attended Dr Barrow's Academy in Soho, later a period at the Royal Academy Schools. He began his career as a serious painter, but living a somewhat dissipated life, with gambling and the rest, he let this lapse and took to drawing. His technique was generally to use a reed pen with an ink composed of vermilion and black, which he would dilute for tones, and finish with local colours. Many of his works were published by Ackermann. 'The Married Man or The Happy Family' (pen and grey ink with water-colour) £1785, 1975, Christie's, London.

RUBENS, SIR PETER PAUL (1577–1640)
Born in Siegen in Westphalia his parents hailed from Antwerp, and the family returned there in 1587. Rubens studied with three teachers one after

'Private Practice Previous to the Ball' by Thomas Rowlandson. (Courtesy Christie's)

the other, Verhaecht, Adam van Noort and Otto van Veen. At 23 he went to Italy and entered the service of the Duke of Mantua. In 1603 he visited Spain and had the chance to view the Royal Collection. Back to Italy and then recalled to Antwerp by news of his mother's illness, he arrived too late, but settled there as Court Painter to the Spanish Governor of the Netherlands. From this time dates the setting up of his immense establishment, which was really a very high-quality picture factory, employing such talented hands as Jan Brueghel, Frans Snyders, Jordaens, Teniers and Van Dyck. Rubens the courtier and travelling diplomat had to do this as he received so many orders from the royalty and noblemen he saw on his trips.

Yet Rubens himself, with his energy and output was a giant among painters, a giant imbued with very considerable talent, and depth of feeling as can be seen in the 'Descent from the Cross' in Antwerp Cathedral, considered as his masterpiece, his magnificently controlled handling has left one of the most dramatically beautiful pictures. Landscape he provides with 'View of the Château Steen'; portraits epitomized by the 'Le Chapeau de Paille'; both of these are in the National Gallery, London. Values today are very high; in 1959 'Adoration of the Kings' went for £275000, in 1806 it brought just £800, in 1783 £700.

RUBIN, REUBEN (1893–1975)
Born in Galatz, Rumania, he went to Israel and studied at the Bezalel School, Jerusalem, and then travelled for further studies to Paris, Rumania and Italy. In 1923 he helped to found the Israeli Association of Painters and Sculptors. His manner had some influence from Modigliani and Matisse. He also designed décor for the theatre and made woodcuts.
'L'Entrée du Verger' $12000, 1975, Sotheby Parke Bernet, New York.

RUBLEV, ANDREY (c 1370–c 1430)
The greatest Russian icon-painter, he is first mentioned in 1405 as working as an assistant to Theophanes the Greek in the Cathedral of the Annunciation in Moscow. Other works ascribed in part or wholly to him are in the Cathedral at Vladimir and entirely his is the icon of the Old Testament Trinity, now in the Tretyakov State Gallery, Moscow.

RUDE, FRANÇOIS (1784–1855)
French sculptor who changed from a style based on the Antique to one embracing Romanticism. Best-known works include the relief in the Arc de Triomphe 'Departure of the Volunteers' and 'Marshal Ney' in the Place de l'Observatoire, Paris.

RUISDAEL, JACOB VAN (c 1628–82)
Born in Haarlem where it is likely he had some tuition from his father Isaac. He was to develop into one of the leading landscape-artists of Holland; his influence was wide, and **among his pupils was Hobbema.**

RUSKIN, JOHN (1819–1900)
English writer on art and social problems, leading art critic of the Victorian times and water-colourist and draughtsman.
'Pillars from Lucerne' £378, 1975, Christie's, London.

RUYSDAEL, SALOMON VAN (c 1600–70)
One of the most prolific Dutch landscape-painters, he was the uncle of Jacob van Ruisdael. His works were of a consistently high quality.
'A view on the Rhine near Arnhem' £36750, 1975, Christie's London.

RYDER, ALBERT PINKHAM (1847–1917)
American painter who worked in the studio of William E. Marshall. Choosing imaginative subjects, he induced into them a quality of mystery which can be noted in that strange picture 'Death on a Pale Horse' in the Cleveland Museum of Art.

RYSBRACK, JOHN MICHAEL (1694–1770)
Flemish sculptor who came and settled in England and by quality was in competition with Roubiliac. He executed a number of tombs in Westminster Abbey including that of Sir Isaac Newton, also the Marlborough Tomb at Blenheim and the excellent equestrian statue of William III in Bristol.

SACHEUSE, JOHN
A pure-blooded Greenland Eskimo who sailed in 1818 with Ross on the voyage to find the elusive North-West Passage. He was to act as an interpreter at £3 a week. He also unofficially recorded some of the happenings in water-colour.

SAENREDAM, PIETER (1597–1665)
Born in Assendelft, he studied with Franz de Grebber. Afterwards he became noted for his delightful church interiors, with carefully worked-out perspective and muted cool colours.

SAINT-GAUDENS, AUGUSTUS (1848–1907)
American sculptor, born in Dublin of a Gascon and an Irish mother, he and his family went to America in the year of his birth. At 13 he was apprenticed to Louis Avet, a cameo-cutter; later he went to the Cooper Union Institute and the National Academy of Design. In 1867 he went to France to attend the École des Beaux-Arts, Paris. Returning to America he was much in demand. Works include: the statue of Lincoln at Lincoln

Park, Chicago, and the Farragut Monument at Madison Square.

SANDBY, PAUL (1725–1809)
Born in Nottingham, he came to London when he was 16 and worked in the draughting office of the Tower. He spent much of his life as a travelling topographer recording in water-colour, or wash and pencil. The first trip was with the Duke of Cumberland to the Highlands of Scotland. Then he worked for some time with his brother Thomas (1721–98), also an artist and an architect, at Windsor. Next he went to Wales with Sir Watkin Williams Wynne. He also worked in aquatint, etching and engraving.
'Part of the hundred steps and Winchester Tower, Windsor Castle' (water-colour with body colour), £1680, 1975, Christie's, London.

SARGENT, JOHN SINGER (1856–1925)
American portrait-painter who attended the Florence Academy and worked in Paris with Carolus Duran. He settled in London in 1884 and had a considerable success with the top society portraits. His style was very much *alla prima*, with a great use of well-loaded brushes with long swirling strokes. Latterly he produced a number of surprisingly sensitive landscapes.
'Les enfants Pailleron' £52 500, 1975, Christie's, London.

SARTO, ANDREA DEL (1486–1531)
One of the most important painters in Florence for his period, he worked in the studio of Piero di Cosimo. He was the first in Florence to rely on 'blocking in' a composition rather than making a careful under-drawing. With fresco he had a particularly accomplished manner; noteworthy examples are the 'Nativity of the Virgin' and the 'Madonna del Sacco', both in SS Annunziata, Florence, and his 'Last Supper' in the Refectory of San Salvi.

SARTORIUS, FRANCIS (c 1775–c 1830)
Horse-painter, whose work can be confused with that of his father. One of his best pictures was of the Marchioness of Salisbury riding at Hatfield, which was hung in the 1806 Academy. He is one who is having a considerable return to popular esteem. In 1874 'Six Huntsmen in an extensive landscape' went for 65 guineas; in 1976 it fetched £5500.

SASSETA, STEFANO DI GIOVANNI (c 1392–1450)
One of the most imaginative and inventive Sienese painters, he took advantage of prevailing fashions and the International Gothic decorative manner; but still held to the underlying ideals of the 14th century in religious art. One of his main works which has been satisfactorily authenticated is the St Francis altarpiece.

SCHLEMMER, OSKAR (1888–1943)
German painter and stage-designer who worked under Adolf Hölzel, he developed a monumental style following on with Runge, Hodler and Puvis de Chavannes. He turned his figures into columns, resolving many of his composition problems by geometry. Schlemmer taught at the Bauhaus for nine years. He executed reliefs in the stairwell of the Art Academy in Weimar, but unfortunately Schulze-Naumburg, the successor of Gropius, had them destroyed.
'Schwarzbezopfte von hinten' (oil and pencil on canvas) £6825, 1975, Christie's, London.

SCHMIDT-ROTTLUFF, KARL (1884–)
He was the youngest member of the Brücke, and early on was heavily influenced by Les Fauves. In 1906 he was painting with his friend Nolde on the island of Alsen, and this did much for the richness of his colour to come. Perhaps in some ways he produced his best work from the wood-block; he adopted a very free style with woodcuts, often working from dark to white, using direct cuts he produced many lively and memorable prints in the manner of 'Head of a Labourer' 1923.
'The Sower' DM 135000, 1975, Hauswedell & Nolte, Hamburg.

SCHONGAUER, MARTIN (c 1430–c 85)
Born at Colmar, he was a painter and particularly an engraver. There is only one picture that is certain to be by him and that is the decorative 'Virgin in a Garden of Roses' (or 'Madonna of the Rosehedge') which is in the Church of St Martin, Colmar. There are more than 100 engravings bearing his initials M.S. with a cross, one arm of which is hooked.
'Christus vor Pilatus' (engraving from 'The Passion') DM 15000, 1975, Karl u. Faber, Munich.

SCHWITTERS, KURT (1887–1948)
German painter and collagist, who studied at the Dresden Academy, and worked his way through many theories and ideas. Influenced at first by Kandinsky he went on to evolve his own idea of Dada, which he named Merz. In his house in Hanover he built a huge Merz-construction 'Cathedral of Erotic Misery' which gradually reached up through all the storeys of the house, and included such as: Nibelung treasures, Goethe's relics, and worn-out writing pencils. He stated 'I simply could not see any reason why old streetcar tickets, driftwood, coat checks, wire and wheel parts, buttons, junk from the attic and heaps of refuse should not be used as material for paintings, any less than colours made in a factory.'

'Mz 223, Heet Water' Sfrs 43 000, 1975, Kornfeld und Klipstein, Berne.

SCOREL, JAN VAN (1495–1562)

Born at Scorel, he studied with James Cornelisz. The year 1519 saw him in Nuremberg visiting Dürer, from there to Venice and here he joined a pilgrimage to Jerusalem, back again to Venice and then to Rome where he painted Adrian VI, the Utrecht Pope, who made him a Canon of Utrecht. He had many of his pictures destroyed by the iconoclasts. Remaining are a number of fine portraits from the 'Jerusalem Pilgrims' to 'Portrait of Agatha Schoonhoven' in the Doria Pamphili Gallery, Rome.

SCOTT, SAMUEL (c 1702–72)

English marine painter in the manner of the Van de Veldes, influenced by Canaletto.
'The River Thames, London' £10 500, 1975, Christie's, London.

SEBASTIANO DEL PIOMBO (c 1485–1547)

Italian painter who probably trained in the studio of Giovanni Bellini, and was strongly affected by the work of Giorgione. In 1511 he journeyed to Rome and met Raphael and Michelangelo. Two portraits in the Uffizi of a Young Woman and Young Man display his fine control of light and modelling. The National Gallery, London, has his 'Raising of Lazarus' which has marks of Michelangelo's influence.

SEGHERS, HERCULES (c 1590–1638)

Dutch landscape-painter, a pupil of Coninxloo. He developed a very original manner with landscapes, often with strange mountains in the background; influences present point to Elsheimer. Rembrandt was known to have had eight of his pictures.

SEQUEIRA, DOMINGOS ANTONIO DE (1768–1837)

Portuguese historical painter and designer born in Lisbon. He studied in Rome with Antonio Cavallucci and on his return to Portugal produced many paintings, also portraits. As proof of his skill as a designer, the silver table service in Apsley House, London, that was given to the Duke of Wellington by the Regent of Portugal, was by him. In 1823 he went to Paris to show his 'Last Moments of the Poet Camoens'. After that he went to Rome and became devout.

SÉRAPHINE DE SENLIS (1864–1934)

(properly Séraphine-Louis). French painter born at Assy in the Oise, she spent her childhood as a shepherdess. Then she worked as a charwoman and somewhere along the line she started to paint with a closely guarded secret of the enamel-like method she used. It was the collector Wilhelm Uhde who discovered her. One day in a café he saw a delightful still life of some apples, on asking who had done it, they told him it was by the servant in the place he was staying. He bought some of her work and then lost touch with her until after the First World War when he again found her 'small and withered, with a fanatical look and livid face framed by discoloured locks' and devoted utterly to her painting.
'Flowers on blue background' F 15 100, 1975, Hôtel Drouot, Paris.

SEROV, VALENTIN ALEXANDROVICH (1865–1911)

Born in St Petersburg, the son of the composer A H Serov, he was a pupil of Repin and at the Academy of Arts under Chistiakov. Painter of portraits, landscapes and historical genre, designer for the theatre and etcher and lithographer. Member of *Mir Iskusstva*. His pupils included, Saryan, Petrov-Vodkin and Yuon.

SEROV, VLADIMIR ALEXANDROVICH (1910–)

Born in Leningrad, he studied at the Academy of Arts under Brodsky and Savinsky. Painter of historical revolutionary subjects and portraits. In the Lenin Museum in Moscow is his 'Delegates from the Villages visiting Lenin during the Revolution'.

SERRES FAMILY

The father, **Dominic the Elder** (1722–93), was born in Auch in Gascony, his parents intended him for the Church; but he ran away to sea. His trading vessel was captured by the English in 1752 and he, being brought to England, set out to become a marine painter, and succeeded so well that soon he was selling freely, and he became a Member of the Royal Academy when it was instituted. He was also appointed Marine Painter to George III. An important work was 'Lord Howe's victory over the combined Fleets of France and Spain.' **John Thomas** (1759–1825) studied with his father and also became a marine painter. He was Drawing Master at the Naval School in Chelsea, and published *The Little Sea Torch for Coasting Ships* with coloured plates and in 1825 his *Liber Nauticus* a handbook for marine painters. Sadly he was ruined by the depravity and high-spending of his wife Olive, who would insist on calling herself the Princess of Cumberland. He became bankrupt and died in prison. **Olive** (1772–1834) wife of John Thomas, she was the daughter of a house-painter named Wilton, although she claimed to be the daughter of the Duke of Cumberland. She painted landscapes. Dominic the Younger was the younger son of Dominic the Elder, and he was a water-colourist and landscape-painter. After a number of years he yielded to melancholia and was supported by his brother John Thomas.

SEURAT, GEORGES (1859–91)
Born in Paris, he had two years at the École Nationale des Beaux-Arts, and after a 12-month period of military training he spent much time studying the pictures of such as Delacroix, Ingres and Veronese in the Louvre and also reading the works of colour theorists, like Charles Blanc, Chevreul, Rood and Sutter. He worked on a principle that form was more important than line. Then he set himself to apply the colours in the manner of Pointillism: pin-head dots of pure colours interspersed so that the eye of the viewer sees them as areas of tone and tint. He was the most disciplined of the Divisionists. There is his 'Une Baignade' in the Tate Gallery, London, 'Young Woman powdering herself' in the Courtauld Collection, London, and another excellent example 'Fishing Fleet at Port-en-Bessin' in the Museum of Modern Art, New York.

SEVERINI, GINO (1883–1966)
Italian painter born at Cortona, he attended evening classes at the Villa Medicis, and then in 1901 met Boccioni and with him worked in Balla's studio on Divisionist principles. Thence to Paris where he met most of the leading figures who were trying out Cubism, Futurism and Synthetic Cubism. After the First World War he went back to a form of Classicism, then turned to religious subjects and finally worked on the patience-demanding art of mosaics.
'Composition 1929' L 7500000, 1975, Finarte, Rome.

SHAHN, BEN (1898–1969)
American painter born in Kaunas, Lithuania, he entered the USA in 1906 and worked as a lithographer. After the First World War he attended New York University, the City College and the National Academy of Design. Early work was harshly critical of the social scene; later he became more withdrawn and was painting imaginative compositions of quality; on the one hand pointing to loneliness and the impact of civilization and on the other painting a still life such as 'Composition with Clarinets and Tin Horn', in the Detroit Institute of Arts, which points to influences from the Bauhaus teachers.

SHEPHERD, DAVID
Animal painter at large today, deeply concerned with Wildlife Conservation. He was awarded the Order of the Golden Ark by HRH the Prince of the Netherlands for his services to conservation in Zambia and to Operation Tiger.

'Charging Elephant' £840, 1975, Christie's, London.

SICKERT, WALTER RICHARD (1860–1942)
English painter born in Munich of an English mother and a Danish father, he attended the Slade School with Alphonse Le Gros and then became a pupil of Whistler's. Later a strong influence on him and a close friend was Degas. **One of the founders of the Camden Town Club**, he was with Wilson Steer one of the most important figures in the vogue of English Impressionism. Always an addict of the theatre, many of his paintings are associated with this subject, put down with warm rich colour and soft outlines, but redolent with atmosphere.
'Street in Dieppe with two figures' ($9\frac{1}{2} \times 7\frac{1}{2}$ in) (241×190.5 mm)) £1050, 1975, Christie's, London.

SIGNAC, PAUL (1863–1935)
French painter born in Paris and a follower of the style of Seurat, he trained at the Académie Libre de Bing under Alexandre Guillemin. He exaggerated Pointillism, applying his colours in quite large spots or rectangles so that an effect somewhat similar to mosaic was achieved.
'Les Andelys, les bains' £50400, 1975, Christie's, London.

SIGNORELLI, LUCA (1441–1523)
Italian painter born at Cortona, and was probably a pupil of Francesca. His manner that developed was Florentine with the accent on outline. His best work is the fresco cycle in the Cappella della Madonna di San Brizio, in the Cathedral of Orvieto; this was begun by Fra Angelico, and Signorelli worked for five years on it. It shows a Dantesque conception of the End of the World, The Coming and Fall of Anti-Christ and the Last Judgement. It is of interest that he does not people his nether world with strange outlandish beasts such as Bosch or Dürer might have conjured up, but rather readily recognizable well-built powerful figures only differing from live mortals by the putrid colour of their flesh.

SIQUEIROS, DAVID ALFARO (1896–   )
Born in Chihuahua, Mexico, his studies included a Government grant to go to Paris. He has always been conscious of the struggles to break through for freedom of expression and personal and group rights. His style that emerged has great power shown by his self-portrait in the Museum of Modern Art, Instituto Nacional de Bellas Artes, Mexico City. This is painted in pyroxylin-bound pigments on celotex and makes considerable use of rough texture and dark warm colours.

SISLEY, ALFRED (1839–99)
Born in Paris, his studies included periods at the École des Beaux-Arts and the Atelier Gleyre. Although one of the most loyal Impressionists he has not had as much publicity as the others, perhaps because he was utterly devoted to the business of painting. He was a man of immense fortitude having to battle continually for the barest necessities to keep himself going. Yet so steadfast and strong was his spirit that no sign of his material need comes through to the harmonious and peaceful beauty he could put down on his canvases. In 'L'Abrevoir' in the Tate Gallery or 'Flood at Port-Marly', in the Louvre, these qualities are evident. His value in his time shows only the oft-repeated indifference or distrust of new talent: in 1890 'Le Verger' £46, in 1906 'View of the Seine' £168. It was not until after the Second World War that his paintings started to be appreciated.
'Sentier dans le jardin de By, Matin de Mai' Sfrs 460000, 1975, Galerie Motte, Geneva.
'La Seine à Bougival' £42000, 1975, Christie's, London.

SLOAN, JOHN (1871–1951)
American painter and engraver who studied at the Pennsylvania Academy of Fine Arts, and received tuition from Robert Henri, he started as an illustrator for newspapers and magazines. A member of The Eight he was a leading figure with the modern Realists.

SLUTER, CLAUS (working about c 1380– c 1406)
One of the greatest sculptors of his time in northern Europe, he worked almost only for Philip the Bold, Duke of Burgundy. His outstanding creation was the 'Puits de Moise' (Well of Moses) at Dijon, a well-head in stone with six full-length Prophets. He achieved an astonishing and sympathetic Realism, which called for a polychrome finish which was often done by Malouel.

SMITH, DAVID (1906–65)
American sculptor, who after initial studies at Ohio University worked in the Studebaker factory as a riveter on a car assembly line. Then he went to the Art Students' League, New York. Early days had him earning a living as a taxi-driver, salesman, carpenter and seaman. Primarily he is a Constructionist and a leading figure in welded metal sculpture.
'Voltri XII' $67000, 1975, Sotheby Parke Bernet, Los Angeles.

SMITH, SIR MATTHEW (1879–1954)
English painter of flowers, landscapes and nudes, very much influenced by Les Fauves. His studies included a time at the Slade School and in the studio of Matisse.

SNYDERS, FRANS (1579–1657)
Born at Antwerp he worked under Pieter Brueghel the Younger and had particular abilities with animals and still life. Rubens often employed him on such subjects in his large compositions.

SODOMA, GIOVANNI ANTONIO BAZZI (1477–1549)
(known as Il Sodoma) Born at Vercelli in northern Italy he studied with Spanzotti and was influenced by the work of Leonardo. He evolved a harmonious poetic treatment for landscape set in the background of his pictures. In San Domenico, Siena is a fine fresco series by him 'The Life of St Catherine'.

SOEST, KONRAD VON (active early part 15th century)
German painter, working in the Dortmund area; his manner was in the so-called Soft Style (the name for the painting in Germany at the end of the 14th and beginning of the 15th century, related to International Gothic it was characterized by flowing rhythms and light gentle sentiment). Best-known and authenticated work is a signed polyptych in the parish church of Niederwildungen.

SOUTINE, CHAIM (1894–1943)
Born at Smilovich, near Minsk, he was tenth in a family of eleven; his father was a poor tailor, but he managed to attend the School of Fine Arts in Vilno. With help from a local doctor who saw his talent, he went on to Paris to the École des Beaux-Arts. His manner expresses an intense involvement with his sitters or with landscapes. The paint is put on with a feeling of urgency and passion. In his lifetime he was a solitary one, passed by, not caring to exhibit but collected by some: in 1923 Doctor Barnes, an American, had bought 100 pictures.
'Maison derrière les arbres' £35700, 1975, Christie's, London.

SPENCER, SIR STANLEY (1891–1959)
Born at Cookham in Berkshire and a pupil at the Slade School, he developed into a mystical religious painter and also a landscape-artist with a meticulous observation and attention to detail. **His murals decorate the War Memorial Chapel at Burghclere in Hampshire.**
'Hilda and I at Burghclere' £12600, 1975, Christie's, London.

STAEL, NICOLAS DE (1914–55)
Russian painter who worked primarily in France; his manner was Abstract and Expressionist with the use of rugged impasto.

'Nature Morte-boite de peinture' £32000, 1975, Sotheby's, London.

STEEN, JAN (1626–79)
Born at Leyden and studied with Jan van Goyen. He painted small genre pieces often with a humorous touch. At one time he leased a brewery in Delft and kept an inn at Lange Brug (Long Bridge) in Leyden. Hard-pressed by creditors, he worked unremittingly to solve his problems and sold his work often at a very low price. A year's rent for his house was only 29 florins, but the landlord demanded three well-painted portraits. 'Peasants playing bowls outside an inn', £27300, 1975, Christie's, London.

STEER, PHILIP WILSON (1860–1942)
English painter who studied first at the Gloucester School of Art and after in Paris at the Académie Julian. **A founder-member of the New English Art Club**, he was influenced by Turner and introduced much of the Impressionists' manner to the English scene.

STEVENS, ALFRED (1817–75)
Born at Blandford Forum in Dorset, he had no academic training, but he was befriended by the Reverend Samuel Best who sent him off on a boat for Italy, but overlooked the fact he would need funds there. Poor Alfred, knowing no Italian, was set on by thieves, drawn into political intrigues, stayed a year and a half in Naples and then walked to Rome, and from there to Florence, where he made some kind of a living copying masters' works for dealers and others. In 1840 he had a time in the studio of Thorvaldsen, the Danish sculptor. Back in England his best work was done as a sculptor, he completed the Wellington Monument for St Paul's, although it was not erected until some 40 years after Stevens's death. He worked not only in marble but also bronze, porcelain and silver. As a painter he is remembered by the sympathetic and sensitive portrait of Mrs Collman in the Tate Gallery, London.

STORCK, ABRAHAM (1630–c 1710)
Born in Amsterdam, his teacher is not known, but there appears to be an influence from Backhuysen. One of his best pictures shows the arrival of the Duke of Marlborough at Amsterdam, with a public procession of ships, barges and yachts decorated over all with flags. He is one who has shown a marked increase today. In 1835 Christie's sold his 'A Cappriccio Mediterranean Harbour Scene' for $7\frac{1}{2}$ guineas; in 1976 it fetched £5250.

STOSS, VEIT (c 1450–1533)
Celebrated German sculptor, who worked on the huge altarpiece in St Mary's Church in Cracow, he also did the red marble tomb for Kasimir IV of Poland. Settled back in Nuremberg he produced among other works the altarpiece for Bamberg.

STROZZI, BERNARDO (1581–1644)
Born in Genoa, he was the pupil of Pietro Sorri, and at an early age he became a Franciscan friar; this led to his nickname of Il Cappuccino. He was considerably influenced by Rubens. In 1610 he left his cloister to help his mother, and remained the rest of his life a painter. In San Domenico, Genoa, there is a powerful and impressive fresco by him of Paradise.

STUART, GILBERT (1755–1828)
American portrait-painter, he travelled to Scotland in 1770, then back to America, and once more over the Atlantic to London to study with Benjamin West. After this to Dublin for six years and once more to America when among many other portraits **he painted George Washington 124 times.**
'Portrait of General William Hull' $8000, 1975, Sotheby Parke Bernet, New York.

STUBBS, GEORGE (1724–1806)
Born in Liverpool, **he was a prodigy, by the age of eight he was studying anatomy**, and soon after entered the studio of a local artist and also was able to see and absorb the great masters in Lord Derby's Collection. He started as a portrait-

'Mrs Richard Yates' by Gilbert Stuart. (Courtesy National Gallery of Art, Washington, DC, Andrew W. Mellon Collection)

painter, then he worked for Wedgwoods. On travels to Italy he accidentally witnessed a lion attacking a horse, which was to be used as the subject for one of his well-known pictures later. Back in England he set about becoming the fine horse-painter he was to be. As preparatory studies he worked for six years on a portfolio of etchings and engravings, which he published under the title of **The Anatomy of the Horse**, which is still counted as the best work of its kind. Most of his dissecting and drawing was done in a lonely Lincolnshire farmhouse, where he could work away at his grim task. His one companion was a Miss Mary Spencer who must have been a remarkable one to have endured the succession of dead horses, each one remaining in the house for up to six or seven weeks. Stubbs himself seemed to have been indifferent to the charnel conditions.
'Portrait of a huntsman, standing by a saddled chestnut hunter' £60000, 1975, Sotheby Parke Bernet.

SULLY, THOMAS (1783–1872)
American portrait-painter born at Horncastle, Lincolnshire, England. He went with his parents to America when only nine. In 1806 he studied with John Trumbull in New York, also Gilbert Stuart. In 1809 he came back to England spending a period in the studio of Benjamin West. Back to America again he set up his studio in Philadelphia. His sitters included: Thomas Jefferson, Fanny Kemble, Lafayette and Washington.

SUTHERLAND, GRAHAM (1903–   )
Born in London, after a projected career as a railway engineer had been abandoned he studied at the Goldsmiths' College School of Art. Since those days long ago he has developed into one of the most powerful portrait-painters of his time. The image he presents is one culled forth from the inner person of his sitter by his perception; often the sitter may be shocked, yet the pictures are documents on those they portray. As well he has painted landscapes in a highly imaginative manner, worked as an Official War Artist in the Second World War and designed the great tapestry for Coventry Cathedral.
'Landscape with stones and grasses' £7800, 1975, Sotheby Parke Bernet.

TAMAYO, RUFINO (1899–   )
Born at Oaxaca, Mexico, of a Zapotecan family. An Expressionist he studied at the San Carlos Academy in Mexico City. At this period he advocated a return to smaller paintings as opposed to the vogue for large frescoes which was sweeping the country. He left Mexico for New York in 1938 and was greatly influenced by the Picasso Exhibition in 1939. His style since has grown increasingly more stark and at times satirical and savage, as with 'The Singer' in the Musée National d'Art Moderne, Paris.
'Hombre contemplando los pajaros' $27000, 1975, Sotheby Parke Bernet, New York.

TANGUY, YVES (1900–55)
Born in Paris, he travelled the world as a merchant seaman, and quite unexpectedly turned to painting after seeing a picture by Giorgio Chirico in the windows of the dealer Paul Guillaume. The art that emerged from him so suddenly, started as simple landscapes, and then rapidly changed to a strange serene Surrealism. In 1926 he did try a somewhat dangerous experiment, by receiving electric shocks he hoped to bring about automatic painting. The year 1939 saw his departure from France for New York and thereafter he lived more or less in artistic isolation.
Untitled' £34000, 1975, Sotheby Parke Bernet.

TATLIN, VLADIMIR (1885–1953)
Russian painter, designer and maker of Abstract constructions, he studied at the Moscow School of Painting, Sculpture and Architecture, and was linked with Anton Pevsner, Naum Gabo and Alexander Rodchenko. In 1919 he erected the model for his famous 'Monument to the Third International'; it consisted of two cylinders and a glass pyramid, revolving at varying speeds, and outside there was an iron spiral. It would have been one of the first Abstract buildings, measuring 1300 ft (396 m) full size.

TCHELITCHEW, PAVEL (1898–1957)
Russian Romantic painter, who was first in Berlin then settled in America. His manner veered from Automatism to Surrealism with a weird distortion of perspective.
'Study for Phenomena', $32000, 1975, Sotheby Parke Bernet, New York.

TENIERS FAMILY
The main ancestor was **Julian I** (c 1558–85) who had two sons, **Julian II** (1572–1615) and **David I** (1582–1649). Julian II had two sons, **Julian III** (Master in 1636) and **Theodore I** (Master in 1636). David I or the Elder painted religious pictures and had four sons, **David II** or the Younger and most famous (1610–90), **Julian IV** (1616–79), **Theodore II** 1619–97) and **Abraham** (1629–70) he learnt from his father and brother, and painted festivals in the style of, but inferior to, David the Younger. David the Younger was a Master at Antwerp by 1632 and worked there until 1651, then moved to Brussels. His work was mainly with genre, although he did some pictures based on the activities of witches, also monkeys and cats dressed up. He was Court Painter to the

Archduke Leopold Wilhelm and also Curator of his collection; of these works he made more than 200 copies and afterwards published engravings from them. David the Younger had a son **David III** (1638–85) who painted in the same manner, but whose work was inferior to that of his father; he also had a son **David IV.** In addition several of the above had more children and nearly all of them painted, so the whole family adds up to one of the biggest puzzles for the art historian.

'Peasants dancing outside a country inn' (David Teniers the Younger) £68 250, 1975, Christie's, London.

'Figures in a Cave' FB 900 000, 1975, Palais des Beaux-Arts, Brussels.

TERBORCH, GERARD (1617–81)
Dutch painter, he was in the studio of Pieter Molijn for about four years. A painter of groups and portraits; in the National Gallery, London, is his 'Peace of Münster, May 15, 1648'.

TERBRUGGHEN, HENDRICK (1588–1629)
Born at Deventer, he became a pupil of Abraham Bloemaert, afterwards going to Italy for ten years, and living in Rome and Naples. When Rubens made his tour through Holland he pronounced Terbrugghen to be one of the ablest painters of his country.

'The Backgammon Players' £199 500, 1977, Christie's, London.

THEODORIC OF PRAGUE
(working between 1348 and 1368)
His name heads the list of the Painters' Guild at Prague in 1348. He was the Court Painter to the Emperor Charles IV and made more than 100 pictures for the Castle of Karlstein in Bohemia, particularly for the Chapel of the Holy Cross in the castle.

THORNHILL, SIR JAMES (1676–1734)
Born at Melcombe Regis, Dorset, he was in the studio of Thomas Highmore. He became the only English painter in the grand Baroque manner. His works include the superb Painted Hall at Greenwich Hospital the oval picture there showing, Triumph of Peace and Liberty (106 × 51 ft 32 × 15·5 m)) is **the largest painting in Britain;** the inside of the dome of St Paul's, here eight grisaille panels show happenings in the Saint's life; and Queen Anne's bedroom at Hampton Court. He had an academy in London, **with Hogarth as the best-known pupil.**

THORVALDSEN, BERTEL (1768–1844)
Danish sculptor who lived and worked in Rome from 1797 to 1838. He studied the Antique and chose Classicism, and he was ranked next to Canova. He was swamped with commissions, and Napoleon had him produce a frieze of Alexander's campaigns. Thorvaldsen did a larger than life statue of Christ, which was **one of the most copied statues of the 19th century.** In Copenhagen there is a Thorvaldsen Museum with a great many of his works or plaster casts of the important ones.

TIDEMAN, ADOLF (1814–76)
Born at Mandal, Norway, he studied at the Copenhagen Academy and then in Düsseldorf with Schadow and Hildebrandt. He was the leading painter in Norway for his time, subject-matter was generally genre with a romantic setting.

TIEPOLO, GIOVANNI BATTISTA
(1696–1770)
Born in Venice and studied with Gregorio Lazzarini and was considerably influenced by Veronese. His style had great flourish and immense skill and represents the flower of the Italian Rococo. His works include 'Gathering of the Manna and Sacrifice of Melchizedek' in the Paris Church, Verolanuova, the Kaisersaal and the Great Staircase in the Palace at Würzburg and fresco cycles at Udine. His son Giandomenico (1727–1804) helped his father with many of his undertakings. In his own work he leaned towards caricature and genre.

'Ariane entourée d'amours' ($10\frac{1}{4} \times 12\frac{3}{8}$ in (260 × 314 mm)) F 55 000, 1975, Palais Galliera, Paris.

'Briseis led to the tent of Agamemnon' (red chalk, pen and brown wash, $9\frac{1}{2} \times 7\frac{3}{4}$ in (241 × 197 mm)) £3360, 1975, Christie's, London.

TINTORETTO, JACOPO (1518–94)
(properly Jacopo Robusti). Born in Venice, his name was derived from his father's profession of dyer (tintore). There are several stories about whether or not he studied with Titian, and if he did as to whether he was thrown out after a few days. However, he certainly had great admiration for the master, as he had written on his studio wall, 'The drawing of Michelangelo and the colour of Titian'. In his art Tintoretto was driven forward fired by an almost pious emotion. It is said he worked out his vast compositions with small clay model figures. In the Doge's Palace in Venice is his magnificent 'Il Paradiso', commissioned after the authorities would have seen his abilities with 'St Mark rescuing the slave', a composition with a large number of figures and clever foreshortening. In the National Gallery, London, is 'St George and the dragon', another courageously handled composition. Brimming with action, it is almost possible to see the princess moving towards the

'Goethe in the Campagna' by Johann Heinrich Wilhelm Tischbein. (Courtesy Städelsches Kunstinstitut, Frankfurt)

foreground. As with many other artists Tintoretto had a large studio workshop, in which apart from others he was helped by his sons Domenico and Marco, also his daughter Marietta. In his handling of light and shade he to a degree was foreshadowing Rembrandt.

'Christ on the Cross', £15000, 1975, Sotheby Parke Bernet.

TISCHBEIN FAMILY

The most important member was **Johann Heinrich Wilhelm**, painter and engraver (1751–1829), who was born at Hayna; he studied with his uncle Johann Henrich Tischbein the Elder and then went to Hamburg and learnt to restore pictures with Johann Jakob Tischbein. After this he began to paint portraits and also travelled to Switzerland and Italy. In 1787 he went to Naples accompanied by Goethe. There is in the Städelsches Kunstinstitut, Frankfurt, Tischbein's painting 'Goethe in the Campagna'. The other members of the family, apart from those already mentioned, include: **Johann Anton** (1720–84) landscape and historical painter, **Johann Friedrich Auguste** (1750–1812), **Johann Heinrich the Younger** (1724–1808) landscapes and portraits and **Karl Ludwig** (1797–1855) portraits and Romantic genre.

TISSOT, JAMES (1836–1902)

Born at Nantes, he studied with Jean Flandrin. He became involved politically at the time of the Commune and he moved to London. Here he painted ladies of fashion and courtly and social scenes such as: 'Ball on Shipboard', 'Walk in the Snow' and 'Meeting of Faust and Marguerite'.

'Seaside' £11550, 1975, Christie's, London.

TITIAN (c 1487–1576)

(properly Tiziano Vecelli). Born at Pieve di Cadore, a village set high in the Venetian Alps, he was in the studio of Giovanni Bellini and later worked with Giorgione, and expanded his theories of colour, light and shade. Titian was the master of rich harmonies of colour, and from him these manners spread across Europe, El Greco taking them to Spain, they were picked up by Rubens, Van Dyck and Rembrandt. The painter prospered: he lived in a fine palace; he was asked for by the highest in the land and sent for by kings from all over Europe. There is the remarkable equestrian portrait of Charles V at the Battle of Mühlberg, in the Prado, Madrid, a fresh way of treating such a subject. But his spread of ideas was amazing, from 'Bacchus and Ariadne', which shows the moment of Bachus arriving in his chariot to find the sorrowing Ariadne deserted by her lover Theseus, in the National Gallery, London, to the delicately treated 'Woman at her Toilet', in the Louvre, and his best work a 'Pietà' in the Accadèmia, Venice, which was intended for his own tomb, but which had to be finished by Palma Giovane, as that curse of the times, the Plague, bore the great Venetian master away.

TOBEY, MARK (1890–1976)

American painter grouped with the Abstract-Expressionists. He started out as an illustrator for a mail order catalogue in Chicago. He left America to study in Europe, also the Orient under Teng Kwei. Back in America he developed his so-called white writing, which appeared as swirling white lines on dark toned and textured backgrounds.

TOULOUSE-LAUTREC, HENRI MARIE RAYMOND DE (1864–1901)

French painter and the son of Count Alphonse de Toulouse-Lautrec Monfa. Serious falls in 1878 and 1879, his legs severely broken, left him after months of pain, a crippled dwarf. He went to Paris and worked in several studios, met Degas who influenced him as to choice of subjects and also with the free use of pastels and gouache. Lautrec, carrying with him the bitterness of his physical state, produced the most amazing series of scorching satire on the social scene set in the cafés and houses of certain repute, the theatre and the Bohemian life of the latter years of the 19th century in Paris. His colour and taste in a strange way had a high degree of fastidiousness, creating beauty from what might have been a squalid situation; this can be seen by 'Seated Girl' in the Louvre, 'Yvette Guilbert', in the Museum of Western Art, Moscow, and 'Au bal du Moulin de la Galette' in the Art Institute, Chicago. 'Fille à l'accroche-cœur' £230000, 1975, Sotheby's, London.

TOWNE, FRANCIS (1740–1816)
English landscape-painter who studied with W. Pars. He worked largely in water-colour and although primarily in the West Country travelled widely in England also Wales, and made a tour to Switzerland and Italy.
'Lake Como, looking towards Monte Lenoni' £380, 1975, Sotheby Parke Bernet.

TROYON, CONSTANT (1810–65)
Born at Sèvres, he worked for a time as a decorator in the porcelain factory; then he developed an aptitude for landscape with animals.

TRUMBULL, JOHN (1756–1843)
American painter who fought in the War of Independence, acting as an aide to George Washington. He studied afterwards with Benjamin West in London. Trumbull painted portraits of Washington and also such subjects as 'The Battle of Bunker Hill' and the 'Signing of the Declaration of Independence'.

TURNER, JOSEPH MALLORD WILLIAM (1775–1851)
The son of a barber in Covent Garden, London, **he was a prodigy, by the age of 14 making his living by the sale of his drawings.** An artistic revolution in himself, probably no painter has ever worked harder than Turner, hardly ever was he idle, his total output of drawings, water-colours and oils numbering thousands. He travelled constantly, seeking fresh subjects all over Britain and much of Europe.

His work to a degree can be divided into three categories. Early pictures such as 'Crossing the Brook', in the National Gallery, London, or landscapes in the tradition of Gainsborough and

'Prancing Horse' by Toulouse-Lautrec. (Courtesy Christie's)

Hobbema. Later after travels to France he was influenced by Poussin, and there came forth mythological subjects such as 'Ulysses deriding Polyphemus' also in the National Gallery, and finally the great masterpieces of his atmospheric period, when the mind of the painter grappled with and caught the furies of elemental forces, the wonders of light and moisture, seen in 'Rain, Steam and Speed, and 'Hastings'; the first in the National Gallery and the second in the Tate Gallery. Yet even this great genius did not reach through to everyone as can be seen when such as his 'Florence from the Boboli Gardens', a water-colour, was sold in 1878 for just nine guineas, in 1975 it fetched £2100, and even nearer to date in 1930 'The Refectory, Kirkstall Abbey, also a water-colour made only 135 guineas, but in 1975 it went for £4725. **In 1976 £340 000 was given for his 'Bridgwater Sea-Piece'. A world record price for a British Painter reached by Christie's.**

TUSSAUD, MARIE (1761–1850)
Born in Strasbourg, France, she was taught wax-modelling by an uncle who made figures for anatomical study. She came to London in 1802 and started her celebrated waxwork museum, showing the likenesses of the famous and infamous.

UCCELLO, PAOLO (c 1396–1475)
Italian painter absorbed in the study of perspective, he did not invent it although he did much to develop the science of vanishing lines. In his three great battle paintings, one of which, the 'Rout of San Romano', is in the National Gallery, London, he makes great play with fallen figures, broken lances and spears. As a young man he had studied geometry with the mathematician Giovanni Moretti. He had a fine feeling for design and also carefully observed nature in his compositions.

UTRILLO, MAURICE (1883–1955)
Born in Montmartre, Paris, he was the illegitimate son of Suzanne Valadon, a travelling acrobat, one-time favourite model of Degas and finally an accomplished painter herself. At first Utrillo was attracted to Impressionism, then Cubism and from then on he developed a highly individual style of his own. Sadly his life was a long struggle with illness, partly brought on by his intemperent habits, yet he worked on. Few can have captured so well the textures of the old buildings of Paris, the flaking plaster, ageing woodwork, old tiles and slates.
'Rue Sainte-Rustique sous la neige' F 139000, 1975, Hôtel Drouot, Paris.
'La rue Sarrette à Paris' L 22000000, 1975, Finarte, Rome.

VALADON, MARIE-CLEMÉNTINE
(1865–1938)
(called Suzanne). Born at Bessines, besides Degas
she also modelled for Renoir, Puvis de Chavannes
and Toulouse-Lautrec. When she started to paint
herself her manner was colourful, with some
allegiance to Gauguin, as with 'Blue Bedroom' in
the Musée d'Art Moderne, Paris.
'Nature Morte aux fleurs et à la chaise' F 28000,
Palais Galliera, Paris.

VALDÉS LEAL, JUAN DE (1622–90)
Spanish painter, son of a Portuguese father, he
worked in Seville and started a painting academy
with Murillo. His work was mainly on religious
subjects with a few examples of still life sometimes
as 'vanity' pieces, with macabre contrasts of skulls
and domestic utensils.

VAN DYCK, SIR ANTHONY (1599–1641)
Born in Antwerp, he was at first in the studio of
Van Balen and then became one of Rubens's main
assistants while still in his teens; an illustration of
his youthful talent is that he was admitted to the
Guild of St Luke at the age of 19. Rubens spared
nothing that the young man should learn all his
secrets. Van Dyck then had a lengthy travel to
Italy, seeking out in particular all the works he
could by Titian. In 1620 he went to England to the
Court of James I, but the trip was a failure, and he
returned to the Continent where he undertook
further travel to Italy in 1621. Between 1628 and
1632 he was in Antwerp. He then was invited
to England by the Court of Charles I where he
remained for the rest of his life. He painted the
intellectual connoisseur Charles many times,
including the triple portrait in the Royal

*'The Three Marys at the Sepulchre' by Jan van
Eyck (possibly with his brother Hubert). (Courtesy
Boymans-van Beuningen Museum, Rotterdam)*

Collection and the elegant equestrian portrait in
the National Gallery, London.

VAN EYCK BROTHERS
Flemish painters, born at Maaseyck, **they were to
be the harbingers of a new way for pictures, a
breaking out from the stylistic bonds of the
Gothic and medieval**. The two brothers were
Hubert (c 1370–1426) and **Jan** (c 1390–1441).
Unfortunately there is little recorded of their lives,
but with their greatest work, the 'Adoration of the
Lamb' there is a statement in Latin on the frame of
the magnificent polyptych, which translated
reads, 'The painter Hubert van Eyck, than whom
none was greater, began it; Jan, second in art,
having completed it at the charge of Jodocus Vyd,
invites you by this verse on the 6th May to
contemplate what has been done.' **This altarpiece must rank as one of the greatest
masterpieces in the world**; it originally
consisted of 12 panels, four of which were painted
on the reverse. The incomparable work has been
subjected to many vicissitudes, panels being
removed as spoils of conquest, but today it is
complete except for one that has disappeared. At
the end of the 16th century the Calvinists took the
whole work, but on this occasion it was soon
returned to Ghent. Two centuries later the
Emperor Joseph II of Austria was shocked by the
nudity of Adam and Eve and these two panels
were removed and stored in an attic. The looting
commissars of Napoleon lifted the four main
central panels and took them to Paris. After
Waterloo General Blücher saw to the return of the
panels to Ghent. In 1816 the Vicar-General Le
Surre, when the bishop Monsignor de Broglie was
away, sold the folding doors of the polyptych,
minus the Adam and Eve panels, to an antique-
dealer from Brussels. This gentleman saw them on
their way to William III of Prussia. In 1895 the side
panels were sawn longitudinally in half so that
both sides could be displayed. Back in Ghent the
rest of the work only just escaped destruction by
fire. At the end of the First World War the
Germans had to give up the panels which were
then in Berlin; thus the work was once again
complete. Peace for a few years followed, but on
the night of 10/11 April 1934, the panel known as
The Upright Judges and the now separated panel
of St John the Baptist vanished. The latter was
eventually returned after a large ransom, but the
whereabouts of the other panel remains one of the
mysteries of the art world. A copy has been made
by J. Vanderveken.
  In the National Gallery, London, are two paintings by Jan Van Eyck, the incomparable 'Jan
Arnolfini and his Wife' and 'Man's portrait', both

of which show the mastery and breakthrough of this the younger brother in a partnership which was to achieve so much.

VAN GOGH, VINCENT (1853–90)

Born at Groot-Zundert, in North Brabant, he was an individual in mental turmoil from his early days. At first he worked as a clerk to a picture-dealer, but left because of an emotional upset occasioned by a love-affair. He tried for the ministry, but was rejected, and so he set himself to preach to the poor downtrodden mining families in Belgium. At about the same time he began to paint, helped by his cousin, the artist Mauve. The time passed quickly as the tempo increased. He went to France, and in Paris came in contact with the Impressionists and their work. In 1886 he attended the École Nationale des Beaux-Arts and met Toulouse-Lautrec, on whose advice he went to Arles in 1888. Here, in Provence, it all seemed right for him: the bright sun-enriched colours, the landscape, the people, the carts, the farm-buildings. His output was amazing, more than 200 pictures in 15 months. Yet in this surge of creativity, his existence was wretched; he sold nothing, had little to eat, his mental state started to crack, hallucinations flooded his mind. There was the pitiful quarrel with Gauguin, after which poor Vincent hacked off his right ear with a razor, then sent it wrapped in a handkerchief as a present to a girl in a brothel. Those around him, terrified of what next he might do, raised a petition to have him put in an asylum. He went to a hospital in Paris. Out again he craved work, but in 1889 he asked himself that he should be placed in the asylum at Saint-Rémy-de-Provence. Working even there his output continued high, 150 paintings and hundreds of drawings in the year he spent in the place. Throughout these times there is the touching friendship of his brother Theo, who did so much to support him. In 1890 Vincent was to meet Dr Gachet, who was to tend him lovingly and he then settled in Auvers, but the final mental storms were gathering and were to blow to tatters his poor ravaged mind.

In July 1890, on the last Sunday of his life, he took himself out into the golden ripening corn-fields, where a few days previously he had painted 'Wheat field with crows', that intense picture with swirling rich corn which is overhung by a dark indigo sky, the whole being imprinted with the ominous black crows; and there he fired a re-volver into his chest, and died two days later. What a legacy this man had left, all colour, texture, excitement, pictures as highly individual as any produced, but that in his time hardly did anything to bring him comfort, but today are as

'The Church at Auvers' by Vincent van Gogh. (Courtesy The Louvre, Jeu de Paume, Paris)

sought after as any. The 'Schoolboy' in blue smock, in the São Paulo Museum, 'Moored Boats' in the Kuntsmuseum in Essen, 'Road with Cypresses' in the Tate Gallery, London; his in-vention and quality seems endless. 'Zinnias dans un Vase' $310000, 1969, Parke Bernet Galleries, New York.

VASARELY, VICTOR (1908–)

Born in Pecs, Hungary, he studied Medicine and then went to the Bauhaus, which was at that time in Budapest under the direction of Moholy-Nagy. In 1930 he moved to Paris. His art underlies the Op Art movement, wherein twisted and swirling geometric line patterns simulate movement; Bridget Riley is the leading protagonist today. 'Celenderis', £4200, 1975, Christie's, London.

VASARI, GIORGIO (1511–74)

Born in Arezzo, he became a minor painter but famous because of his writings with the *Lives of the most eminent Painters, Sculptors and Architects*, first published in 1550 and an enlarged edition in 1568.

VELÁZQUEZ, DIEGO RODRIGUEZ DE SILVA (1599–1660)

Born in Seville, he studied first with Francisco

Herrera, but the master's brutal ways scared the young painter away and to the studio of Francisco Pacheco, whose daughter he ultimately married. After this on a visit to Madrid he met Count Olivares who introduced him to Philip IV, with which monarch the painter was to have a long and profitable friendship. Velázquez had great quality and taste, which gave his portraits, such as 'Lady with a Fan' in the National Gallery, London, a sensitive and searching atmosphere. In contrast are 'The Surrender of Breda' and 'The Tapestry Weavers', both in the Prado, Madrid. A major work of his had not been on the market since before the First World War, when a portrait of Philip IV made £82000, until **27 November 1970, when the 'Portrait of Juan de Pareja', also known as The Slave of Velázquez**, made the **world record figure for a painting of** £2310000 at Christie's, London.

VELDE, VAN DE, FAMILY

The father was **Willem the Elder** (1610–93) who was born in Leyden; he started life as a sailor, and then commenced painting and drawing, whereupon he acquired such skill that the Dutch authorities provided him with a small vessel from which to sketch sea battles. Charles II invited him to England in 1675, and he stayed there for the rest of his time. He had two sons, **Willem the Younger** (1633–1707) and **Adriaen** (1636–72). Willem the Younger, **the finest marine painter produced by the Dutch**, studied with his father and Simon de Vlieger. He went to England in 1674 and by records was extraordinarily productive; it was said that between 1778 and 1780 about 8000 pictures were sold at public auction. Apparently he could fill a quire of paper with sketches in an evening. Adriaen studied with his father and Wynants and Potter; he painted landscapes and battles, also he worked figures into

compositions at times for Hobbema, Wynants, and Verboom. Esaias (c 1519–1630) was probably the brother of Willem the Elder, also a landscape- and battle-painter; his best-known pupil was Jan van Goyen.

'Fisherman beaching a fishing-smack' (Willem the Younger) £26250, 1975, Christie's, London. In 1947 it made only £462.

'A Winter Scene' (Adriaen) £7200, 1975, Sotheby Parke Bernet.

'Winter landscape with figures skating' (Esaias) DM 40000. Kunsthaus Lempertz, Cologne.

VERMEER, JAN (1632–75)

Born in Delft, he may have studied with Carel Fabritius. He became the painter of harmony, whether with landscape or an interior with such as 'A Woman at the Virginals', in the National Gallery, London. His art is one of observation for the effects of light on flesh, textiles, polished surfaces and as coming through a window. He was a very slow worker, and there are probably less than 40 paintings than can be unquestionably credited to him. The noting of his prices from the 18th century through to the 20th is of interest. 'The Woman with the Pitcher' in 1719 brought £10 10s, in 1907 the Rijksmuseum gave £50000 for it. In 1810 'The Singing Lesson' made £51. In 1837 'Lady and Servant' went for £16. In 1931, Andrew Mellon gave in excess of £100000 for the 'Girl with the Red Hat'; it is now in the National Gallery, Washington, DC. In 1959 'Girl's Head' brought about £400000, in 1816 in Holland it had made just three florins.

VERNET, CLAUDE-JOSEPH AND FAMILY (c 1712–89)

French painter, born at Avignon, he studied with his father. Memorable works are seaports, of which he painted a series for Louis XV. He had a son, **Antoine Horace** (1758–1836) well known for pictures of battles and horses. Antoine had a son **Émile Jean Horace** (1789–1863), also a battle-painter.

VERONESE, PAOLO CALIARI (1528–88)

From Verona came this buoyant character, who studied with his uncle Antonio Bodile, and who rose quickly to fame. He produced a number of paintings of great beauty, although at times it was felt he approached religious subjects too light-heartedly; in fact because of the 'Feast in the House of Levi', now in the Accadèmia, Venice, he was hauled before the Inquisition.

VERROCCHIO, ANDREA DEL (1435–88)

Florentine painter, sculptor, goldsmith and architect, **he ran the largest art workshop in Florence**. The leading pupil was Leonardo, and also attending were Lorenzo di Credi and possibly

'Skirmishing Cavalry' by Esaias van de Velde. (Photo by the Author)

for a short time Botticelli. Verrocchio is best remembered by the magnificently fierce equestrian bronze statue of the *condottieri* Bartolomeo Colleoni in Venice.

VIEIRA DA SILVA, MARIA HELENA (1908–)

Born in Portugal, she studied in Paris and stayed on; part of her tuition was from Roger Bissière (1886–1965) who taught at the Académie Ranson. She married the Hungarian Abstract painter Arpad Szenes. Her expression is in the Abstract; as well as painting she has designed tapestries. 'La garde des anges' £13000, 1975, Sotheby Parke Bernet.

VIGÉE-LEBRUN, LOUISE ÉLIZABETH (1755–1842)

French portrait-painter who studied with her father a pastellist, Vigée. She was influenced by Greuze. Her sitters included Lady Emma Hamilton, Lord Byron, the Prince of Wales and also more than 20 portraits of Marie-Antoinette.

VIGELAND, GUSTAV (1869–1943)

Born at Mandal, way up in the north of Norway, he was carving wood when but a child. He was a pupil of the sculptor Bergslien, in Oslo, later with Bissen in Copenhagen and lastly he went to Paris in 1892 when he was a pupil in the studio of Rodin. At the the start of his career he was carving Gothic figures for the Trondheim Old Cathedral. His memorial is really the Park of Sculpture at Frognor, Oslo, **one of the greatest demonstrations of working in granite anywhere. Most striking work there is the Obelisk with about 100 intertwined figures.**

VLAMINCK, MAURICE DE (1876–1958)

A Fauve, and an individual who liked to boast he had never been in the Louvre, who played the violin, was a racing-cyclist, later a motoring enthusiast who went for speed and the open roads and loved crowds. Any discipline fretted him; priests, police, teachers were avoided as they could fetter his freedom. His pictures are broadly painted, with great excitement in the use of colour with heavy impasto. 'Le pécheur à la ligne' £118000, 1975, Sotheby Parke Bernet.

VOS, DE, FAMILY AND OTHERS OF THE NAME

Painters of this name have probably caused more confusion for those rustling through the records to find a provenance than most. The interwoven story starts in 1585 with Cornelis de Vos. Then in the 17th century there was another Cornelis, who was no relation to the first one. In the 16th century Lambertus de Vos from Meehlin was working,

'La Route avec Deux Personnes' by Maurice de Vlaminck. (Courtesy Christie's)

and a little earlier than him in the same century comes Marten de Vos from Antwerp, the son of Pieter de Vos. Marten de Vos had a son Marten, he also had a brother Pieter, who had a son Willem. Then there was Paulus the brother of the first-mentioned Cornelis. Again in the 17th century there was Jan de Vos and Simon de Vos, and there are some writers who claim there are also another Pieter and Willem and Hendrik as well as even more Christian names. In a recent sale there was a 'Madonna and Child with Saint Joseph, in a wooded landscape' on a panel by C. de Vos; was this one of the Cornelises or was it another one of them emerging to fame? What really complicates the scene is that they all painted portraits, landscapes and animals.

VUILLARD, EDOUARD (1868–1940)

Born at Cuiseaux in Saône-et-Loire, he was a member of the Nabis and was influenced by Sérusier, and interested in Intimism. A painter of intimate scenes in the streets of Paris, and in the houses. A soft harmonious colourist. 'La Mère de l'artiste à la fenêtre. Rue de la Tour' £15750, 1975, Christie's, London.

WALLIS, ALFRED (1855–1942)

Cornish fisherman living in St Ives. He started to paint in 1928, pure Primitive scenes that he knew, mostly associated with the sea and ships; he was encouraged by Ben Nicholson and Christopher Wood.

WARD, JAMES (1769–1859)

Born in London, he studied with John Raphael Smith and became one of the leading animal painters; his style to a degree was influenced by his brother-in-law, George Morland. In the Nottingham Gallery there is a magnificent bull, cow and calf in a landscape.

'Mameluke, a bay racehorse' £29400, 1975, Christie's, London.

WATTEAU, ANTOINE (1684–1721)

Born in Valenciennes, the son of the village carpenter. He studied first with Guerin and then went to Paris and worked with Gillot on theatrical scenery, also grotesque effects for carnivals. He developed into the master of Fêtes compositions that became a fashion for collectors, semi-wild landscapes with beautiful people, often in costume

'Indian in Body Paint' is a water-colour by John White, Sir Walter Raleigh's artist-recorder (done in America c. 1577–90). It bears the inscription, 'The manner of their attire and painting them Selves when they goe to their generall huntings, or at theire Solemne feasts.' (The work may possibly have been inspired by a water-colour of Jacques Le Moyne de Morgues). (Courtesy of the Trustees of the British Museum)

occupied with masques and music. The 'Champs Elysées' in the Wallace Collection, London, is in line with this approach. The style also embraced mythological subjects such as 'Embarkation for Cythera' in the Louvre.

WATTS, GEORGE FREDERICK (1817–1904)

Victorian portrait, historical and mythological painter, he studied at the Royal Academy Schools and in Florence. His sitters included: Cardinal Manning, in the National Portrait Gallery, London, Tennyson, Browning, Swinburne, Rossetti and William Morris.

WEST, BENJAMIN (1738–1820)

American-born painter, who may have had some rudimentary instruction from the Cherokee Indians. He went to Italy in 1760 and then to London three years later, where he settled. A painter of portraits, historical and religious themes, his studio became a centre for visiting American artists. **He succeeded Sir Joshua Reynolds as President of the Royal Academy.**

WESTMACOTT, SIR RICHARD (1775–1856)

English sculptor who worked in the studio of Canova in Rome, on his return to London he was in demand, works include the Monument to Charles James Fox in Westminster Abbey and the statue of Achilles in Hyde Park, a memorial to Wellington. His son Richard (1799–1872) followed his father and is recalled by the pediment on the Royal Exchange.

WEYDEN, ROGIER VAN DER (c 1399–1464)

The leading Flemish painter of the first half of the 15th century; he was born at Tournai and became a pupil of Robert Campin. A painter moved by a deep religious conviction. He had the technical ability of the Van Eyck brothers, but his colour is cooler; and later he was to an extent influenced by the Renaissance when he visited Italy in 1450. This can be noted in the 'Entombment' in the Uffizi in Florence; other works include portraits such as 'Francisque d'Este' in the Metropolitan Museum of Art, New York.

WHEATLEY, FRANCIS (1747–1801)

English portrait and 'conversation piece' painter, he probably studied with Zoffany. He worked in London, and for a few years in Dublin, where among other pictures he painted a large composition 'The Irish House of Commons' into which he put portraits of all the leading Irish politicians.

'The Garden Party: the Oliver and the Ward families grouped in a garden' £55000, 1975, Sotheby's, London.

WHISTLER, JAMES MCNEILL
(1834–1903)
Born at Lowell, Massachusetts, his father a military engineer. The boy spent several years in St Petersburg where his father was consulting engineer of the Moscow and St Petersburg Railway. Later he was to train in Paris in the Atelier Gleyre. He was a strange figure with a small goatee beard, patent-leather boots, dark hair in ringlets and liking for black straw hats; he also had a prickly nature always ready for a duel of wit and words. He moved to London in 1859. Whistler was a sensitive painter with a liking for evening or night scenes, and subtle portraits; an exquisite colourist; also he etched some very fine plates which put him about third in line behind Rembrandt and Goya. He endured the notorious lawsuit, which followed on Ruskin's derogation of his 'Nocturne in Black and Gold: The Falling Rocket' and the accusatory remarks made by the critic: 'I have seen and heard much of Cockney impudence before now, but never expected to hear a coxcomb ask two hundred guineas for flinging a pot of paint in the public face.' Whistler won the law case with damages of one farthing but no costs, which bankrupted him. 'Miss Cecily Alexander', in the National Gallery, London, demonstrates his taste and quality. Latterly he substituted a butterfly for his signature.
'Chelsea houses' ($5\frac{1}{4} \times 9\frac{1}{4}$ in (133×235 mm)) $4250, 1975, Sotheby Parke Bernet, New York.

WHITELEY, BRETT (1939–)
Born in Sydney, Australia, he studied in night classes at the Julian Ashton School, Sydney. He has lived and worked since 1960 mostly in London. Preferring large canvases, he forms his compositions from broken and twisted nudes and suggestive writhing limbs.
'Homage' (oil, collage and black ink) £680, 1975, Sotheby Parke Bernet.

WIERTZ, ANTOINE JOSEPH (1806–65)
Belgian painter, he studied at the Antwerp Academy and later at the Louvre. He worked on portraits, and religious subjects, which at times contained some way-out erotic and macabre features, inspiration for Surrealists to come. There is a Wiertz Museum in Brussels in what was once his studio.

WILKIE, SIR DAVID (1785–1841)
Scottish painter of genre who studied at the Edinburgh School of Art. He was influenced by the Dutch painters Ostade and Teniers, subjects included such as 'The Village Festival' in the Tate Gallery, London, 'The Blind Fiddler', 'The Reading of the Will' and 'Queen Victoria's First Council'.
'Distraining for Rent' £50400, 1975, Christie's, London.

WILSON, RICHARD (1714–82)
Born at Pinegas in Montgomeryshire, Wales, he was first put in the studio of a little-known portrait-painter in London, Thomas Wright. After this for some years he managed to make a living practising this art, then having saved, he set out on a trip to Italy. He met Zuccarelli and the two of them worked side by side for some years. On his return to Britain he turned away from doing further portraits and set out to paint pure landscapes, not the romanticized ruin-filled fantasies by those he had seen in Italy. But success was to elude him as fashion was calling for just those pictures he rejected the thought of doing, also it was a vogue to patronize foreign artists. Wilson had as a further handicap a boorish manner; but paint on he did, leaving more remarkably fine pictures of the British landscape. **He was a founder-member of the Royal Academy.** Many of his best works are still in private collections, but noteworthy in public hands are 'Snowdon' in the Walker Art Gallery, Liverpool, 'Okehampton Castle' in the City Art Gallery, Birmingham, and 'On Hounslow Heath' in the Tate Gallery, London.

WINTERHALTER, FRANZ XAVIER
(1806–73)
German painter of royalty, he studied at Munich and later set himself up at Karlsruhe. There, a successful portrait of the Grand Duke Leopold had set him on the road to considerable success and he moved to Paris, but was often travelling to the various Courts of Europe. His sitters included: Queen Victoria and her family, the Empress of Russia, Louis-Philippe and his Queen, Queen Isabella of Spain, Napoleon III, Prince Metternich, the Emperor Franz-Josef and the Empress Eugénie.

WITZ, CONRAD (c 1400–46)
Among the leading painters of the period he was born at Rottweil, Swabia, but worked most of his life in Switzerland. His style displays an acquaintance with the work of his contemporaries, the Van Eycks, but at the same time has a simplification and monumental quality more associated with the Italians. His finest work is 'The Miraculous Draught of Fish' in the Musée d'Art et d'Histoire, in Geneva. **It is the first landscape in European art with a recognizable view** and also it is dated 1444 and signed. He was an artist of

remarkable talents, an accurate observer of nature, but at the same time able to sift out the essential components.

WOLGEMUT, MICHAEL (1434–1519)
Born in Nuremberg, a painter and woodcutter who did much to advance the technique of the latter. **His most famous pupil was Albrecht Dürer.**

WOOTTON, JOHN (d 1756)
A pupil of Jan Wijck, and one who made a considerable name as a painter of horses, often on very large canvases. he also produced a number of landscapes in the classic style of Claude and battle scenes.
'Little David' £13 000, 1975 Sotheby Parke Bernet.

WOTRUBA, FRITZ (1907–75)
Born in Vienna, he became Austria's leading sculptor. He studied engraving first, and then became a pupil of Anton Hanak. He produced a number of bronzes, but first he is a worker in stone, his figures massive, responding to the hardness of the material, but containing a feeling of rhythm.

WOUWERMAN, PHILIPS (1619–68)
Born in Haarlem, he was in the studio of Jan Wynants, developing into a painter of genre scenes with skirmishes, battles, hunting scenes set out in the pictures with style, clarity and a sensitive feeling for landscape. He was prolific, with something like 1200 paintings being actually known. Today his values are increasing; for example, 'An Italianate River Landscape, sold with Christie's in 1889 for 460 guineas; in 1975 it made £16 000.

WRIGHT, JOSEPH (1734–97)
(known as Wright of Derby). Born in Derby, he studied in London with Thomas Hudson. He went to Italy in 1774 and here must have absorbed the influence of Caravaggio. On his return he set up as a portrait-painter, but more important were his experiments in unusual light effects as with 'The Experiment with the Air Pump' in the Tate Gallery, London.
'Portrait of Harry Peckham' £13 650, 1975, Christie's, London

YEATS, JACK (1871–1957)
Born in London, the brother of the poet W B Yeats, he studied at the Westminster School of Art under Fred Brown. He lived and worked in Ireland and was one of the most remarkable colourists, with a highly individual method of applying his paint. His pictures capture the very feeling of the scenes he worked on, the characters, movement and the landscape.
'The Engineer' £5 775, 1975, Christie's, London.

ZADKINE, OSSIP (1890–1967)
Born in Smolensk, Russia, the sculptor came to England and attended the London Polytechnic in 1907 and then went to Paris in 1909. He went through a period of Cubism but later changed, bringing in a sense of movement and his figures echoing some tenets of the Expressionists.
'Adam and Eve' (bronze) DM 16 000, 1975, Galerie Wolfgang Ketterer, Munich.

ZOFFANY, JOHANN (c 1734–1810)
German painter of portraits and 'conversation pieces' from Frankfurt am Main. He studied in Italy and worked mostly in England. **A founder-member of the Royal Academy**, he painted many theatrical scenes generally showing an actual moment in the play. David Garrick featured in a number of them.

ZUCCARELLI, FRANCESCO (1702–88)
Italian painter of landscapes with ruins, peasants and animals, he had a great vogue in England, which he visited twice staying for several years each time. **He was a founder-member of the Royal Academy.**
'Italian Landscapes' (a pair) £14 700, 1975, Christie's, London.

ZURBARAN, FRANCISCO DE (1598–1664)
Born near Bádajos, he specialized in studies of monks using the manner of Caravaggio with light and shade, as with 'St Francis in meditation' in the National Gallery, London. Although filled with pious emotion, it is too studied, and fails to light up the imagination in the manner of El Greco.

Currency abbreviations are: $Austr – Australian dollar, Asch – Austrian Schilling, FB – Belgian Franc, Dkr – Danish Kronen, F – French Franc, DM – Deutsch Mark, Fl – Dutch Florin, L – Italian Lire, Rand – S. Africa, Skr – Swedish Kronen, Sfrs – Swiss Francs, Yen – Japan.

Exchange rates with £ as at December 31st 1977:

Austria	28.25 Asch
Belgium	62.25 FB
Denmark	10.92 Kr
France	8.88 Fr
W. Germany	3.97 DM
Holland	4.31 Guilders
Italy	1,650 L
Japan	460 Yen
S. Africa	1.82 Rand
Sweden	8.85 Kr
Switzerland	3.79 Fr
U.S.	1.9075 dollars

Chapter Seven

Collectors and Keepers

The collecting of works of art started long before galleries and museums were founded. It was a pursuit for the wealthy and powerful, which grew from a love of beautiful things, an appreciation of craftsmanship and a feeling that values could rise with time. Undoubtedly also a number gained a sense of greater power by the ownership of superb works. Some objects might have been acquired because of religious, political or social associations, but generally these factors did not seem to influence the collector. He was and is, however, affected by slumps, violence and wars, and with the last come instances of the predatory collector.

A supreme example of such a one in recent times is Adolf Hitler. He gave orders to party officials working under the *Einsatzstab* Rosenberg, with the encouragement of Göring, to get together by threat, requisitioning or outright theft an accumulation of first quality art treasures that could have more than filled a very large museum. An extract from an *Einsatzstab* listing for 1 April 1943 shows among many other items:

5225 paintings and 297 sculptures

Just over a year later, in July, records showed categories that included:

5281 drawings, paintings, pastels and water-colours
684 books, enamels, glass, manuscripts and miniatures
583 medallions, plaques, sculptures and terra-cottas
1286 objects of Oriental origin, included: bronzes, carvings, pictures, porcelain, screens, weapons
259 antiquities which included: bronzes, cut gems, dishes, jewellery, terracottas and vases.

The private collector first comes to prominence during the Hellenistic dynasty, known as the Attalids, who ruled from 283 to 133 BC. King Attalus II enlarged the empire of Pergamum, and in so doing made an impressive collection of books, jewels, paintings and sculpture; he also must have been one of the first to catalogue his acquisitions, which included, besides originals, copies of those objects which it had been impossible to get hold of.

With the rise of the Roman Empire, so was there a wave of searching for and taking Greek antiquities. It became the fashionable thing for the leading figures to possess the finest and rarest. Men of the stature of Appius Claudius, Lucullus, Pompey, Julius Caesar, Nero and Hadrian became collectors. Galleries for showing the treasures were added to the luxurious villas. At times some of these men must have been deceived, as often when there is an acceleration of the collecting desire, a spate of copies and forgeries arrives on the market to supply the demand.

It is of interest to note that at this time there was raised **the first voice stating the importance of works of art as a cultural benefit and the right that all should have to see them**. Marcus Agrippa (63–12 BC) made this announcement and advocated that the treasures of the wealthy should be on public display, not shut away in private galleries.

During the medieval period collecting rested largely with the Church. There was the Abbot Suger of the Abbey of Saint-Denis near Paris who gathered to his care many fine jewels, enamels, rare vases, textiles and stained glass. The treasures of the ecclesiastical establishments were added to by rich gifts of paintings, sculpture, vestments, gold and silver plate, illuminated manuscripts; and as well as this artists and craftsmen lived around the abbeys, churches and cathedrals, to be employed on adding enrichment. But with all this the accent was mostly on the rarity and value and not on the aesthetic quality. This was a reawakening to come with the Renaissance, and the flowering of Humanism. It was the time of such as the Medicis, especially Lorenzo the Magnificent, the Gonzaga, the Este and many others. Lorenzo who may have been on one hand a brutal oppressor was on the other **among the first of the truly great collectors**, a true connoisseur (from the Italian *conoscitore* and derived from the Latin *cognascere*). Under his patronage many of the greatest names of the Renaissance received encouragement. Artists emerged from the obscurity of anonymity and were sought after for their individual style and genius.

In the 17th century the English awoke to the quality of European art, in particular painting, and a small group of connoisseurs who included Prince Henry, his younger brother Charles I, Thomas, 2nd Earl of Arundel in the Howard line and the Duke of Buckingham, began to collect in a grand manner. The greatest of these three was Charles I, a highly enlightened intelligent man, he brought to England what was **one of the finest collections of paintings of any time**. His most celebrated acquisition was the greater part of the Collection of the Duke of Mantua. The taste of the King was wide and covered not just the Italian school, but artists as diverse as Dürer, Geertgen and Rubens. At the peak his collection numbered some 1387 fine paintings. After the execution of Charles I, the Cromwell régime committed an act of great ignorance: they ordered the sale of the Royal Collections, part retribution perhaps, partly to raise funds; but all the same a savage loss to the art heritage of the country. Buyers from Europe bought in large numbers, and nearly all the paintings left the country. Agents for such as Philip IV of Spain, the Archduke Leopold Wilhelm and Mazarin were successful with their bids and galleries such as the Louvre, the Prado and the Kunsthistorisches Museum, Vienna added some of the finest works to their catalogues.

A thought might be given to the risks that attended this moving about of large works, notably paintings on canvas, which are prone to damage. They were probably crated in some way and then would be humped across the map on the backs of mules and in springless wagons. One report even states that there had to be man-handling over the worst passages of the Alps. Weather could have given rise to worry, as certainly passages by sea would, in small rolling sailing-ships, which did at times founder with their precious cargoes.

The return of the Monarchy in England brought a slight impulse for collectors, but it was not until the 18th century that the pace again quickened. The father of Horace Walpole got together a collection of paintings at Houghton, that was acknowledged as **the finest of the century**. When it was sold complete to Catherine the Great, Parliament was besought by Wilkes to intercede and keep it as a foundation for a National Gallery, but his plea was in vain and the Hermitage in Leningrad benefited. During the first half of the century the owners of such as Blenheim, Chatsworth and Holkham covered their walls largely with the Italian school and there was also a vogue for the work of such as Poussin and Claude.

This predilection for the work of foreign artists by English collectors to a high degree lasted right through the 18th century and well into the 19th. Native painters of the quality of Richard Wilson were neglected, William Hogarth turned the acid of his satire on the English aristocratic collector when he produced a print of a monkey watering dead plants, the plants intended to be the works of the Old Masters. But the fashion continued even if it meant the noble walls had to be covered with copies of the works of the great, as there were not enough originals to go round.

The Grand Tour complex had a firm grip on the nobility, it was as much a part of education as learning to read and write. Seldom did the travellers return empty-handed. It is recorded that Lord Burlington came back from his sojourn in Europe, particularly Italy, **with 900 packing-cases holding works of art**.

Taste began to change in the 19th century, and collectors like the Prince Regent, the 3rd Marquess of Hertford and Sir Robert Peel turned their attention to the Dutch painters. Small landscapes, genre and still lifes were sought after. The taste for small conversation pieces and scenes of everyday life brought a demand for English painters, not Constable or Turner, but rather Wilkie, Mulready and Leslie.

The French Revolution made available large numbers of works of art on the English market. **One of the most notable collections bought in entirety was that of the Duc d'Orleans**; in 1792 the Marquess of Stafford, the Earl of Carlisle and the Duke of Bridgwater got together and settled the deal. The great houses were being filled with treasures by their discerning owners; Baron Ferdinand de Rothschild at Waddesdon Manor, the 4th Lord Hertford, the 10th Earl of Northumberland, and across the country stately homes became individual repositories of outstanding collections, not only pictures and sculpture, but fine ceramics, furniture, silver, bronzes, tapestries and objects of great beauty and rarity.

The late 19th and early 20th century not only brought a change in painting and sculpture styles, but also heralded yet another removal for the works of art; this time there was to be a sensational drain of Europe's cultural treasures across the Atlantic to where fortunes of vast size were being made from oil, coke, manufacturing, banking and the rest. Under the guidance of such as Lord Duveen the new connoisseurs swept the masterpieces away, inflating the prices to three or four times. Such a one was John Pierpont Morgan (1837–1913), the international banker, so rich that he could twice save America from insolvency. **He was the biggest and most dominating figure in this art-buying bonanza.** He bought art collections as avidly as he did ships or railways. His billions brooked no denial, as the multi-nought cheques were signed. In about twenty years he spent $60000000; on objects which have since gone to enrich public collections in America, notably the Metropolitan Museum of Art in New York. The list is long of those Americans who have made it, spent it and then willed it for the enjoyment and benefit of the people. Isabella Stewart Gardner, Henry Clay Frick, Robert Lehman, Charles Lang Freer, Samuel Marx, Harry Lewis Winston, J. Paul Getty, Forsyth Wickes, Katherine Sophie Dreier, Albert and Mary Lasker, Joseph H. Hirschhorn, the uranium king, and the truly fabulous Rockefellers, all were capable of spending on art sums that even today have not all been topped.

Back in Europe there still remain, however, many strongholds where collections old and new are breathtaking with their sheer quality. There are the Italian paintings of Count Vittorio Cini. This collection was begun between 1910 and 1915, and many of the works of art are listed as national treasures. They are housed in the Castello di Monselice and the Palazzo Loredan in Venice. Among them are the superb 'Aurora' by Guardi, the vital, alive 'Madonna and Child with Two Angels' by Piero di Cosimo and the 'Two Friends' by Pontormo. At Lugano, Switzerland, the treasures of Baron H H Thyssen-Bornemisza: Tiepolo's moving 'The Death of Hyacinth', Holbein's 'King Henry VIII', the extraordinary 'Christ Disputing with the Doctors' by Dürer, the fresh sparkling 'Family Group' by Frans Hals and Fragonard's 'See-saw'.

There is Emil G. Bührle in Zürich with an outstanding collection of French 19th-century painting from 'The Death of Hassan' by Delacroix to 'Bordeaux Harbour' by Manet and the powerful 'Sower' by Van Gogh; also in Switzerland Oskar Reinhart with Manet, Sisley, Renoir, Courbet, Chardin and the fine portrait of

Dr Johannes Cuspinian by Lucas Cranach. To the south, in Venice, is Peggy Guggenheim with the only collection in Europe of 20th-century works put together on a systematic historical basis. Represented are examples from all the major art movements since 1910. As a study ground for the student it is unique as there are rare early examples from such as the Dada and Suprematist movements, comprehensive works of the de Stijl group and many others looking well in the rather unexpected habitat of the Palazzo Venier dei Leoni. Painters whose work is there include: Picasso, Braque, Severini, Kandinsky, Malevich, Chagall and Matta; and sculptors with them are: Kemeny, Paolozzi, Moore, Armitage and Giacometti.

A last example of a private collector is Hugh Lane who brought together a set of pictures, matched with considerable taste of French 19th-century painting. He, sadly, went down with the *Lusitania* in 1915, and, also rather sadly, at the moment a number of his fine pictures, because of a legal argument, are doomed to sail the Irish Sea every so many years as they commute between the Municipal Gallery, Dublin and the National Gallery, London.

GALLERIES AND MUSEUMS

With today's figures, major galleries and museums and their contents must be easily the most valuable buildings there are. In many cases it would be almost impossible to work out just what the collections are worth. Out of the thousands of such institutions, it is only possible to bring in a few; these can perhaps point to how such great places come into being and how long ago it was that some of them were opened.

Italy probably has more museums and galleries per head of her population than anywhere else. What better place to start with than Florence, that is in reality one great big museum itself, so steeped is the old town in the arts; indeed the Renaissance seems very close in among the palaces, the churches and the narrow streets. Here is one of the outstanding galleries in the world, the Uffizi. The pictures are in the upper floor of the Uffizi Palace which was designed by Vasari and built between 1560 and 1574. The collection owes much to the acquisitions made by the Medicis, and has been added to later by Ferdinando I and II and the della Rovere family, Cosimo III added 17th-century Dutch paintings and there is the famous collection of artists' self-portraits. In 1565 Vasari had built to his plans the corridor over the Ponte Vecchio and connecting with the Pitti Palace, which stands as the second most important gallery in the city. This palace was begun in 1440 by Luca Pitti, a rich

Historisches Museum. (Courtesy Landeshauptstadt Hannover)

competitor of the Medicis. In 1549 Cosimo I made it a Medici palace. The galleries on the upper floor have 500 fine paintings from the Medici collections, also there is an admirable set of Gobelin tapestries depicting the story of Esther. In Florence too is the Bargello, originally the home of the chief magistrate, the Palazzo del Podestà, later it was made into a prison, and it was not until 1865 that it became a museum, which holds among other great works **Donatello's 'David'** and also Verrocchio's, Cellini's bronzes and excellent Della Robbias; many of the exhibits were unfortunately damaged in the 1966 Flood, those in the Uffizi and the Pitti escaping damage because they were on upper floors. There is one other place that should not be left out as it has one of the greatest sculptural treasures in the world, that is the Accadèmia di Belle Arti. **There is Michelangelo's 'David' in all its perfection.**

In Milan is the Brera, which was first opened to the public in 1809 in the palace of that name that had belonged to the Jesuit Order. It houses a fine collection of the Italian school. When in this city, a visit should be made to the refectory of the Santa Maria delle Grazie, to take a lingering look at what remains of **Leonardo's 'Last Supper'.** To the South, in Naples, is the Capodimonte with a large collection of pictures from the 13th to the 19th century.

To Rome which, like Florence, is almost a museum in itself with the modern city interspersed with fine old buildings and ruins. The Galleria Borghese, founded about 1616, houses a

collection built up from works assembled by Cardinal Scipio and his descendant Camillo, and **ranks high as a group of works brought together by one family.** At the Musèo Capitolino is an important group of classic sculpture including: 'The Dying Gaul', 'Capitoline Venus', a copy after Praxiteles and 'Apollo of Omphalos'. The Palazzo Barberini has outstanding Italian school paintings also works by Holbein and El Greco. **But for most the Sistine Chapel stands as one of the great creative wonders in art.** Close by there are the superb frescoes of Raphael and a number of galleries in the Vatican complex. The Pinacoteca Vaticana, inaugurated in 1932 by Pope Pius XI, holds work by Leonardo, Fra Angelico, Raphael and Caravaggio. The Musèo Gregoriano Etrusco, founded by Pope Gregory XVI in 1837, has on display much material from the Etruscan tombs. There is also the celebrated Bibliotèca Apostòlica Vaticana with its famous art collections which include the **Aldobrandini Wedding (a fresco)**, Gothic ivory-carvings, early icons, and beautiful early Syrian woven fabrics with subjects based on the Annunciation and the Nativity.

Venice has the Accadèmia di Belle Arti with outstanding collections of the works of Titian, Tintoretto, Giorgione and the Bellinis, and also the Palace of the Doges with Tintoretto's immense 'Il Paradiso'.

Germany has always been very collection and museum minded. In Munich are the two great collections in the Alte Pinakothek, devoted to Old Masters and the Neue Pinakothek to 19th- and 20th-century works. Apart from paintings the Alte Pinakothek has a collection of **20 000 drawings and 300 000 engravings.** In the city there is also the Glyptothek founded by Ludwig I of Bavaria in 1816 which houses ancient and modern sculpture, including the 'Barberini Fawn'.

Austria has, in the Albertina Collection in Vienna, **one of the outstanding collections of drawings.** The building was formerly the Palace of the Archduke Frederick of Austria and was built by Montoyer between 1801 and 1804. The collection is particularly strong in the 15th- and 16th-century Germans such as Dürer.

Russia with the State Hermitage in Leningrad, **the largest public museum and art gallery in the Soviet Union, has one of the most important museums in the world.** It was opened in 1764 as a Court museum and to the public in 1852. It takes its name from the retreat, where Catherine the Great held a private exhibition of works which she jealously guarded from the gaze of the unchosen. In its history the collection has been several times eroded by sales.

Nicholas I sold 1219 pictures in 1853, and earlier this century a number of outstanding works of such as Van Eyck, Rembrandt, Watteau, Rubens, Dirck Bouts and Guardi went by private transaction to the Mellon Collection and the Gulbenkian Foundation's Museum in Lisbon; at the same time more pictures went in public sales in Berlin and Leipzig, and in 1928 a Berlin gallery had on offer from the Hermitage a number of pictures by such as Boucher, Canaletto, Rubens, Tintoretto and Teniers. During the last war a large part of the collections was evacuated by train almost under the muzzles of the German guns, to a safe place at Sverdlovsk on the eastern side of the Urals. **Today the Museum houses more than 2 500 000 objects, 40 000 drawings and 500 000 engravings.**

In Moscow itself there is the State Pushkin Museum of Fine Arts with 500 000 items of ancient Eastern, Graeco-Roman, Byzantine, European and American art and also the State Tretyakov Gallery, which has **one of the finest showings of icons** and works of Russian painters, sculptors and graphic artists, some 40 000 exhibits in all.

Often, to fully appreciate some particular painter's work, it is best done in his native light. Rembrandt is one of those who is seen to perfection in Holland, either in the Rijksmuseum, Amsterdam or the Mauritshuis, The Hague. The former was founded in 1808 as the Royal Museum, being built by Louis-Napoleon as King of Holland. It contains among many masterpieces Rembrandt's magnificent great canvas 'The Night Watch', which was stupidly attacked by a man with a knife in 1976.

Belgium has Bruges where the arts have been entwined in the old city with its canals and cobbles. The Groeningemuseum contains an interesting and well-balanced collection of the early Flemish painters including Jan van Eyck, Memline, Bosch and Gerard David. In the Notre-Dame is a white marble 'Virgin and Child' an early work by Michelangelo.

In France resides the world's best-known painting, **the 'Mona Lisa' by Leonardo da Vinci.** The painter, although an Italian, died at Clos-Lucé, just outside Amboise on the Loire, whither he had come to spend his last years under the gentle understanding patronage of François I. The great museum the Louvre has the 'Mona Lisa', the most copied picture, and also probably one which has more tales of forgery and theft around it than any other. The Louvre has had a somewhat harsh career. The present buildings stand on the site of a fortress erected by Philip-Augustus about 1190. It was reconstructed in the

Museo del Prado, Gallería central. (Courtesy Spanish National Tourist Office)

13th century by Charles V. The whole site on the Right Bank of the Seine occupies 45 acres (18 ha). The first part of the modern structure was built in 1541 to the plan of Lescot. Louis XIV had the main part of the great square erected; his Minister of Finance, Colbert, saw to it that treasures poured in to suit the whim of the 'Sun King', choice paintings arrived from England as the Cromwellian auctioneers dispersed Charles I's Collection. The building and enrichment with sculpture went on until 1678, then interest palled, the King had gone to Versailles.

Dreadful things happened, rooms were invaded by the poor, businessmen traded from the entrances. In the middle of the 18th century someone thought of making a proper museum there. Just then the Revolution struck. But a short time later with the encouragement of Jacques Louis David the doors of this museum were opened to all in 1793. Further revolutions of 1830 and 1848 left the great complex unharmed, although the Communards in 1871 did fire the Library and the Tuileries, but the blaze was quenched before serious damage occurred. **Today the Louvre lays claim to being the greatest over-all treasure-house** with its varied collections that include: paintings, sculpture, engravings, drawings, tapestries, ivories, porcelain, gems, terracottas, bronzes from all periods.

To study the genius of Spanish painters the Prado, Madrid, must be visited. This fine gallery is the work of Juan de Villanueva and was started in 1785 and completed in 1830. The 2500 paintings have come largely from the Royal Collections. Apart from El Greco, Velázquez, Ribera,

Zurbáran, Murillo and Goya there are excellent examples of Dürer, Bosch, Tintoretto, Rubens, Cranach and 20 pictures by Titian.

In Geneva is the Musée d'Art et d'Histoire which holds an absorbing collection of early German and Flemish Primitives as well as Swiss painters, in particular an imposing collection of work by Hodler. Specialist showings of works by one artist can be absorbing, being given the chance to study the progress of a creative mind. In Oslo,

Victoria and Albert Museum – Raphael Cartoon Gallery. Conservation requirements are met by automatic control of the roof-light venetian blinds and the artificial lighting system. The average illuminations on the Cartoons, produced by 750 watt tungsten-halogen lamps in louvred directional reflectors, is 150 lux. (Crown Copyright. Victoria and Albert Museum)

The Metropolitan Museum of Art's Store-room 1, where new objects to the collections are registered and examined upon arrival. (Courtesy The Metropolitan Museum of Art)

for instance, there is the Munch Museum, in Bruges the Brangwyn, near Flensburg the Nolde, and many others in the different countries.

The third claimant to the largest and widest collection is the British Museum (founded in 1753). It is, of course, **the largest museum in the United Kingdom.** The main building in Bloomsbury, London, was built in 1823 and has a total floor area of 17·57 acres (7·11 ha). The genesis of this vast establishment came from the legacy of his collection to the nation by Sir Hans Sloane (1660–1735); to this was added the library of George II. It was opened to the public in 1759, under the name of the British Museum and in the building known as Montague House in Bloomsbury. The art sections contain the famous collections of Prints and Drawings, and antiquities from all over the Middle East. Here are the **famous Elgin Marbles.** In England there is the Ashmolean Museum in Oxford, built in 1679. This august institution had a strange start. The story began with a father and son, both named John Tradescant, who were naturalists and horticulturists. John the Younger left their gleanings of 'twelve cartloads of curiosities' to one Elias Ashmole, who after an interval gave them to Oxford University who then had a building designed by Thomas Wood, into which the exhibits could go. Today there are notable examples of antiquities, ivories, sculpture, and an interesting selection of the caricatures by Max Beerbohm.

The National Gallery, London has possibly the finest over-all collection from the different schools and countries. It was founded in 1824 and has been in the present building designed by William Wilkins since 1838.

In the past century more museums and galleries must have been opened in America than anywhere else. Works of art, antiquity and interest have flowed across the Atlantic and Pacific oceans to find new homes and new eyes to view them. **In New York is yet another entrant for the largest and finest collection in the world. It is the Metropolitan Museum of Art which was founded in 1871.** In a way it is similar to the Louvre as it embraces practically every category of collecting. The picture galleries have a very comprehensive selection from all schools. There is also **one of the world's leading collections of medieval art** originally made by George Grey Barnard and given to the museum by John D. Rockefeller Junior.

The National Gallery of America is in Washington, DC and was established in 1937, but the doors did not open to the public until 1941. Andrew William Mellon (born 1855), an

American financier, gave the gallery which is **one of the largest white marble buildings in the world, being some 785 ft (239 m) long.** The large majority of the pictures were donated by philanthropists including: Paul Mellon, 150 paintings with examples by Rembrandt, Botticelli, Van Eyck, Raphael and Vermeer; the Kress Collection of nearly 400 paintings and sculpture, Italian school with Duccio, Giotto, Bellini, Giorgione, Titian and Tintoretto; the Widener Collection with El Greco, Vermeer, Rembrandt, Turner, Constable, Manet, Renoir and Degas; Chester Dale who sent Impressionists and earlier French painters of the 19th century, and El Greco, Chardin, Rubens and Boucher; and Lessing J. Rosenwald who gave over 8000 drawings and prints.

Around the world there are new concepts being tried out for gallery and museum display. One of the first of these was the Guggenheim Museum in New York, designed by Frank Lloyd Wright; it is a spiral-shaped gallery of unique design; visitors start by taking a lift to the top and then walk down the gently sloping ramp while viewing the paintings. New techniques with lighting have been set up as with the National Gallery, Melbourne, and the Great Hall at the Victoria Arts centre also in Melbourne, where there is a coloured transparent ceiling that gives a great sensation of space.

In Paris, in 1975, there opened the huge and ambitious Centre National d'Art et de Culture Georges-Pompidou. This combines in one a museum of National Modern Art, which includes not only painting and sculpture but also drawings, prints, photography and films; there is the centre for all forms of design and a research unit for the visual, musical, acoustics and voice. In Genoa is the reconstructed Palazzo Bianco Museum, which has been brilliantly carried out by the architect Franco Albini. In Brazil there arose the São Paulo Museum, a brave showing of old and new objects together, with an avowed policy to bring the people into closer contact with and to have a better understanding of the arts. Sweden has the Kunstgalerie, Lund, with massive sloping skylights that fill the galleries with a high quality of illumination. At Saint-Paul-de-Vence, France, is the Maeght Foundation Museum, again a new idea; it consists of a number of separate buildings, grouped rather like a small village on a hill overlooking the Bay of Antibes. In Tokyo is the National Museum for Western Art designed by Le Corbusier, planned so that the available space can be used in many ways. For a feeling of sheer spaciousness the Museum of Modern Art, Rio de Janeiro, would be hard to better. The architect,

Out-door sculpture related to urban environment, becoming a part of a street scene. A work by Karl Prantl. (1971.) (Courtesy Hauptamt für Hochbauwesen Nürnberg)

Henie-Onstad Art Centre, Hovikodden, near Oslo. (Courtesy of the Director)

Affonso-Eduardo Reidy, has given maximum use to large areas of glass and exploited the possibilities of reinforced concrete. At Pforzheim, Germany there is the Reuchlinhaus, named after a 15th-century humanist; here the architect Manfred Lehmbruck has produced a highly flexible arrangement which allows the spaces for exhibition to be altered as desired.

Another fashion for the showing of sculpture is the park, where the exhibits are set outside and in and among natural features, such as rocks, hills, trees and vegetation. They are proving highly successful. Examples include the Kröller-Müller in Holland, which combines the park with a gallery; the Shaw Park in St Louis, Missouri; Louisiana, also with a gallery, at Humblebæk, Denmark; the Vigeland-Parken in Oslo; the Henie-Onstad Art Centre at Hovikodden in Baerum, near Oslo; and one of the newest at West Bretton, near Wakefield, Yorkshire, which was opened in September 1977.

Chapter Eight

The Darker Side

The turnover of the world art market in 1965 was around £125 000 000; an opinion has been expressed that some £12 000 000 of that sum came from forgeries, either outright or from copies that had got into the wrong hands and have been 'assisted' with floating signatures and carefully fabricated provenances. In 1976 the sales-rooms of the world pushed the over-all figure up to and in excess of £300 000 000. If the ratio for fakes remains constant, it would indicate a rather horrific figure of somewhere over £30 000 000 finding its way, with deductions for the 'helpmates' *en route*, into the forger's wallet.

The easy money pulls them in, but these operators in the shade are shrewd and watch markets and fashions. When they see a supply drying up for a particular category, artist or craftsman they can be quick to satisfy the demand.

There is a strange legal anomaly with regard to forgery. It is not a crime to privately make a fake,

Modern forgery of El Greco with part of top paint layer removed to show earlier picture underneath.
(Courtesy Doerner-Institut, Munich)

or to copy or imitate a work of art. The crime of forgery is only perpetrated when the forger or his agents or other interested parties utter the forgery, which is the legal jargon, for selling or making an attempt to sell the forgery as a genuine object.

How long ago did the practice start? Certainly well before the Renaissance. Metal and stone objects that have proved to be fakes have been excavated from Egyptian tombs that date some appreciable time before the Christian era. There is some evidence that the forging of works of art begins at approximately the same time as collecting, which comes back again to the case of limited supply for an increasing demand, that nearly always will be met by illegal productions.

Among the first serious collectors on a large scale were the Romans with their treasure-filled villas and well-stocked galleries; they paralleled many of the great private collectors of this century. Records show how in the times of the Emperors lesser contemporary artists were wont to sign their work with a famous name; **paintings would be labelled by Zeuxis, sculpture by Praxiteles.**

During the Renaissance there was a heavy demand for the Antique, particularly in Italy. When the young Michelangelo was working under the patronage of Lorenzo the Magnificent he produced a marble 'Sleeping Cupid' in the classic style. Lorenzo saw it and admired the talent of the younger man, but as he did not want it himself he put it to Michelangelo that it could have some market value if made to look old. This the young sculptor achieved by burying the 'Sleeping Cupid' in a patch of damp sour ground for a period. This gave the white marble convincing age-stains. The figure was sent to an art-dealer in Rome, Baldassare de Milanese, who sold it to Cardinal Riario of San Giorgio for 200 gold ducats and incidentally gave Michelangelo only 30 ducats. Later the Cardinal found out the deception and the dealer had to refund the money. Although Michelangelo's original 'Sleeping Cupid' has disappeared there have been a number of forgeries pretending they are the little figure.

One artist to suffer from plagiarization and the forging of his signature or monogram was Albrecht Dürer. His simple and famous AD mark was too tempting for the producers of lesser prints. In 1498 Dürer published his series based on the Apocalypse, and when he brought out a 3rd edition in 1511 he saw that it was necessary to warn would-be predators that his patron the Emperor Maximilian forbade anyone to copy the cuts or to sell within the limits of the Empire fake prints pretending they were by Dürer; if they did, they would risk confiscation of their goods and other punishments. In 1502, nine years previously, an artist from Frankfurt, Hieronymous Greff, had brought out an aped Apocalypse. So bad did

'Mars or an Etruscan Warrior', 20th century AD in style of 5th century BC. (Courtesy The Metropolitan Museum of Art)

A recent fake of a Lucas Cranach by Franz Wolfgang Rohrich. (Courtesy Doerner-Institut, Munich)

matters get that by 1512 one brazen character had the cheek to stand outside the Rathaus in Nuremberg selling forged Dürer prints complete with the AD monogram. The most accomplished Dürer forger was Marcantonio Raimondi (c 1480–1534), an Italian who must have had some considerable skill because earlier he had also faked Raphael. Raimondi produced copies of Dürer's 'Little Passion' and 17 of the cuts of his 'Life of the Virgin', all within a very short time of publication by the artist. Following these up with 'Adam and Eve', he even in this instance took the pains to fill out the tablet exactly as the original 'Albert Dvrer Noricos Faciebat 1504'. Actually Raimondi was an artist of quality in his own right, and in some of his copies it can be seen how he brings in effects of his own which makes them less faithful than others.

Dürer must have attracted more forgers and copyists than any other artist, with the possible exception of Leonardo and his 'Mona Lisa'. There are listed more than 300 persons who worked 'after' or in the manner of Dürer, all probably trying to cash in on his great success as a master of the print. There were the youthful Hieronymus Wierx, who made an excellent version of the 'Knight, Death and the Devil' when he was only 15; Virgil Solis, the skilled Nuremberg engraver, who had the grace to sign his imitations VS; others included Erhard Schön, J C Vischer, Joh. van Goosen, Hieronymus and Lambert Hopfer, Ulrich Kraus and Martin Rota.

Copyists either innocently or by design have caused much trouble and confusion. One of the best examples of a painter being freely copied is Raphael. There was Giulio dei Giannuzzi, at times called Giulio Pippi but more often Giulio Romano, who was born in Rome in 1492 and went to work in Raphael's studio; he assisted the master in his work at the Vatican and is considered one of his leading pupils. It is known that he copied a number of Raphael's Madonnas. When the master died he left all his studio equipment to Giulio Romano and Gianfrancesco Penni, who were his executors, and they were asked to finish his incomplete frescoes in the Sala di Constantino in the Vatican. After this there comes quite a company of copyists. The Flemish painter Jan Gillis Delcour, born about 1632 near Liège, went to Rome and worked under Andrea Sacchi and Carlo Maratti. In Spain there was Juan Bautista del Mazo, born in Madrid in 1610; he was equally adept at copying not only Raphael but also Tintoretto, Titian and Veronese. Philip IV had him copy the Venetian pictures in his collection, which he did so well that it was only with great difficulty that the originals could be picked out. From Denmark came Nikolaj Abraham Abilgaard, mentioned in Chapter Six. On account of the fire at the Christianburg Palace, where he stored his paintings, some of which were destroyed, records are vague as to whether his copies of Raphael, Titian and Michelangelo survived or not.

The talented copyist appear all over Europe. In the 17th century there was Adriaan Hanneman from The Hague who came to England in the steps of Van Dyck and spent his time up till the Civil War producing telling portraits in the manner of that master. Van Dyck was also well copied by Adriaan de Lelie, who was born at Tilburgen in 1755 and Jean Eugène Charles Alberti, born in Amsterdam in 1781. Frans Hals the Younger copied and imitated his famous father. In the case of Pieter de Ryng, the Flemish painter of still life of high quality in the manner of de Heem, dealers found the paintings would not sell under Pieter de Ryng's name in England so they promptly changed the signature to de Heem.

An example of totally innocent copying is concerned with Turner, who was one of the most prolific of artists. John Ruskin with good intent encouraged several artists to copy Turner's work so as to help them make progress, he thought. Such as Arthur Severn, H B Brabazon and William Hackstown did this so well that their results have often been accepted as original Turners, and the truth has only been discovered when they have been checked against the lists in the British Museum of the Turner Bequest.

Peter Thompson working in the 19th century thought up a rather ingenious idea. **He brought into being not only fake paintings but also a fake artist who was supposed to have done them.** Little is known of Thompson but he gave his spurious artist, Captain John Eyre, a full biography, which included that he was 'born' in Bakewell, Derbyshire on 6 October 1604, and he was a 'descendant' of Simon Eyre, the shoemaker, Lord Mayor of London. There were details of his 'service' with the Royalist cause until the 'trial' of John Hampden, after which he 'went' along with the Roundheads, and he had 'met' and 'painted' Cromwell. He was 'wounded' at Marston Moor and 'died' at Bakewell in 1644. Then happily some 300 pictures by him were found in his house.

Thompson made his created artist produce drawings of the fortifications of London in pen and ink in the manner of Wenceslaus Hollar. The drawings were made on old paper with weak ink that gave the impression that it had faded, also added were comments in what appeared to be

17th-century script. To launch his artist Thompson produced etched facsimiles of the fortification drawings marked at five and ten shillings each; his subscribers' list was headed by Prince Albert. It is quite possible still to come across an Eyre Hollar.

Among the dubious practices worked on pictures is the one of painting out or adding details. This may be because of prudery, as with wisps of drapery being trailed across nudes, or with painters like Lely who presented the ladies of their time with very low *décolletée*, it became the habit of later owners to have an artist add long thick locks of hair in the best position to be concealing. Figures have been added, expressions altered, or flags changed on ships, decorations exchanged, whole figures taken out as with the case of Reynolds's picture showing the Misses Payne at the harpsichord with their mother. Later someone must have objected to Mrs Payne as they had her hidden by a patch of sky and some shrubs. The mother was eventually saved from oblivion by the Curator of the Lady Lever Art Gallery and the canvas restored to the original composition.

Pastiches can often cause trouble. A painter deliberately takes features from two or more pictures and fuses them together on one canvas. This can be especially confusing when the painting is made at the same period as those from which it borrows. Another trap can be when a forger produces his fake and then deliberately damages it, perhaps even tearing the canvas. He then sets about restoring it, working on the theory that, the fact that it has been restored will convince the would-be buyer that it must be genuine. Where the forger has difficulty in obtaining the old period canvas he needs, he will line his canvas on which he has already painted with another canvas on the back which will tend to conceal the defect.

Art historians, critics and scholars usually when they find themselves with divergent opinions keep the squabbling behind doors. But now and then matters get out of hand. In September 1909 the Kaiser Friedrich Museum in Berlin took the wraps off a new and wonderful acquisition. It was a wax bust of Flora by no less than Leonardo da Vinci, they said. It had been purchased at the behest of Dr Wilhelm von Bode, the Director-General of Prussian Museums, one of the most awe-filling figures in the international art scene.

Suddenly, towards the end of October, rumours came from England that the 'Flora' was not by Leonardo but by some little-known English sculptor. Had the German tax-payers put up 160000 marks for a waxen fake? Very soon battle lines were formed. On one side there were Dr Bode, Edmund Hildebrandt and the art critic

from the *Berliner Tageblatt*, Adolph Donath. One of those crying forgery on the other side was Gustav Pauli of the Kunsthalle, Hamburg. He said the 'Flora' stylistically was wrong and she looked like a bust of Queen Victoria. *The Times* printed a letter from Charles Cooksey, a Southampton auctioneer, that suggested the 'Flora' had been produced by a minor English sculptor Richard Cockle Lucas, and that the son, A D Lucas, could come forward and confirm that he assisted his father with the bust. It got more and more complicated when the English art historian, E V Lucas, felt the 'Flora' must be Cockle Lucas's, then 14 years later wrote to *The Times* stating it was after all an authentic work by Leonardo. The matter progressed to more complications after scientific examinations which showed that archil was present, a dye in use in Leonardo's time. Then Augusto Jandolo made it known that another 'Flora' had come to light, only this time it was a marble one. The dealer, Alfredo Barsanti had bought it for 50 lire. He quite quickly sold it to the Museum of Fine Arts in Boston for 48000 lire. This unexpected fortune must have over-excited Barsanti as he found yet another 'Flora', this time in Florence and accredited it to Verrocchio, in whose studio Leonardo had worked. He paid 2000 lire for it and hoped to repeat matters. But experts playing cautious told him it was a poor forgery. The dealer himself, somewhat unwisely, cleaned the marble bust, but he was again in luck as the Italian art historian Ruddioro Schiff said that it was good and quite authentic. In the end Lord Duveen bought it for onward transit to one of his clients.

Such blatant contradictions as those that the wax 'Flora' provoked can have a somewhat shaking effect on the collector's judgement, just whom does he appeal to for help.

A second quake shook the art world in 1918, as the market was flooding with works of considerable importance. Pictures, sculpture and precious objects came up for sale from many different and at times untraceable origins. Among this treasure trove some superb sculptures began to appear in Paris, their origin was apparently Italy. Who was doing the off-loading, some impoverished noble family or was the Vatican quietly disposing of some of its great hoard? The jarring truth when it became known was that these fine objects were forgeries.

The creator was Alceo Dossena (1878–1937); he was a totally humble man with his stonemason's workshops alongside the Tiber in Rome; a craftsman of high talent. These fine works in the style of the early Greeks and Romans, Gothic and Renaissance periods which he produced were not intended by him to be presented as forgeries. He never tried to pretend that they were by anyone else but himself; it was a chance meeting with a greedy, unscrupulous dealer that caused him to receive the label of forger.

On Christmas Eve 1916 Dossena bound for some seasonal leave left his army unit at Poggio Mirteto and headed for Rome. Wrapped up in paper he had with him a small figure of the Madonna in wood he had carved. In a café he ran into the dealer Alfredo Fasoli, who, when he was shown the Madonna thought at first it was old and that perhaps the soldier had picked it up somewhere. He handed over 100 lire and when he got it home and took a close look he knew it was a modern work.

Fasoli schemed how he could exploit this simple talented man. After some time he met Dossena again and saw more of his amazing versatility; he suggested how Dossena might give his works the impression of age. Then from the workshop began to come works in stone, terracotta and wood in varying styles but all quite convincing. Fasoli ordered from Dossena an Early Renaissance tomb in the style of Mino da Fiesole, for which he found an eager buyer at 6 000 000 lire and gave the sculptor 25 000 lire. Fasoli made fantastic profits for himself. Of course he had his expenses: the histories of these masterpieces had to be invented, faked certificates of authenticity provided, provenance proved. But still there were profits galore.

Gradually Dossena heard whispers of the vast sums of money being paid for his works. He tried at one time to get an advance against his next orders, but the answer was no. The callous nature of this reply is probably what brought Dossena to make his own exposure. But the art world was not to be turned upside-down so easily again. Involved were museum directors who had spent large sums of other people's money, critics, connoisseurs and those who know all about such things. They scoffed at his, what they thought, conceited claim. Then they weakened slightly and said that if he had done these things they were obviously copies, but no one could produce the originals from which he was supposed to have worked. To prove his case Dossena eventually allowed himself to be filmed while he actually modelled. But he could not win, for now the experts started to point out flaws in his work that apparently they had not seen before. Dossena tried to sue Fasoli for fraud, only to find the dealer attacked him as a political agitator. His last days were spent sadly in Rome's pauper hospital.

One of the most ambitious schemes for large-scale forgeries was thought out by Lothar Malskat and Dietrich Fey; it involved the faking of murals in the Marienkirche, Port of Lübeck in West Germany. The church had been set on fire during a Royal Air Force raid in the last war and been severely damaged. It was known that the interior had been originally decorated with frescoes, also that these had been covered over with whitewash by iconoclasts in the 15th century. The fire and attentions of the fire-brigade with their hoses served to uncover enough of the paintings to show their quality; also the number of them and the large areas they covered. Unfortunately, the gutted church was at the mercy of the weather for several years before the war ceased, and the paintings deteriorated considerably, leaving but faint traces.

Restoration work was set in hand under the supervision of a committee, finances coming from the government and public sources. Dietrich Fey was the appointed restorer and he was to be assisted by a young painter of considerable skill, Lothar Malskat, who had also some experience in restoration work. The work progressed high up inside the church on scaffolding and screened from the public view. By September 1951 the task had been completed, and at a viewing many top art experts attended also Chancellor Adenauer. Fey was rewarded with an honour, a special stamp showed a detail from the restoration. But Malskat got neither plaudits nor apparently did Fey pay him his share of the fee.

The young painter must have suffered his chagrin in silence during the winter until in the spring he could no longer contain himself. On 9 May 1952 he issued a declaration that the frescoes in the Marienkirche were not restorations at all but they were forgeries by himself and Fey. As with Dossena earlier the pundits of the art world were not going to believe it; they knew the two restorers had parted their association and felt this was just a ruse by Malskat to hurt Fey. Various top art historians voiced opinions as to the extraordinary importance of these frescoes, the finest examples of intact work from the 13th and 14th centuries in western Europe.

To lend weight to his announcement Malskat was obliged to elucidate further. He showed how he had copied some details for the frescoes from Bernath's *History of Fresco Painting*, also portraits of Coptic Saints in the Kaiser Friedrich Museum in Berlin. Models he had used included a German film actress, his father, members of the church staff, he had even in one part painted in the head of Rasputin with a halo. Later in 1952 Malskat gave up further secrets of his clandestine work when he admitted to forging more paintings on the walls of Schleswig Cathedral. In part of these there was a frieze of medallions containing animals, some imaginary some real, including turkeys. The last brought out into the open a strange argument. Some quite rightly pointed out these birds just were not around when the frescoes were thought to have been painted in the Early Gothic time; turkeys having been brought back from the Americas in about 1550. Anachronisms such as this have often provided incriminating evidence against the forger.

In October Malskat continued his self-incrimination and came up with the admission that he had made some 600 forgeries or imitations of subjects that ranged from Indian miniatures to paintings by such as Chagall, Corot, Degas, Gauguin, Hodler, Matisse, Rembrandt and Watteau.

The legal wheels of investigation gradually brought Malskat and Fey to trial in August 1954; the proceedings ground on for 66 days and the verdict in the end was 18 months for Malskat and 20 months for Fey. Malskat had been the main producer of the fakes and Fey had issued them.

The old story about Corot, that he painted 700 pictures of which 8000 are in America, has more than a handful of truth with it. Actually, the total of his works supposed to be in America has now multiplied many times; for some reason he remains one of the favourite painters in that country. **The artist in his lifetime started the rot by signing works in his style by his pupils, thereby under the law actually becoming a forger himself; in that he was signing a picture actually by someone else.** He was also prone to working into his students' paintings, and confused the legal consultants in later years; they had to use such terms as a 'quarter forgery' or a 'half forgery'.

When Corot died a wholesale business started up to supply the enormous demand for his pictures, some of the practitioners being so sure of themselves that spurious Corots were even produced in media that the master had never used in all his painting life. An official and authoritative catalogue of the artist's work was produced in 1905 by Alfred Robaut, and it can assist in verification, but the flood of copies and fakes went on and on. Fortunately for the forger there seems to be a large number of collectors who are only too ready to buy on only the slenderest of provenances and the slightest of advice. Sometimes they will ignore readily available source material, and once the purchase of a fake has been made and they discover they have been cheated, vanity can prevent them disclosing the fact.

Corot's fakers were not just confined to France. In 1888 more than 200 pictures pertaining to be by him were sent to France from a Belgian

A pseudo period scene, painted with acrylic colours, and emanating from a studio in Spain.

A spurious horse painting, carried out in acrylic, from Spain. With acrylic colours a slight rubbery feel is generally a lasting feature of the paint film.

A modern marine painting in acrylic, from Spain, with a false appearance of age.

workshop. Collectors fell around trying to get hold of the master's works. There was the pathetic Dr Jousseaume, who spent most of his life buying up Corots, none of which were ever in Robaut's catalogue. For the fraudulent dealers that brought round the miserable specimens, he only had one condition, and that was that a painting or a drawing should cost no more than 100 francs. When he died **he was found to have bought 2414 paintings, water-colours and drawings.** The crafty vendors had apparently got round the awkward matter of Robaut's catalogue by saying they had come across a secret hoard of Corot's works.

Many of the French painters from 1850 and well into the 20th century have suffered from the forger's zeal. Faked Douanier Rousseaus were produced that in the early 1920s fetched up to 25 000 francs. There was Mme Claude Latour who was convicted in 1947 for mass-producing Picassos and Utrillos; much of her work being shifted by a 22-year-old dealer Jacques Marisse, who paid her between 100 and 1000 francs and was then selling the fakes as originals for up to 70 000. Mme Latour must have had some skill, because when Utrillo himself was confronted with her work, he admitted that he could not point to the fake and original in each case. His work has been forged probably nearly as much as Corot's, and known workshop areas include Zürich and Germany, although these may be blind leads and only be used as issuing points.

In Spain somewhere there is a studio workshop which is busy producing fakes carried out in acrylic colours, which will dry through quickly, on canvas, and then lining them for reasons mentioned earlier. The subjects include 18th-

'Procuress' by Han van Meegeren, which purports to be a copy of a picture by Theodor van Baburen (1570–1624) that appeared in a painting by Vermeer. (Courtesy Courtauld Institute Gallery)

century sailing-ships, horses, small genre pieces, not signed by any particular artist but quite cleverly done to suggest a particular school or painter. These pictures can crop up in house sales, small auction-rooms or with lesser dealers.

Most professions have an acknowledged master, and as far as the forging of paintings is concerned the uncrowned prince must be the late Han van Meegeren. After all, he did fool in 1937 the leading critic Abraham Bredius who went into a eulogy when he saw van Meegeren's 'Christ's Supper at Emmaus' and raised no protest when the Rembrandt Society and the State bought it as a genuine Vermeer for the Boymans Museum in Rotterdam for £58 000. In this instance only Duveen's agent in Paris apparently saw through the deceit when he proclaimed the picture '. . . a rotten fake'.

Van Meegeren was one step at least ahead of his brother fakers. He realized experts might be expecting to find religious subjects by Jan Vermeer, the great 17th-century Dutch painter, so he fabricated not only the master's style but also his subjects. He went to considerable pains to procure the right age canvas; choosing his pigments at first with research and knowledge, he obtained genuine ultramarine, which comes from lapis lazuli, from Winsor & Newton in London. Later perhaps success blurred his intelligence, for he used cobalt, a pigment not known in Vermeer's time. To get the paints to set quickly and hard he mixed them with phenol and formaldehyde. He heated the result, rolled the canvas to produce cracking and over the years between 1937 and 1943 made a handsome income.

But when fate caught up with him he was not arrested in the first instance as a forger but for 'collaboration with the enemy'. The reason, he had sold 'Christ and the Adulteress' to Hermann Göring, and the Dutch authorities believed he had got hold of a previously unknown work by Vermeer and had let it go to one of the country's arch enemies. Van Meegeren languished in prison for some time and then he told his captors the truth, which was that he had only sold the obese Marshal a fake. His trial produced another long argument of the experts. But conclusive evidence was produced by the principal scientific investigator, the late Dr Paul Coremans, who also gave as his opinion that Han van Meegeren was the greatest forger of all time. As such he had reaped a reward suitable to the title. For six faked Vermeers and two faked de Hoochs he had grossed £821 000.

Yet, was he the greatest? There are signs that another who may have surpassed him could one day be exposed.

'*Madonna of the Veil*', *a forgery in the manner of Sandro Botticelli, painted on a poplar panel with tempera largely mixed with egg white (c 1920–30). (Courtesy Courtauld Institute Gallery)*

Index

(Text references in Roman, illustrations in *italics*)